Music Engineering

Second edition

Richard Brice

Newnes

OXFORD   BOSTON   JOHANNESBURG   MELBOURNE   NEW DELHI   SINGAPORE

Newnes
An imprint of Butterworth-Heinemann
Linacre House, Jordan Hill, Oxford OX2 8DP
225 Wildwood Avenue, Woburn, MA 01801-2041
A division of Reed Educational and Professional Publishing Ltd

ℛ A member of the Reed Elsevier plc group

First published 1998
Second edition 2001

British Library Cataloguing in Publication Data
A catalogue record for this book is available from the British Library

ISBN 0 7506 5040 0

Library of Congress Cataloguing in Publication Data
A catalogue record for this book is available from the Library of Congress

Composition by Genesis Typesetting, Rochester, Kent
Printed and bound in Great Britain by Biddles Ltd, *www.biddles.co.uk*

Contents

Preface to the second edition

The point of any second edition is to revise and bring up to date the text of the first. This I have aimed to do and you will find herein much more detail on newer sound-synthesis techniques; like granular synthesis and physical modelling as well as much more technical detail concerning digital signal processing. There is also additional content for the computer based musician; more on using a PC, both as a recording device, and as a signal-processor. You will also find, paradoxically, further historical information; included here because of the on-going retro-fascination with valve, signal-processing technology and analogue synthesisers. But the biggest modification of all, is the inclusion of new Fact Sheets (fourteen in all) detailing a great deal of background information to the understanding of the body of the text, as well as information which may be useful for reference and further study. Why this change?

In the Preface to the first edition I wrote, *Music Engineering* does not require a deep knowledge of complex electronic concepts. However . . . something has to be assumed and, in this case, it is a knowledge of basic electronics.' In short, I think I made a mistake in this assumption! It is my belief that my failure to address the fundamental information alienated a great many potential readers of the original *Music Engineering*. It is particularly this omission that I have aimed to redress in this new edition and the first seven Fact Sheets were written to give a 'whirlwind' background in analogue and digital electronics. Clearly, not everything is covered and further study will be required to master electronics, but my hope is that these small, relatively undemanding, lessons will give sufficient information to enrich the main text and to whet the appetite for further and wider reading. Fact Sheets 8 and onwards are different: these extend the scope of the text to include practical designs and reference material. I hope, therefore, that – if you aspire to play or work somewhere in the increasingly technical music-media industry – you will find much of interest in these pages, irrespective of your current level of knowledge.

Acknowledgements

In addition to those who are mentioned in the Preface to the first edition, and whom I should wish to thank again – their inspiration and assistance being as much in evidence here as in the previous edition, I should also wish to thank Martin Eccles, editor of *Electronics World*, who gave permission for several diagrams from the vast archive of the erstwhile *Wireless World* magazine. Finally, I should like to thank Claire, Marion, Stanislas and David who are my inspiration in all things.

Richard Brice
Ave d'Iena, Paris 16eme

Preface to the first edition

Although technical in content, *Music Engineering* does not require a deep knowledge of complex electronics concepts. However, as with all books which cover a wide range of subject material, something has to be assumed and, in this case, it is a knowledge of basic electronics. This I did for fear that 'pulling off the road every five minutes' to explain another basic concept would interrupt our journey unacceptably. Moreover in 'dumbing-down' the text too much, I worried I would run the risk of irritating more knowledgeable readers. In spite of this, and although there are some points of interest on our journey which will require a technical background, there is much here for the novice too. Broad technical issues, where they are necessary for comprehension, are introduced as and when they are required. For instance, a technical understanding of amplitude distortion effects and mechanisms is required for Chapter 4, so this precedes the description of valve circuitry contained within the chapter. In this way I hope the following will appeal to a wide audience and will be as much at home on a bookshelf next door to books on MIDI, Home Recording and Orchestration as alongside the heavy tomes in an engineer's bookcase.

One warning – there is some mathematics early on! Now equations are great if you like them and 'bad news' if you don't. Where I have used them (and it's only very rarely), they're for the benefit of the former, who want to get to the 'nuts and bolts' of nature's mechanisms. The latter can simply skip ahead. If you don't like maths and have no feel for it, you'll lose nothing by pushing onwards.

Acknowledgements

When I was six years old my father built me a little transistor radio with earphones. I was thrilled. Some nights after it was finished my mother discovered me sitting up late at night in my bed crying. When she asked what was the matter, I said, 'Mimi's dead'. I had listened to the whole of Puccini's *La Bohème*! It's not too simplistic to say that this book owes its genesis to my father's inspiration. First with that little radio and many times thereafter – like the time he helped me design and build my first guitar amp. That I do it now, and here, may compensate a little for my omission to thank him before it was too late. Thanks are also due to my friend Simon Nield. Our late-night, alchohol-fuelled musings on many of the subjects considered hereafter benefited my knowledge and understanding as much as did harm to my liver! Thanks, too, to those who supplied pictures and information. Particularly Roland UK, Brad Coates of Melbourne Music Centre, Jim Dowler of ADT and Andy Smith of BASF. And to the staff of Butterworth-Heinemann for their assistance in preparing this book.

Technology note

Finally, you may notice, flicking through the pages, that many of the circuit examples are analogue rather than digital. There are a number of reasons for this. First, especially in the home and project studio market, most equipment is still analogue (except the tape or disk recorders). Second, and this is overridingly the most important reason, digital circuits of the signal processing devices described simply do not exist! Which is to say, the functionality is performed within Digital Signal Processing (DSP) integrated circuits which are programmed like microprocessors to perform the operations required. The hardware configuration of a DSP solution thereby elucidates the function of circuit not at all. Of course, the important point is to understand the technique and function of the particular audio process; whether it be flanging, phasing, mixing or compression and so on. So, where I hoped an analogue circuit might aid that understanding, I have chosen it. In other cases I have used block diagrams which may be regarded as hardware blocks or software processes.

Richard Brice
Ave d'Iena, Paris
1998

1
Jilted Generation – Science and sensibility

Who this book is for

In 1837, a Dr Page of Massachusetts created a ringing tone using an apparatus involving a horseshoe magnet and a coil of copper wire. He called his creation 'galvanic music'. Although his contribution to the world of electronic music is neither noteworthy as the invention of a serious instrument or as a work of musical art, it does demonstrate an original mind at work. One that sought to break the barriers of conventional thinking and indeed of conventional culture. A mind that sought to cross the desert that exists between the arts and the sciences. Page started down the long road which, starting as a dirt track, led to the Theremin and the loudspeaker. A track which had turned into a 'two-lane black-top' by the time of the invention of the electric guitar and the Hammond organ, and had become an Interstate by the time it reached the multi-track tape recorder and the MINIMOOG synthesiser. Even to today, when music and electronics race along together on an eight-lane freeway.

Each step along this route betrays yet another restive mind at work and interestingly, neither the arts or the sciences appear to have the monopoly on restlessness! No better example exists of this than the two men who are immortalised in the names of the world's first two electric guitars: Leo Fender was an electronics technician who turned his skills to musical ends, inventing the Telecaster; Les Paul was a musician who turned his prolific mind to technology. Same motivation, very different men, very different guitars. This book is full of the inventions of fertile, enterprising minds and I hope that it will be of interest to electronics engineers who, like Leo Fender, have acquired an interest in music and for musicians who, like Les Paul, have become fascinated in the technology of electric and electronic music making and who wish to learn more. For all these individuals, I have adopted the collective term musician-engineer, the two

1

parts of which define the ends of the spectrum of people to whom I hope the book will appeal.

Music and the twentieth century

As I write this second edition of *Music Engineering*, the twentieth century has passed. Who knows, or can know, what the future will bring? But looking back, any author choosing to write a history of the world would have to devote a long chapter to the last one-hundred years. It would not make easy reading either. Within fourteen years of the beginning of the twentieth century, the mechanical genius of the previous hundred years had been turned upon ourselves in a war of unspeakable horror. A war that lasted (according to the theory that the Second World War was a re-kindling of the First) over thirty years. It was a century in which we came face to face with the dark-side of ourselves. From mustard gas to work-camps, from ethnic-cleansing to the atomic bomb, the 'post-Hiroshima' citizens of the world have a unique – and uncomfortable – vision of what it is to be human. The twentieth century was the century during which, to quote W.H. Auden, 'the Devil broke parole'. It was also the century during which, due to a kind of intellectual 'trickle-down effect', the philosophical certainties which underpinned society for hundreds of generations evaporated. The widespread beliefs in God and in the immutability of His creation were gone. In an age during which politics summoned the power to smash whole cities and the families that lived in them, is it any wonder that the belief systems which had underpinned stable societies should have been smashed as well? And throughout the troubled century, the science of electronics graced us with ambiguous blessings, like an ambivalent angel. The technology of the twentieth century (our century) is the technology of electronics. From the field-telephone to the smart bomb, electronics has been our dark companion. From television to the X-ray, our friend. It has also made each of us a part of a far bigger world. The global village is far too cosy a name for the factionalised, polluted, half-starving world community to which we are all members, but who cannot now not be aware of what it is to be part of this community with all the benefits and disadvantages that brings?

In order to depict in music a unique vision of this most unconventional time, composers sought ways of breaking the conventional bounds of music. The twentieth century was a century of 'movements' as groups of musicians struggled to express the bewildering litany of new experiences the age brought. They were legion, especially after the end of the 1939–1946 war. Integral serialism (a movement which attracted composers such as Boulez, Berio and Nono) sought to break the mould of traditional musical ideas and associations by taking the ideas of serialism (see Fact Sheet #6), developed by Schoenberg, Berg and Webern, and applying these concepts to all elements of a musical structure; to rhythm, dynamics and so on. Other groups, the aleatorists and the followers of Fluxus, sought – by introducing indeterminacy and chance into their compositions – to redefine what is art and what is not. Indeed the slogan of Fluxus was, 'Art is Life and Life is Art'. Even to the extent, as in the case of John Cage and La Monte Young, of seeming to resign from the process of composition altogether! Others, whilst retaining conventional musical instruments, sought to capitalise on bizarre instrumental uses; encouraging the exercise of split-notes in the brass and the woodwinds or have emphasised ancillary instrumental sounds, key noise for instance. Fortunately for us – and for future generations of concert goers – not all composers succumbed to the philosophy of Adorno (1958) who believed, 'Only out of its own confusion can art deal with a confused society'. They took a different path – the one that concerns us here. They opted to exploit the new 'sound-world' fostered by electronics, to explore and explain the Zeitgeist.

Electronics

Lest I give the impression that this book will concentrate on, so-called, art-music. Let me 'come-clean' at once as to my own preferences: I believe there can hardly be a better sound-world than that produced by electronics to describe our unparalleled age. Listen to that film-soundtrack, doesn't the drone of the engines of a squadron of bombers sound like an electronic tone? And what better aural sensation better symbolises our fragmented cosmic life-raft than the swoosh, splash, crackle cacophony of a tuning short-wave receiver? But these sounds are not the sole province of high-art. One of the frustrations of the impossible 'discipline' of aesthetics is that the learned opinions of one generation are often laughable in the next. The artists, writers and musicians who are lauded in their own day often slip unnoticed into history leaving subsequent generations to discover intellects ignored and misunderstood in their own time. Never is this inability accurately to judge the 'half-life' of music more true than in cases where music appears, on the surface, to 'slip the bonds' of its own conception[1]. In our own time, popular, rock and (particularly) dance music – which often appears to ignore our own time – may ultimately most eloquently express it. History proves that art does not have to prick our consciences to speak to our hearts.

Ironically, the examples of the electronic sound-world discovered by the post war avant-garde composers remain largely unknown and un-liked. Possibly history will judge these composers differently, but I doubt it. They probably tried too hard to depict the harsh, exigent realities of our times in music which is, itself, too abrasive and demanding. But – and this is the crucial point – their legacy has found a vital place in all sectors of today's music industry and, most importantly of all, via this, in our collective consciousness; precisely because of the resonances electronic music finds in our hearts and souls. History, as always, will be the judge but, I believe the 25th century Mars-dweller is more likely to regard the characteristic sound of our age as an analogue synthesiser patch or a coruscating electric guitar, than that most noble invention of the nineteenth century – the orchestra – being asked to tap, rasp and otherwise abuse their instruments in a grotesque parody of orchestral technique! Couple with that the undeniable fact that electronics, in the form of recording and reproduction technology, has brought about a revolution in the dissemination of music and its cultural significance is almost beyond comprehension. For electronics is not just the messenger, it's the message itself.

The messenger and the message

This idea that electronics is both messenger and message is more than a mere poetic conceit. For it implies, with some degree of accuracy, that this book is two books in one. A book about recording (and by association, reproduction) technology and a book about electronic musical instruments and effect technology. Clearly a well-informed recording engineer, design engineer or musician-engineer must be aware of both and both are included for completeness. Furthermore, it is not strictly necessary to distinguish a pattern separating the two disciplines. However, the unmistakable motif which emerges between these two strands of technology illuminates the subject in a fresh and distinctive way and it also sheds light on debates ranging from valve amplifiers versus transistor amplifiers to that 'hardy perennial' – the analogue versus digital controversy. It explains why musicians treasure ancient, flagging Hammond organs rather than opt for modern, sampling 'equivalents' and other Objectivist versus Subjectivist conundrums where 'worse' is considered 'better'. In short, it explains why musicians hear success where engineers often see failure!

This pattern concerns the adoption or rejection of traditional values of engineering excellence in electrical circuits. For instance, electronics for recording has traditionally placed a high degree of emphasis on the inherent linearity of the transfer function. Not so musical electronics where, for creative and aesthetic reasons, circuits which either inadvertently or deliberately introduce some form of non-linearity are often preferred and sought after. The most obvious example is the guitar distortion-pedal whose only duty is to introduce gross non-linearity into the system transfer function. But there are many less conspicuous examples. The rearguard action against digital processing, and the recent renaissance in valve equipment, both bear witness to this real distinction between the electronics of the message bearer and the message creation itself. Virtually all the debates in audio arise as a consequence of failing adequately to distinguish between these two opposing philosophies.

But which is right? The answer is neither – but a change is at hand. Traditional wisdom has it that the message bearer must 'cosset' a musical signal, it must do it justice – think of the term fidelity in high-fidelity. The same wisdom never constrained the instrumental amplifier and loud-speaker designer. She has always had the freedom to tailor performance parameters to her taste in the hope of winning sales from like-minded (or like-eared!) musicians. Interestingly, recent work in audio electronics and perceptual psychology suggests that the age of a slavish commitment to 'pure engineering' specifications may not be what is ultimately required. Not because these are not valuable in themselves but because human perception may require different cues and a more 'open-minded' philosophy. If there is a lesson, it is that engineers need to listen. (For this reason, the psychology of auditory perception figures in the following; especially in Chapters 2 and 11.)

A perfect example of this more open-minded approach, concerns left–right channel crosstalk. For years the reduction of this effect has been a primary aim of recording system's designers and one of the 'triumphs' of digital audio mooted as its elimination. Unfortunately we now know that there exist beneficial effects of left–right crosstalk. So, crosstalk, long a feature of analogue recording systems, has been largely misunderstood and the deleterious results from its elimination in digital systems similarly mistaken. This has nurtured much of the debate between analogue and digital recording systems and explains many of the apparent 'short-comings' of digital recording. (This particular issue is discussed at length in Chapter 11.)

New threats . . . new opportunities

Being a time of great change, the twentieth century demanded a new mutable art to describe it. But, from the agricultural revolution to the Pentium chip, change brings new opportunities and new threats. As yet, we have no reason to believe that this new century will be more stable than the last. We live in the middle of a time of undeniable sociological and demographical transformation. Digital technology is altering our lives in untold ways. Both in our leisure time; wherein exploding markets for television, computer games and the Internet compete for our attention. And in our work; where new methodologies must evolve in the face of new, global pressures. As so often in a technological revolution, the paradigm shift always originates with an outsider; Faraday was a lab-technician, Einstein was a patent clerk. So it is with music and recording technology; we must look outside the industry to see the forces which will shape this next millennium. Central to the future, is the concept (dubbed by MIT's Nicholas Negroponte, as 'convergence') as digital technology destroys the boundaries between computing and traditional media. The dramatic growth of the world-wide entertainment market that these changes foretell, presents real challenges for musicians and engineers alike. Above all, one technology – as yet, but a few years old – is altering the business of music worldwide. That is the Internet and the MPEG coding of audio information for download as MP3 files. That this technology, with its inherent undermining of traditional streams of revenue, will change the whole sociology and politics of music production is obvious. The twentieth century, with the development of recording, offered a few musicians untold wealth whilst depriving many of their livelihoods: for every Irvin Berlin, a hundred thousand cinema piano players went on the dole. Recording technology, disc production and distribution has always been sufficiently expensive to ensure heavily capitalised, centralised music making. And, like any autocracy, it formed an elite. Many see the advent of MP3 files of music on the Internet as 'the end of the road' for the music business and, it's true that the democratising effect of this new technology will undermine the ability of a few to make fabulous wealth from a handful of songs. But is that so bad? The twentieth century was the exception: the great composers of the past weren't millionaires and their ambition wasn't to be millionaires. If they were fortunate, they lived comfortable, but far from extravagant, lives. They were driven by the desire to make music: not to drive Rolls Royce cars into swimming pools! MP3 files and ever faster Internet connections will reverse the anomalous trend of the twentieth century. It likely that the future will be more like the past; with many more musicians making music for many forms of entertainment and for smaller audiences.

That's where this book aims to help. To introduce the broad range of skills the world-wide, multi-media driven music-scene will demand. To that end, in the following, you will find sections on all the traditional knowledge-base demanded of the musician-engineer; the technology of microphones (Chapter 3), of electric and electronic instruments (Chapters 5 and 7) of electronic effects (Chapter 6) of consoles, amplifiers and loudspeakers (Chapters 12 through 14) and sound recording (Chapter 9). As well as sections on the new technology of audio workstations and hard disk editing, and an extended section on digital concepts and digital interfaces (Chapter 10), as well as MIDI and multi-channel and multi-dimensional audio (Chapters 8 and 11 respectively) and a crucial last chapter on the world of television, television concepts and synchronisation issues. In addition, you will find Fact Sheets which aim to give a background in analogue and digital electronics and to include some practical designs and reference material.

The music industry is just like any other, if you want to get on, you have to rely on new skills. The good news is, with the right skill-set, these changes mean more opportunities for us all.

Reference
Adorno, T.W. (1958) Philosophy of New Music (Philosophie der neuen Musik), Ullstein Bucher.

Notes

1 Think of Beethoven. His music speaks so eloquently in works which so heroically transcend his own misery.

Fact Sheet #1: Foundations

- A simple circuit – Ohm's law
- Alternating current
- Magnetism and electricity
- Generators
- Electrical signals

A simple circuit – Ohm's law

Figure F1.1 shows a simple electrical circuit, showing a battery, connecting wire and an electrical resistance. Notice that electrical symbols are highly stylised and only look approximately like their real-life counterparts. Actually this circuit does nothing interesting at all except turn the electrical energy

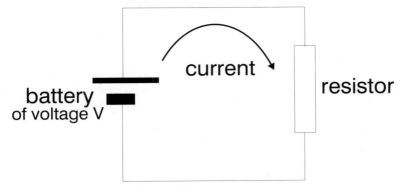

Figure F1.1 *A simple circuit*

of the battery into heat energy by means of the resistor. But it does illustrate the most important relation in electro-technical technology. The relation known as Ohm's law.

Ohm's Law states that the current (*I*) through a given resistor (*R*) is proportional to the voltage across its terminals (*V*). Equally it allows you to calculate an unknown resistance if you know the current flowing through it and the voltage across it. Or, what the voltage will be across a resistor when a certain current flows through it. Mathematically speaking we can say,

$V = I . R$ or,

$I = V/R$ or,

$R = V/I$

all of which are exactly the same.

Alternating current

The current from a battery (as we saw) is known as direct current because it always flows in one direction. Interestingly convention has it that the current flows from the positive terminal of the battery to the negative terminal but in fact it doesn't! Physical current actually flows from the negative terminal to the positive! Whoops! This was a mistake made in the early days of the science of electricity – before the discovery of the electron; the negatively charged particle which is responsible for the propagation of electricity in a conductor. Trouble is, the convention has stuck. To distinguish the conventional from the real the terms conventional-current and electron-current are sometimes used.

However, there is a very important form of current which alternately flows in either direction, such a current is not obtainable from a battery but is easily obtainable from a rotating machine known as an electricity generator. It is known as alternating current or AC. This is the type of current which is supplied to your home which will be between 100 V and 240 V and will alternate its direction 50 or 60 times a second (50/60 Hz) depending on where you live. For power distribution AC has a number of very distinct advantages; the most important of which is that it may 'stepped-up' and 'stepped-down' in voltage for distribution purposes using a device known as a transformer.

The wires that connect your home to the electricity generating station are very long, perhaps hundreds, even thousands, of miles! Now the metal chosen for the wires to cover these distances are deliberately chosen because it is a good conductor of electricity. But the wires are not perfect. In other words they have resistance. From what we know about Ohm's law, we know that some of the current through these wires will be turned into heat, resulting in a voltage (known in this case as volt-drop) appearing across the length of the wire in the proportion,

$V = I/R$

where V is the volt-drop, I is the current you are using in your home and R is the resistance of the wires.

Imagine if you tried to supply one light bulb in the arrangement shown in Figure F1.2a. The current would be in the region of 1/2 amp, and the resistance of the cables between the power station and your home in the region of 200 ohms. Applying Ohm's law we can calculate that the volt-drop across the cable will be $1/2 \times 200 = 100$ V, leaving just 120 V to light the lamp. Unfortunately that means the bulb will only burn very dimly. In fact the system simply isn't practical.

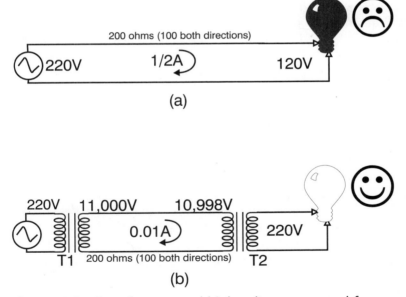

Figure F1.2 *Transformers and high voltages are used for power distribution as at (b)*

Instead transformers are used, as shown in Figure F1.2b. A transformer is made by winding two lengths of wire on a common iron frame. The world's first transformer which was invented by Michael Faraday. He discovered the relationship between magnetism and electricity which other scientists had suspected but been unable to pin down.

Magnetism and electricity – the dynamic duo

It had been known for a long time that if electricity flowed in a wire it created a magnetic field, which could be detected by a small magnetic compass. Faraday reasoned that if electricity could cause magnetism, then it ought to be possible to transform the magnetism back into electricity again. He used a primitive transformer and connected a battery to one of the windings (now known as the primary winding) and used a galvanometer (what we would call a volt-meter today) to look at the current in the second (or secondary) winding. He noticed that when he connected and disconnected the battery, a current flowed in the 'unconnected' secondary winding. The story has it that he realised this as he was putting his equipment away one night – in disgust! Having stared at the equipment throughout a whole, fruitless day, it was only as he disconnected the battery, he saw the galvonometer swing. And that's the crucial point. Electricity will create magnetism but only changing magnetism will make electricity. That's the reason AC (changing current) has to be used for power distribution because a DC transformer does not (and cannot) exist.

Generators

Before we return to practical transformers it's worth noting that the generator drawn in Figure F1.2 uses the same principle for the generation of electricity. Inside a generator a large magnetic assembly is turned near a wire winding. It's the changing magnetic field inside the generator (as the magnet moves) which causes the curent to flow in the winding.

A practical power distribution system using AC current and transformers

Transformers transform the relationship of voltage and current in exact proportion to their winding ratio, as shown in Figure

F1.2b. So, the load of 1/2 amp at 220 V present in the home is transformed into a load 1/100 amp at 11 000 V. Calculating using Ohm's law once again, we can work out the volt-drop,

Volt-drop $= 0.01 \times 200 = 2$ volts

and that 2 volts in 11 000 volts! This almost seems like magic and it demonstrates at a stroke why AC is used for power transmission. Always remember, however, that a transformer is not magic and the power in all parts of the system is the same. Power is measured in watts and is defined as the product of volts times current, so that,

$11\,000 \times 0.01 = 220 \times 0.5 = 110$ watts

Transformers can also play a role in audio as you'll see in later chapters.

Electrical signals

For power distribution an alternating current in the form of a sine wave at 50 Hz is utilised. But alternating currents of greater complexity form the basis of electrical signals. These range from audio signals of as low as a few cycles per second to radio or data signals of millions of cycles per second. The range of audio signals is in the region of approximately 5 Hz to 20 000 Hz. Before we leave this subject, note that the relationship between electricity and magnetism discovered by Faraday is even more profound than he supposed and accounts for phenomenon from radio to heat, to light!

2
Good Vibrations – The nature of sound

The physics of sound

Sound waves are pressure variations in the physical atmosphere. When I speak, the sound I make creates a series of compressions and expansions in the air immediately around me. These travel away at about 300 metres per second (700 mph) in the form of waves which spread out like ripples on a pond. If I'm talking to myself – a pastime hardly to be recommended – these waves collide with the walls, chairs, tables – whatever – and make them move ever so slightly. The waves are thus turned into heat and 'disappear'. If there is anyone there to hear me, that part of the wave which impinges on their external ear will travel down their auditory canal and cause their eardrum to move slightly. What happens after that is a subject we'll look at later in the chapter. All that matters now is that the sound, once received, is heard by the listener.

When we hear a voice of a friend, we recognise it instantly. Similarly if we're musical, we can recognise the sound of a particular musical instrument. (Some people are even able to recognise the identity of an instrumentalist by the sound alone.) So it's clear that the sound of these people and instruments must be different. There are plenty of lay terms to describe the tone of someone's voice: rich, reedy, discordant, syrupy, seductive and so on. Musicians too have their adjectives, but these are poetic rather than precise. Fortunately for the engineer, physicists and mathematicians have provided a precise way of characterising any sound – whenever or however it is produced.

It is a demonstrable property of all musical sound sources that they oscillate: an oboe reed vibrates minutely back and forth when it is blown; the air inside a flute swells and compresses by an equal and opposite amount as it is played; a guitar string twangs back and forth. Each vibration is termed a cycle. Any repeating pattern can be thought of as a cycle and any musical sound, or indeed any musical signal, comprises many of these cycles. Each starts from nothing and grows in one direction, reverses towards the rest position and grows in the opposite direction, reaches a peak and dies away again to the rest position before commencing the entire sequence again. Engineers often term the alternating directions the positive and negative directions to distinguish them. The simplest musical sound is elicited when a tone-producing object vibrates backwards and forwards exhibiting what physicists call simple harmonic motion. When an object vibrates in this way it follows the path traced out in Figure 2.1. Such a motion is known as sinusoidal and the trace is known as a sine wave. Because the illustration of the wave follows the shape or form of the sound we can refer to this type of illustration as a waveform. Just as a mighty palace may be built from little bricks, so the whole science of sound waves – and of electrical signals in general – is built from these little bricks we call sine waves.

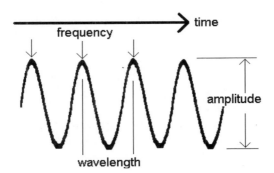

Figure 2.1 *A sine wave*

Such a pure tone, as illustrated, actually sounds rather dull and characterless. But we can still vary such a sound in two important ways. First, we can vary the number of cycles of oscillation which take place per second. Musicians refer to this variable as pitch – physicists call it frequency. The frequency variable is referred to in hertz (Hz) meaning the number of cycles which occur per second. Second, we can alter its loudness; this is related to the size, rather than the rapidity, of the oscillation. In broad principle, things which oscillate violently produce loud sounds. This variable is known as the amplitude of the wave.

Unfortunately, it would be pretty boring music which was made up solely of sine tones despite being able to vary their pitch and loudness. The waveform of a guitar sound is shown in Figure 2.2.

Figure 2.2 *A guitar waveform*

As you can see, the guitar waveform has a fundamental periodicity like the sine wave but much more is going on. If we were to play and record the waveform of other instruments each playing the same pitch note, we would notice a similar but different pattern; the periodicity would remain the same but the extra small, superimposed movements would be different. The term we use to describe the character of the sound is called timbre and the timbre of a sound relates to these extra movements which superimpose themselves upon the fundamental sinusoidal movement which determines the fundamental pitch of the musical note. Fortunately these extra movements are amenable to analysis too, in a quite remarkable way.

Fourier

In the eighteenth century, J.B. Fourier – the son of a poor tailor who rose ultimately to be scientific adviser to Napoleon – showed that any signal that can be generated, can be alternatively expressed as a sum of sinusoids of various frequencies. With this deduction, he gave the world a whole new way of comprehending waveforms. Previously only comprehensible as a time-based phenomena, Fourier gave us new eyes to see with. Instead of thinking of waveforms in the time base (or the time domain) as we see them displayed on an oscilloscope, we may think of them in the frequency base (or the frequency domain) comprised of the sum of various sine waves of different amplitudes and phase.[1] In time, engineers have given us the tools to 'see' waveforms expressed in the frequency domain. These are known as spectrum analysers or, eponymously, as Fourier analysers (see Figure 2.3). The subject of the Fourier transform, which bestows the ability to translate between these two modes of description, is so significant in many of the applications considered hereafter, that a quick revision of the Fourier transform may be of use to some readers. This is necessarily a little mathematical. But, don't worry if you don't like maths. If you skip the next section, the book will still make sense!

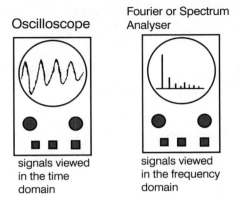

Figure 2.3 *A Fourier analyser and oscilloscope compared*

The Fourier transform

The Fourier transform exists because an electrical signal may be described just as accurately in two different ways: in the frequency domain and in the time domain. In many practical fields, signals are more often thought of in the time base rather than in the frequency base – sound signals and television signals are both good examples. But how do we get from the time domain description (a sine function, for example) to a frequency domain description? In fact the process is very simple, if a little labour intensive. The easiest way to imagine the process is to consider the way a spectrum analyser works.

Consider the simplest input waveform of all – a single, pure sine wave. When this signal is input to a spectrum analyser, it is multiplied within the unit by another variable frequency sine-wave signal. This second signal is internally generated within the spectrum analyser and is known as the basis function. As a result of the multiplication, new waveforms are generated. Some of these waveforms are illustrated in Figure 2.4.

The resulting signals are subsequently low-pass filtered (note that this is the same as saying the time integral is calculated) and the resulting steady voltage is used to drive some form of display device. The principle involved may be appreciated without recourse to difficult mathematics but one axiom must be understood: when two sine waves are multiplied together, the result contains only the sum and difference frequencies of the two original waves. Or mathematically put:

$$\sin A \times \sin B = 1/2\{\cos(A - B) - \cos(A + B)\}$$

Inside the spectrum analyser, because the output of the multiplier stage is low pass filtered, at all practical frequencies, the sum frequencies

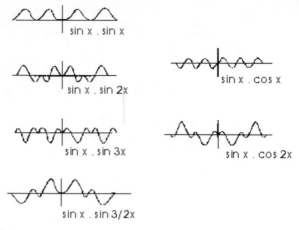

Figure 2.4

disappear leaving only the difference frequency. And this will only be a steady DC signal ($A - B = 0\,\text{Hz}$) when the two frequencies (the input signal – or target signal and the search frequency – or basis function) are exactly the same. Figure 2.4 illustrates this; only ($\sin x$) $\times$ ($\sin x$) results in a waveform which is asymmetrical about the x axis. In this manner, the single, component sine wave frequencies within a complex input waveform may be 'sought out' by selecting different frequency basis functions and noting the DC voltage resulting from the multiplication voltage followed by low-pass filtering.[2]

Actually the scheme described so far, though it often represents the complete system within a spectrum analyser, doesn't give us all the information we need to know about a waveform in the frequency domain. So far we only have a measure of the magnitudes of each of the sine-wave frequencies present within the input signal. In order to reconstruct a waveform from the frequency domain description we need to know the phases of each of the frequencies present within the original signal. It would be quite possible to do this by constructing some form of calibrated phase-shifting arrangement and adjusting for maximum $0\,\text{Hz}$ output, once this had been found, using the technique above. Noting the phase value would yield all the information required for a complete frequency domain description. But this isn't done. Instead a technique is used whereby the input signal is multiplied by a sine basis function and a cosine basis function. These two functions can be thought of as separated by 90° in phase. If you look at Figure 2.4 again, you'll notice that ($\sin x$) $\times$ ($\sin x$) produces a continuous offset whereas ($\sin x$) $\times$ ($\cos x$) does not. Whatever the phase of the input signal, it will generate a result from one or other (or both) of the multiplication processes. By knowing the magnitude of both the sine and cosine multiplications, it is possible to calculate the true magnitude of the original frequency component signal by calculating the square root of the sum of the squares of the two results and its phase, because the tangent of the phase angle = result1/result2. With that intuitive explanation under our belts, let's look at the maths!

The frequency domain description of the signal completely specifies the signal in terms of the amplitude and phases of the various sinusoidal frequency components.

Any signal $x(t)$, expressed as a function of time – in the so-called time domain can instead be expressed in the frequency domain $x(\omega)$, in terms of its frequency spectrum. The continuous-time Fourier integral provides the means for obtaining the frequency-domain representation of a signal from its time-domain representation and vice versa. They are often written like this:

Fourier Transform

$$x(\omega) = \int_{-\infty}^{+\infty} x(t)e^{-j\omega t}\,dt$$

Inverse Fourier Transform

$$x(t) = \frac{1}{2\pi} \int_{-\infty}^{+\infty} x(\omega)e^{-j\omega t}\,d\omega$$

where $x(t)$ is a time-domain signal, $x(\omega)$, is the complex Fourier spectrum of the signal and (ω) is the frequency variable. This may look pretty scary but take heart. The quoted exponential form of Fourier's integral is derived from Euler's Formula:

$$\exp jA = \cos A + j \sin A,$$

the alternative form is found by replacing A by $-A$,

$$\exp(-jA) = \cos(-A) + j \sin(-A)$$

which is equivalent to

$$\exp(-jA) = \cos A - j \sin A$$

so it is really just a shorthand way of writing the process of multiplication by both sine and cosine basis functions and performing the integration which represents the equivalent of the subsequent low-pass filtering mentioned above.

An important difference between analogue and digital implementations is that in the former, the search (basis) function is usually in the form of a frequency sweep. The digital version, because it is a sampled system, only requires that the basis function operates at a number of discrete frequencies; it is therefore known as the discrete Fourier transform or DFT. The fast Fourier transform is just a shortcut version of the full DFT. An incidental benefit of the discrete Fourier transform is that the maths is a bit easier to grasp. Because time is considered to be discrete (on non-continuous) in a digital system, the complication of conceptualising integrals which go from the beginning to the end of time or from infinitely negative frequencies to infinitely high frequency can be dispensed with! Instead the process becomes one of discrete summation. So, if we have a digital signal and we wish to discover the amplitudes and phases of the frequencies which comprise a time-domain signal $x[n]$, we can calculate it thus:

$$a_k = \sum_{N=0}^{N-1} x[n] e^{(-j2\pi kn/N)}$$

where a_k represents the kth spectral component and N is the number of sample values in each period.

Transients

The way a musical note starts is of particular importance in our ability to recognise the instrument on which it is played. The more characteristic and sharply defined the beginning of a note, the more rapidly we are able to determine the instrument from which it is elicited. This bias towards transient information is even evident in spoken English where we use about 16 long sounds (known as phonemes) against about 27 short phonemes. Consider the transient information considered in a vocalised list of words which end the same way: coat, boat, dote, throat, note, wrote, tote and vote for instance! Importantly, transients, too, can be analysed in terms of a combination of sinusoids of differing amplitudes and phases using the Fourier integral as described above.

Musical ranges

The relationship between written musical pitch (on a stave) and frequency (in hertz) is illustrated in Figure 2.5. Note also the annotations on this diagram which indicate the lower and upper frequency extremes of particular instruments. Remember that the frequency components of the sound produced by each of these instruments extends very much

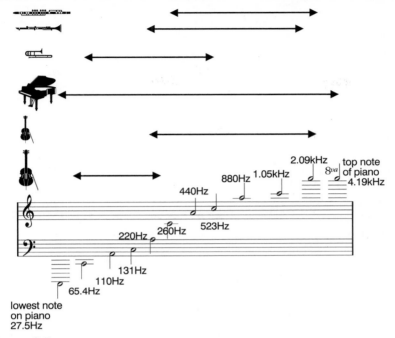

Figure 2.5

higher than the fundamental tone. Take for instance the highest note on a grand piano. Its fundamental is about 4.19 kHz but the fourth harmonic of this note, which is certainly seen to be present if the sound of this tone is analysed on a spectrum analyser, is well above 16 kHz. The frequency ranges of various instruments are tabulated in Table 2.1.

Musical scales and temperaments

If a note of particular pitch is chosen at random, other notes may be picked out by ear as being in some way intuitively related to the original note. The relationship between these 'kindred' notes was first studied by Pythagoras around 500 BC. He demonstrated, by experiment, that notes which sounded related were also associated by a simple ratio of two small whole numbers; the simplest being the octave in which one note is related to the other in the ratio 2:1. A pair of notes associated in this manner are termed a musical interval. Other musical intervals are defined in terms of their simple numerical ratio in Table 2.2.

When the musical notes in these simple ratios are sounded simultaneously, the sound is always pleasant – or consonant. Relationships which

Table 2.1

Instrument	Lower frequency limit (Hz)	Upper frequency limit (Hz)
Choir	82**	880**
Piano	27.5	4190
Violin	196	2600
Viola	131	1050
Cello	65.4	780
Flute	260	2090
Piccolo	532	4190
Guitar (electric)	82.4	2090*
Guitar (bass)	41.2	220*

* Approximate figure; depends on type of instrument.
** This range may be extended by individual singers.

depart from these integral ratios tend to sound harsh or dissonant. Why do we consider some combinations of tones pleasant and others unpleasant? While conditioning when young inevitably plays a role, there is also some evidence that we prefer pairs of tones for which there is a similarity in the time patterns of the neural discharge of auditory neurones (Moore 1989). In other words, the basis of consonance and dissonance appears to have some physiological foundation and is not acquired solely by exposure and by learning. Certainly Pythagoras's own work suggests that the perception must have preceded the theory.

Table 2.2

Interval	Frequency ratio
Octave	2:1
Fifth	3:2 (1.5)
Fourth	4:3 (1.333)
Major third	5:4 (1.25)
Minor third	6:5 (1.2)
Major sixth	5:3 (1.667)
Minor sixth	8:5 (1.6)

Such is the power of Pythagoras' influence that the musical scales which comprise modern music are constructed from the intervals derived from his investigations into consonant intervals. His legacy may be most easily understood by considering the white notes of a piano. Having established the small whole number relationships between each of these notes, the Greek musical system reached full maturity when seven different natural scales or modes were 'mapped out' from these seven basic notes (and their octave transpositions). Each musical mode starting on a different note thus:

C-C   Lydian
D-D   Phrygian
E-E   Dorian
F-F   Hypolydian
G-G   Hypophrygian
A-A   Hypodorian or Aeolian
B-B   Mixolydian

The melodic and expressive qualities obtainable from the Greek system of modes was considerable, indeed the Greeks set great store in the ennobling effect of musical study and performance. And these modes (and many more still more complex – involving half and quarter tones) persist in the music of non-Western cultures; Indian and Persian musicians having developed particularly complex systems. However, in Christendom the situation began to crystallise over time, so that by the fourth century AD only four of the original seven Greek modes were recognised and codified by the Christian church (Wood 1944).

The rise of polyphony (the sounding of two simultaneous melodies sung together) and the ensuing rise of harmony, hardened the situation further so that by the end of the Renaissance only two modes remained in common usage, thereafter forming the basis of the Western system of musical scales. These last remaining two modes – the Greek Lydian and the Greek Hypodorian or Aeolian – are equivalent to the modern C major and A minor scales respectively. Melodies and harmonies constructed using these scales have distinctly different characters; they are said to be composed within a major or minor key.

In order that composers could best contrast the nature of the two moods elicited by these two scales, many wished to construct a minor scale which started with C as its root note (rather than A). In so doing, a musical piece might be able smoothly (and thus evocatively) to move from the darker atmosphere of the minor scale to the lighter, more robust nature of the major key.

Unfortunately, such a process presents a problem because the third note of the minor scale (the interval of the minor third) does not appear in a seven-note scale constructed with C as its root without the addition of an extra note just below E; the note that we now know as E-flat (E♭). By similar reasoning the notes of A-flat and B-flat need to be added too. This process of mapping the seven notes of the minor and major scales onto different root notes (a process referred to as transposition) also resulted in the invention of the extra notes D-flat and G-flat. The corresponding process of transposing the major scale to start from another root note – for instance from A – results in it being necessary to invent three more notes just above C (known as C-sharp or C♯), just above F and above G (known as F-sharp and G-sharp respectively).

Unfortunately, the note created just above G (G-sharp) by this process is not equivalent to the note produced just below A (A-flat) which was created in the practice of transposing the minor scale to begin on C as a root note! The same is true for C-sharp and D-flat, for F-sharp and G-flat and so on. In each case these notes are similar but not identical. This complication produced a very enigmatic situation for 300 years and one which, in some degree, persists to this day. Its (partial) resolution involved the radical step of treating each of the non-equivalent note pairs as equal and abandoning a scale based upon natural ratios except for the relationship of the octave. Thus 12 new notes were created in the series as shown in Table 2.3.

This new series is known as the equal-tempered, chromatic scale. Each note in the series being approximately 1.0595 times the frequency of the

Table 2.3

Note	Interval	Frequency (non-standard)	Ratio of a	Ratio of b
C	unison	100	1.00a	0.5b
C♯ (D♭)	minor second	105.95	1.0595a	0.52975b
D	major second	112.25	1.1225a	0.56125b
D♯ (E♭)	minor third	118.92	1.1892a	0.5946b
E	major third	125.99	1.25599a	0.62995b
F	perfect fourth	133.48	1.3348a	0.6674b
F♯ (G♭)	augmented fourth	141.42	1.4142a	0.7071b
G	perfect fifth	149.83	1.4983a	0.74915b
G♯ (A♭)	minor sixth	158.74	1.5874a	0.7937b
A	major sixth	168.18	1.6818a	0.8409b
A♯ (B♭)	minor seventh	178.18	1.7818a	0.8909b
B	major seventh	188.77	1.8877a	0.94385b
C	octave	200.00	2.00a	1.00b

note below. If you compare the values in this table with those in Table 2.2, these 'new' notes enjoy interesting relationships with the notes of the natural scale codified by Pythagoras. Note particularly that the perfect fourth and the perfect fifth are only in 0.1% error with their natural counterparts – a fact that has great significance in tonal music because of the close relationship these notes possess. The minor and major thirds, however, and the major sixth and minor seventh are all nearly 1% in error!

In spite of the tuning compromises, the development of this method of scales and tuning permitted composers to exploit all 24 minor and major keys (Aeolian and Lydian modes) starting on any note; a reality which was exploited by J.S. Bach in his set of preludes and fugues known as the 'Well Tempered Clavier' (24 in all – one in each of every possible key). The invention of equal-tempered tuning having been made, the scene was set for the flowering of classical Western art-music; a form which sets great store in the ability to move between different keys for its expressive and dramatic potential.

It is noteworthy that Pythagoras' natural scales and modes are still with us in the music of jazz, rock and in twentieth-century classical music. Furthermore, while the employ of the equal-tempered, chromatic tuning is mandatory for pianos, organs and other polyphonic instruments, it is not for instruments which cannot produce simultaneous notes of different pitch (so-called monophonic instruments) and for instruments which may vary their pitch continuously under the influence of the player. All the orchestral string family are of this type, as is the trombone and many others in which – even though they have keys – the player is able to produce a band of pitch possibilities around a particular note. Despite the convenience which equal-temperament confers, it may be observed, for instance, that singers and musicians playing instruments of continuously variable pitch consistently play the minor sixth in the key of C minor (the note A-flat) higher (sharper) than they pitch the major third in the key of E major (the note G-sharp). In other words, they tend to revert to natural scales – these feeling more intuitively 'right'. The Western system of music and musical notation acknowledges this phenomenon in its perpetuation of the system of sharps and flats – a procedure which appears to attribute two names to the same note, and one which has perplexed and flummoxed many a novice music student!

The harmonic series

Figure 2.6a illustrates a stretched string fixed firmly at either end. If this string is plucked at its centre point it will vibrate predominantly as shown in 2.6b. If the string is touched gently at its centre point and plucked

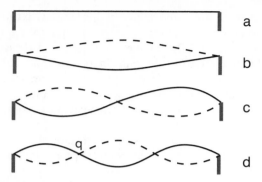

Figure 2.6   *Harmonics in a string*

toward one end or the other, it will tend to vibrate as shown at 2.6c. And if it is touched at point q, and again plucked towards the end, it will vibrate as shown at 2.6d. This pattern is theoretically infinite; each time the wave pattern on the string gaining an extra half-wavelength. Suppose that the vibration of the open string (as shown at 2.6b), produced the musical note C three octaves below middle C, the subsequent notes follow a distinct and repeatable pattern of musical intervals above the note of the open string. These notes are shown in Table 2.4.

Or, in musical notation:

Figure 2.7   *The harmonic series*

And they are termed the harmonic series. (Indeed it is the presence of these extra harmonics which distinguished the guitar waveform from the pure sine tone illustrated above. For it is impossible, when plucking a string, not to elicit some vibrational modes of the harmonics as well as the mode of the fundamental.) A similar pattern is obtainable from exciting the air within a tube, either by 'blowing a raspberry' into one end of the tube; as is done in the case of most brass instruments, or by exciting the air within the tube by means of a slit which produces turbulence, as is the

Table 2.4

Harmonic	Musical note	Comment
Fundamental	C	
2nd (1st overtone)	c	octave
3rd	g	twelfth (octave + fifth)
4th	c'	fifteenth (two octaves)
5th	e'	seventeenth (two octaves + major third)
6th	g'	nineteenth (two octaves + perfect fifth)
7th	b-flat' (nearest note)	dissonant; not in natural scale
8th	c"	three octaves
9th	d"	major 23rd (three octaves + second)
10th	e"	major 24th (three octaves + third)
11th	f" (nearest note)	dissonant; not in natural scale
12th	g"	major 26th (three octaves + fifth)
13th	a" (nearest note)	dissonant; not in natural scale
14th	b-flat"	dissonant; not in natural scale
15th	b"	major 28th
16th	c'''	four octaves
17th	C#'''	dissonant; not in natural scale
18th	d'''	major 30th
19th	d#'''	dissonant; not in natural scale
20th	e'''	major 31st

case with a pipe organ or recorder, or by means of a vibrating reed, as is the case with the oboe or the clarinet. In each case the physics of the tone production is different, but the harmonic series is always the same. Take time to look at the harmonic series and try to get a feeling for the musical relationships between the fundamental note and its harmonics. If you have a piano, play the series to yourself to familiarise yourself with the sound or feel of each of the harmonics and their relationship with the fundamental. So much of what follows relates time and time again to this series and to its physical, acoustical and psychological aspects that one can hardly overstate its importance to our craft. Indeed you might say this series is to the engineer-musician what the periodic table is to the chemist! There is, however, one word of warning, and that concerns the different nomenclature adopted by electronic engineers and musicians in describing the terms of this series. We shall see later on that this same series is produced as a result of amplitude non-linearity in electrical circuits; so-called harmonic distortion. But the engineer refers to the second note of the series as the second harmonic. For instance, an engineer will say its presence betrays second-harmonic distortion. Whereas the musician will refer to this note as the first harmonic in the series; the first note being referred to as the fundamental.

Measuring sound

The decibel

The decibel is very widely used in electrical and electronic engineering. Essentially the decibel or dB (named after Alexander Graham Bell – the inventor of the telephone) is used to express a voltage or current ratio or power ratio. For example, two AC signal voltages V_a and V_b are related in amplitude like this:

$20.\log(V_a/V_b)$ (dB)

Two AC signal currents I_a and I_b are related in amplitude like this:

$20.\log(I_a/I_b)$ (dB)

And two powers (P_a, P_b) are related thus:

$10.\log(P_a/P_b)$ (dB)

The essential thing to remember about the use of decibels is that they may only be used when they express a ratio. It's simply meaningless, as one sometimes sees in the popular press, to refer to a sound source as 100 dB. In fact, whenever you see dB, it's good practice to ask yourself, 'dB relative to what?' In the context of this book, decibel (dB) ratios are common in relation to a fairly limited set of standard quantities. These are listed below. Note that it is becoming good practice to append a letter (or letters) to the term dB to indicate the standard to which the ratio is made.

dBSPL: Power ratio used to express sound level relative to 10^{-12} watts per square metre (W/m^2) or a pressure ratio relative to the equivalent pressure of $20\,\mu Pa$ (20 micropascals); this power (and pressure) being taken as the absolute lower threshold of hearing for a 1 kHz sine-wave sound. In other words, a 0 dBSPL signal is just audible.

dBFS: Voltage level (or numerical relationship) of a signal intended for conversion to, or from, the digital domain relative to the maximum code or full scale (FS) value.

dBm: Power ratio relative to 1 milliwatt. This term is very often misused, for it is often quoted to describe a voltage in relation to 0.775 V RMS; in which case the term dBu should be used. The confusion arises because of the gradual abandonment of 600 Ω terminating impedances for audio lines. Originally signal levels were quoted relative to 1 milliwatt. And a signal of 0.775 V RMS does indeed produce 1 milliwatt of dissipation in a 600 Ω load. A signal of twice this value (in voltage terms) would produce four times the power in a standard load, hence it would be described as

$10.\log(4/1)$

or 6 dB more powerful than the standard signal, i.e. +6 dBm. With the gradual abandonment of 600 Ω lines, signals are more often compared in voltage level terms, but the term dBm is still wrongly applied when it simply relates to a signal voltage comparison.

dBu: Voltage ratio relative to 0.775 V RMS (see dBm).

dBV: Voltage ratio expressed in relation to 1 V RMS.

dB0VU: Current ratio relative to the current required to maintain a steady zero reading on a VU level meter (see Chapter 12 for more information on VU meters and reference levels).

The phon
The phon is used to express loudness, that is the perceived intensity of any sound – however complex. This is achieved by referring the sound in question to a sine-wave source of known intensity. The procedure is this: the sound is presented alternately with a 1 kHz sine-wave source. The listener is required to adjust the level of the sine-wave source until the perceptible intensity is the same as that for the complex sound. The value of the intensity of the 1 kHz tone is then used directly as the value in phons. So that if, for instance, the sound of a passing car is judged to peak at an equivalent level as an 80 dBSPL tone, the sound is said to have a loudness of 80 phons. Some common noise levels are quoted in Table 2.5.

Table 2.5

Phons	Noise sources
140	Gunshot at close range
120	Loud rock concert, jet aircraft taking off
100	Shouting at close range, very busy street
90	Busy city street
70	Average conversation
60	Typical small office or restaurant
50	Average living room; quiet conversation
40	Quiet living room, recording studio
30	Quiet house in country
20	Country area at night
0	Reference level

The mel

The mel is the unit of pitch. The pitch of a 1 kHz sine-wave tone (presented at 40 dBSPL) is 1000 mels. The pitch of any sound judged to be n times a 1 mel tone is termed n mels.

Radiation of sound

The comparison between sound waves spreading out in air and the ripples moving on the surface of water is frequently drawn. With good reason, for the similarity is very close. The essential similarity lies in that the condition (motion, elevation and depression of the wave) is transmitted from point to point by a medium which is comprised of particles (air or water molecules) which oscillate but ultimately do not move from their initial position. As Wood (1944) says:

> The illusion that the water is travelling with the waves may be very strong as we watch them, but every sea-swimmer knows that while the water at the crest is travelling with the waves, the water at the trough is travelling against the waves and there is no forward movement of the water as a whole.

Furthermore, sound waves exhibit all the properties easily observed in water waves; diffraction, interference, reflection and refraction and so on. Each of these properties is not so easy to observe in the case of sound waves. For a start, the medium in which they move is invisible! Second, the wavelengths involved are often very large. However, we are often sensible of each of these phenomena by way of a common observation which we may not intuitively attribute to wave motion. The exception, of course, is echo, which is a pretty obvious form of reflection; one in which the reflecting surface is a considerable distance away. However, reflection happens, to a greater or lesser extent, when a sound wave reaches any surface (Figure 2.8). As a general rule, hard, smooth surfaces reflect the

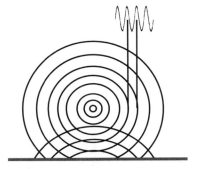

Figure 2.8   *Reflection of sound waves*

most, whereas soft, porous materials have the least reflecting power. Because the design of concert halls and recording studios often has to account for, or deal with, troublesome reflections, these latter surfaces are often employed deliberately – in which case they are referred to as good absorbers of sound. With the exception of outdoors, most everyday situations take place in rooms – large and small – where there is often a significant degree and number of acoustic reflections. These acoustic environments are classified as reverberant and a room with more reflections possesses more reverberation than a room with fewer reflections. The time it takes for the reverberation in a room to die away to one-millionth of the intensity of the original sound (-60 dB) is known as the reverberation time.

Refraction of sound is apparent in the phenomenon that sounds carry better in the quiet of the evening than in the heat of the day. This is because in the evening, when the air near the ground is cooler than the air above it, the sound waves are bent towards the earth (because sound travels faster in warm air). In contrast, in the day, when the air near the ground is hotter than the air above it, the sound waves are bent upwards away from our ears.

Constructive and destructive interference is apparent to us in the phenomenon known as beats wherein two notes of nearly equal pitch are sounded simultaneously. Because the alternate compressions and rarefactions from each of these sound sources gradually move in time relation (phase) with one another, the sound waves periodically add (reinforce) and subtract (destructively interfere) in the air, so we become aware of a slow alteration of sound and silence.

Diffraction (Figure 2.9) is such a common phenomenon that most of us would only be aware of it if sound were not to diffract! The easiest way to illustrate diffraction in everyday experience is to try to imagine a world in which this was not a common experience. (This would be the case if the sounds we experience were of a much higher frequency – and therefore of shorter wavelength.) In such a world, sound would travel in straight lines

Figure 2.9　*Diffraction of sound waves*

and not be diffracted by objects of everyday size (just like light), so, if you opened your front door, the sound from outside would stream in but would not 'spread out' to fill the room. Controlled diffraction from a slot is sometimes employed in the design of loudspeakers as we shall see in Chapter 14. In fact my imaginary world isn't actually all that different because it is the case that low-frequency sounds are much more easily diffracted than high-frequency sounds. This is due to their longer wavelengths. High-frequency sounds do have a tendency to be directional which is why loudspeakers seem to have less treble when you leave the room.

The near and far field

In likening a sound wave to a ripple spreading out on a lake we have to make a distinction between the area immediately surrounding a sound source, the so-called near-field, and the far-field well away from the sound source. In the near-field, sound pressures may not be directly proportional to sound intensity and the particle velocity of the air molecules is not always aligned with the direction of propagation of the wave. In short, the sound source itself disturbs the nature of sound propagation in this proximate region. It is only in the far-field, well away from the sound source, that one can speak of the radiation of a point sound source (like a loudspeaker) as spherical waves of sound emanating into free space. In the far-field particle velocity is always in the direction of propagation (assuming the absence of interfering objects).

The distinction between the near-field and the far-field is important in the context of monitoring loudspeakers and microphones. However, in this context the meaning is slightly different. Here it is used to distinguish the usage of loudspeakers and microphones in reverberant acoustic conditions. When people listen to a loudspeaker in a room they usually listen at a distance of several metres. Because of this they hear a combination of sounds; the sound directly from the loudspeaker and reflections from walls, windows, furniture and so on. This is illustrated in Figure 2.10. This situation has many benefits because the reflections add to produce a greater sound intensity for the listener. However, these

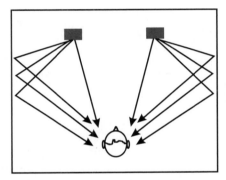

Figure 2.10 *Free-field or far-field sound source*

reflections, despite their superficial, constructive qualities, have the effect of 'smearing' or 'muddying' the sounds, because they arrive later than the original sound. This is inevitable unless the room is very carefully designed to have a high degree of sound absorption and therefore a short reverberation time. The modification or design of a room to possess these characteristics is inevitably expensive. However the sound engineer has no option but to choose the listening conditions which ensure the clearest possible reproduction in order to make consistent technical and artistic judgements. So, when large loudspeakers are required for monitoring, recording studio control rooms are carefully constructed so that the reflections do not 'interfere' with the sounds from the remote loudspeakers themselves. There is, however, a cheaper alternative – known as near-field monitoring – where the loudspeakers are sited close to the listening position; often on the console meter-bridge. The idea being that the relatively intense signal directly from the loudspeaker predominates over the reflected sound which is forced, by careful choice of loudspeaker placement, to have a very indirect route to the listening position. This is illustrated in Figure 2.11.

Near-field monitoring is gaining acceptance as the preferred recording technique firstly because it is cheaper than the alternative, but also, due to limited housing space and shorter reverberation times (because of the presence of fitted carpets and soft furnishings), listening conditions in the home more nearly approximate the near-field situation.

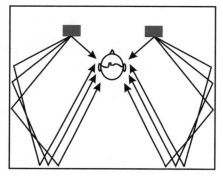

Figure 2.11   *Near-field conditions*

The corollary of near-field monitoring is near-field microphone technique. The idea being, once again, to suppress the effect of the acoustics of the room on the signal: in this case, the signal received by the microphone. This is accomplished by placing the microphone as close as possible to the original sound source. Classical microphone technique has 'shied away' from this method and it is for this reason that recording venues for classical music must be more carefully selected than those for rock and pop recording where near-field (or close) microphone technique is the norm.

The physiology of hearing

Studies of the physiology of the ear reveal that the process of Fourier analysis, referred to earlier, is more than a mere mathematical conception. Anatomical and psychophysiological studies have revealed that the ear executes something very close to a mechanical Fourier analysis on the sounds it collects and passes a frequency domain representation of those sounds onto higher neural centres. An illustration of the human ear is given in Figure 2.12.

internal and middle ear

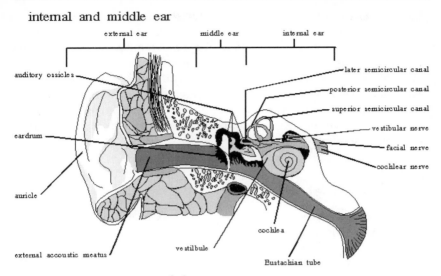

Figure 2.12 *Anatomy of the ear*

After first interacting with the auricle or pinna, sound waves travel down the auditory canal to the eardrum. The position of the eardrum marks the boundary between the external ear and the middle ear. The middle ear is an air-filled cavity housing three tiny bones: the hammer, the anvil and the stirrup. These three bones communicate the vibrations of the eardrum to the oval window on the surface of the inner ear. The manner in which these bones are pivoted, and because the base of the hammer is broader than the base of the stirrup, there exists a considerable mechanical advantage from eardrum to inner ear. A tube runs from the base of the middle ear to the throat, this is known as the Eustachian tube. Its action is to ensure equal pressure exists on either side of the eardrum and that it is open when swallowing. The inner ear is formed in two sections, the cochlea (the spiral structure which looks like a snail's shell) and the three semicircular canals. These later structures are involved with the sense of balance and motion.

The stirrup is firmly attached to the membrane which covers the oval window aperture of the cochlea. The cochlea is full of fluid and is divided along its entire length by the Reissner's membrane and the basilar membrane upon which rests the organ of Corti. When the stirrup moves, it acts like a piston at the oval window and this sets the fluid within the cochlea into motion. This motion, trapped within the enclosed cochlea, creates a standing wave pattern – and therefore a distortion – in the basilar membrane. Importantly, the mechanical properties of the basilar membrane change considerably along its length. As a result the position of the peak in the pattern of vibration varies depending on the frequency of stimulation. The cochlea and its components thus work as a frequency-to-position translation device. Where the basilar membrane is deflected most, there fire the hair cells of the organ of Corti, which interface the afferent neurones that carry signals to the higher levels of the auditory system. The signals leaving the ear are therefore in the form of a frequency domain representation. The intensity of each frequency range (the exact nature and extent of these ranges is considered later) is coded by means of a pulse rate modulation scheme.

The psychology of hearing

Psychoacoustics is the study of the psychology of hearing. Look again at Table 2.5. It tells us a remarkable story. We can hear, without damage, a ratio of sound intensities of about 1:1000 000 000 000. The quietest whisper we can hear is a billionth of the intensity of the sound of a jet aircraft taking off heard at close range. In engineering terms you could say human audition is equivalent to a true 20-bit system – 16 times better than the signal processing inside a compact disc player! Interestingly, the tiniest sound we can hear occurs when our eardrums move less than the diameter of a single atom of hydrogen. Any more sensitive and we would be kept awake at night by the 'sound' of the random movement of the nitrogen molecules within the air around us. In other words, the dynamic range of hearing is so wide as to be up against fundamental physical limitations. It tells us too that we can perceive the direction of a sound within about 1° of arc (for sounds in the lateral plane in the forward direction). Not as good as sight then (which can resolve about one minute of arc) but better than the illusion obtainable from two-loudspeaker stereophony. Psychoacoustics also informs us that two loudspeaker stereophony can be improved, and thereby give a more accurate spatial representation, by stereophonic enhancement techniques. (We shall see how this can be achieved in Chapter 11.)

Frequency masking

The cochlea and its components work as a frequency-to-position translation device, the position of the peak in the pattern of vibration on the basilar membrane depending on the frequency of stimulation. It goes without saying that the position of this deflection cannot be vanishingly small, it has to have some dimension. This might lead us to expect that there must be a degree of uncertainty in pitch perception and indeed there is, although it is very small indeed, especially at low frequencies. This is because the afferent neurones which carry signals to the higher levels of the auditory system 'lock on' and fire together at a particular point in the deflection cycle (the peak). In other words a phase detection, frequency discriminator is at work. This is a truly wonderful system, but is has one drawback. Due to the phase-locking effect, louder signals will predominate over smaller ones, masking a quieter sound in the same frequency range. (Exactly the same thing happens in FM radio where this phenomenon is known as capture effect.) The range of frequencies over which one sound can mask another is known as a critical band, a concept due to Fletcher (quoted in Moore 1989). Masking is very familiar to us in our daily lives. For instance, it accounts for why we cannot hear someone whisper when someone else is shouting. The masking effect of a pure tone gives us a clearer idea about what is going on. Figure 2.13 illustrates the unusual curve which delineates the masking level in the presence of an 85 dBSPL tone. All sounds underneath the curve are effectively

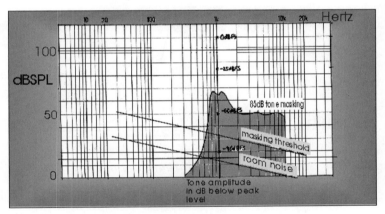

Figure 2.13   *The phenomenon of masking*

inaudible when the tone is present! Notice that a loud, pure sound only masks a quieter one when the louder sound is lower in frequency than the quieter, and only then when both signals are relatively close in frequency. Wideband sounds have a correspondingly wide masking effect. This too is illustrated in Figure 2.13 where you'll notice the lower curve indicates the room noise in dBSPL in relation to frequency for an average room-noise figure of 45 dBSPL. (Notice that the noise level is predominantly low frequency, a sign that the majority of the noise in modern life is mechanical in origin.) The nearly parallel line above this room-noise curve indicates masking threshold. Essentially this illustrates the intensity level, in dBSPL, to which a tone of the indicated frequency would need to be raised in order to become audible.

Temporal masking

Virtually all references in the engineering literature refer cheerfully to an effect known as temporal masking in which a sound of sufficient amplitude will mask sounds immediately preceding or following it in time; as illustrated in Figure 2.14. When sound is masked by a subsequent signal the phenomenon is known as backward masking and typical quoted figures for masking are in the range of 5 to 20 mS. The masking effect which follows a sound is referred to as forward masking may last as long as 50 to 200 mS depending on the level of the masker and the masked stimulus.

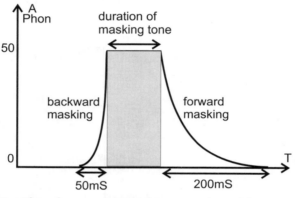

Figure 2.14   *The phenomenon of temporal masking*

Unfortunately, the real situation with temporal masking is more complicated and a review of the psychological literature reveals that experiments to investigate backward masking in particular depend strongly on how much practice the subjects have received – with highly practised subjects showing little or no backward masking (Moore 1989). Forward masking is however well defined (although the nature of the underlying process is still not understood) and can be substantial even with well practised subjects.

The phenomenon of masking is important to digital audio compression as we shall see in Chapter 10.

References

Moore, B.C.J. (1989) *An Introduction to the Psychology of Hearing* (3rd edition). *Academic Press.*

Wood, A. (1944) *The Physics of Music.* Methuen & Co. Ltd.

Notes

1 The phase of a sine-wave oscillation relates to its position with reference to some point in time. Because we can think of waves as cycles, we can express the various points on the wave in terms of an angle relative to the beginning of the sine-wave (at 0°). The positive zero crossing is therefore at 0°, the first peak at 90° etc.
2 Although I have explained this in terms of voltages, clearly the same principle obtains in a digital system for symbolic numerical magnitudes.

Fact Sheet #2: AC circuits

● Capacitors
● Inductors
● LC circuits

Capacitors

The simplest form of a capacitor is two metal plates separated by empty space. In fact the symbol for a capacitor is intended to depict just that, see Figure F2.1a. However a practical capacitor made like this would have a very low value unless the plates were the size of a house, so practical capacitors are often made by means of plates wound up together and separated by an insulator known as a dielectric. You might well ask, what use is a component with two conductors separated by an insulator? How can a current pass through such a device?

We've already seen (in Fact Sheet 1) how electrical current can 'jump across space' in a transformer by means of a magnetic field. A very similar effect happens in a capacitor; except the transfer of current transpires due to an electrical field which, once again, must be changing. In fact in a capacitor, the faster the current changes (i.e. the higher the frequency), the better the transfer of current. You might say that a capacitor is a device that changes its resistance to current on the basis of how fast that current changes but in fact resistance would be the wrong word. Instead you have to use the word reactance or impedance.

This is the crucial property which makes capacitors useful, they can be used to select (or reject – depending on their circuit position) a range of frequencies. A simple example is a tweeter crossover circuit, which can be as simple as a single capacitor in

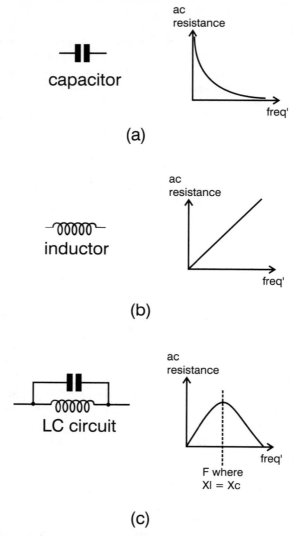

Figure F2.1 *The phenomenon of impedance*

series with the loudspeaker as shown in Figure F2.2. Here the capacitor presents a high impedance to the flow of current at low (bass) frequencies; thereby preventing these signals from reaching and damaging the tweeter unit. But at high frequencies the capacitor presents very little impedance to the flow of current and thereby allows these components to excite the speaker unit.

Inductors

It's a common enough observation that it takes a lot of energy to get something going and a lot of energy to stop it once it's got rolling. Physicists refer to this property as inertia. It turns out that inductors are much the same.

An inductor is superficially similar to a transformer, but it has only one winding (hence the schematic symbol Figure F2.1b) and its circuit application is therfore completely different. Essentially its role in a circuit concerns the property mentioned above which might be expressed thus: when a current tries to flow in an inductor, there is an initial reluctance to this happening. Similarly once current is flowing in an inductor, it's hard to stop it! This effect is actually referred to as reluctance.

This effect is especially useful in filtering applications because, if an AC current changes direction relatively slowly, the inductor presents no impedance to the flow of electricity. If however the AC current changes direction rapidly, the reluctance of the inductor will prevent the current from passing through the winding. In this respect the inductor should be compared with a capacitor, which demonstrates an exactly reciprocal effect – allowing high frequencies to pass but prohibiting the passage of low-frequency alternating currents. With this in mind look again at Figure F2.2, here you will see that the bass loudspeaker is fed from the power amplifier via an inductor which obstructs the flow of high-frequency signals.

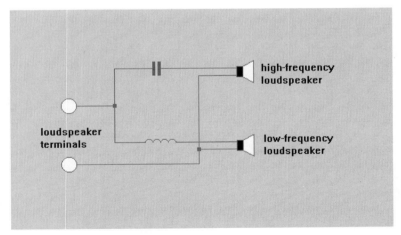

Figure F2.2 *Loudspeaker crossover circuit*

LC circuits

We've seen how the impedance (AC resistance) of a capacior and inductor varies with frequency. The impedance of a capacitor falls with frequency as shown in Figure F2.1a; that's why it was used to block low frequencies in the crossover circuit of Figure F2.2. On the other hand, an inductor's impedance rises with frequency; as shown in Figure F2.1b too. This is why the inductor was put in series with the bass loudspeaker (the woofer) in Figure F2.2; to block the high frequencies from reaching the bass loudspeaker.

Now something really interesting happens when an inductor and capacitor are used together. At the frequency at which their impedances are the same, they resonate and can produce a very large output on the basis of a small input (Figure F2.1c). LC circuits are used in audio equalisers where they are set to resonate over a particular band of frequencies; other circuit elements being so arranged as to cause the LC circuit to subtract or boost a particular range of frequencies.

3
Stand By Me – Microphones and their applications

Principles

A microphone is a device for turning acoustic energy, in the form of sound, into electrical energy, a similar function to that performed by the ear. Fortunately, the process is far simpler than that described for the ear because, instead of transforming the sound vibrations into the frequency domain and coding the intensity and low-frequency phase information into a pulse-density bitstream, the microphone simply converts the vibrations in the air into analogous electrical vibrations within an electric circuit. The signal is thus an analogy of the original sound, and circuits which continue to treat the signal in this manner are known as analogue circuits. There are, as yet, no direct digital transduction processes between acoustic and electric energy so analogue circuitry and analogue processes still play an important part in the amplification and transmission of audio information.

Microphones may be catalogued in a number of ways. First, by way of the physical quantity which they transduce; one type converts the pressure of a sound wave into electrical voltage and the second type converts the sound wave's velocity into electrical voltage. Second they may be characterised by the manner in which they convert one type of energy to another. In this category there exist two important classes: moving coil and electrostatic.

Pressure microphones

A pressure microphone converts acoustic pressure to electrical energy. Because pressure is a scalar quantity, a pressure microphone is theoretically non-directional (meaning that it is equally sensitive to sounds falling

41

upon it from any direction.) The pressure microphone is normally fashioned so that a pressure-sensitive element is coupled to a diaphragm which is open to the air on one side and sealed on the other. In this manner, it most clearly mimics the mechanical arrangement of the human ear. Indeed the ear too is a pressure sensitive transducer. In practice it is not only unnecessary to transduce the overall barometric pressure (i.e. the 0 Hz signal) but essential for the correct operation of the microphone that the ambient pressure on both sides of the microphone's diaphragm remains the same. A normalising vent is therefore always included behind the diaphragm of such a microphone. Essentially the human hearing system functions the same way by means of the Eustachian tube, by which means pressure can be equalised either side of the eardrum. Like the ear the theoretically non-directional response of any practical microphone is modified by the interaction of incident sound with the physical mounting. Nature has sought to maximise this interaction by means of the external ear or pinna whereas most microphone designers seek to reduce it to a minimum. Nevertheless a pressure microphone is characterised by an omnidirectional response which is nearly perfect at low frequencies and which deteriorates to become more directionally sensitive at high frequencies.

Pressure gradient or velocity microphones

The diaphragm of a pressure gradient or velocity microphone is open to the air on both sides so the force upon it is directly proportional to pressure difference. Sounds falling on the diaphragm from directly in front and directly behind generate the largest electrical signal. Sounds which fall on the microphone from angles away from this axis elicit gradually decreasing electrical signals from the microphone until a point where, if a sound arrives at the diaphragm from an angle which is on a line with the axis of the diaphragm itself, no electrical signal is generated at all. This double-lobed characteristic is known as a figure-of-eight polar response.

Despite the fact that the omnidirectional, pressure-sensitive microphone and the bidirectional, velocity-sensitive microphone offer a choice of directional characteristics, in many applications neither is entirely appropriate. Often what is required is a microphone with a unidirectional response which may be positioned in favour of transducing one particular and desired sound source while discriminating against others. Just such a response is obtainable by combining two microphones within the same housing, one pressure sensitive and the other velocity sensitive. By electrically summing their outputs a unidirectional or cardioid response is obtained. Professional microphones

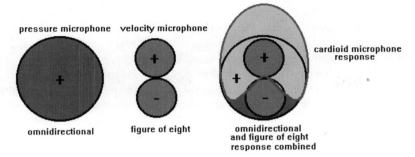

Figure 3.1   *Microphone polar patterns*

used in television and multimedia are often of this type. Figure 3.1 illustrates the manner in which a microphone's directional characteristic may be specified in terms of a polar diagram and also the way in which a unidirectional response may be obtained by a combination of omnidirectional and figure-of-eight response.

Transduction method

The most common type of microphone is the moving coil type which works just like a small loudspeaker in reverse (see Figure 3.2). Inside a moving coil microphone, sound waves mechanically excite a small diaphragm and a coil of wire – which is mechanically coupled to the diaphragm. The coil of wire is contrived to sit in a magnetic field so, once it is set into motion, a small voltage is generated across the coil. This tiny voltage generator has the advantage that it has a very low impedance (of

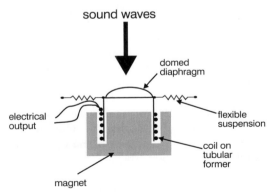

Figure 3.2   *Moving coil microphone*

a few turns of wire at audio frequencies) so it is 'stepped-up' using a transformer housed within the microphone case. Final output impedance is in the region of 600 Ω. Moving coil microphones are very robust and are excellent for general purpose and hand-held usage.

The second most widespread microphone type is the electrostatic type, often simply dubbed capacitor microphone or, rather anachronistically, condenser type. In this variety, the diaphragm – upon which falls the incident sound – is actually one plate of a capacitor (see Figure 3.3). By allowing one plate to move very easily, the capacitance value may be altered by the action of sound waves. Relying on the relationship $Q = CV$ which may be re-written $V = Q/C$, a voltage change may be detected across the plates of a changing capacitance provided a mechanism is in place to keep the charge constant. This requirement is usually fulfilled by

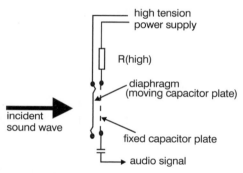

Figure 3.3   *Electrostatic microphone*

means of a constant high tension polarising voltage which is supplied across the plates of the capacitor via a high value resistor. The small alternating signal voltage is taken from one or other of the metallic faces of the element via an isolating capacitor. Unlike the moving coil type, the voltage leaving the plates is not only very small but is also generated across an extremely high (theoretically infinite) impedance. This naturally precludes the use of a transformer to step up the voltage to a level suitable for transferring to the microphone mixer via a length of cable. Instead a small voltage amplifier must be included within the microphone case – usually with an FET (or occasionally vacuum tube) input stage providing the necessary very high input impedance. The two requirements – for a high-tension supply to provide the capacitor-element's polarisation voltage and for power, to energise the FET pre-amplifier – explain why capacitor microphones demand rather more of ancillary equipment than do moving-coil microphones. Nevertheless despite this and their other

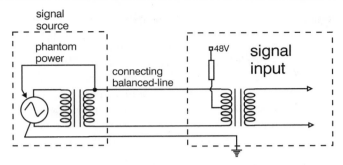

Figure 3.4 *Phantom power arrangement*

drawbacks, such as mechanical fragility, capacitor microphones dominate the professional recording scene because of their generally superior qualities. Professional recording mixers provide the necessary energy to capacitor microphones in the form of a 48 V, phantom supply which runs to the microphone via the signal cable isolated from the signal electronics via capacitors or signal transformers as illustrated in Figure 3.4.

Practical cardioid microphones

Although it is perfectly possible to create a cardioid microphone using a pressure and velocity microphone combined (and several commercial microphones have employed this technique), practical designers of cardioid microphones more often achieve a unidirectional response by applying a technique known as acoustic delay. This method relies on sound reaching the back of the microphone diaphragm after it reaches the front. Sometimes this is as simple as an aperture towards the end of the microphone body as shown in Figure 3.5. Other designs employ acoustic 'windows' covered with material of carefully selected mass and compliance to create an acoustic delay line. However, in either case the operation is the same and this is also illustrated in Figure 3.5. Essentially

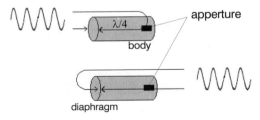

Figure 3.5 *Delay derived cardioid microphones*

mid-frequency sounds which arrive from the front are phase reversed, by way of their journey along the outside and back along the inside of the microphone body, and so reinforce the acoustic signal present at the diaphragm. On the other hand, mid-frequency sounds which fall upon the microphone from the back, travel similar distances to both the front and rear of the diaphragm, and so cancel, resulting in little or no output. Such a device cannot be employed to obtain a unidirectional response at high frequencies but, fortunately, the interaction of these small-wavelength sounds with the microphone housing itself is enough to obtain a directional response in this response range. At low frequencies, the response of such a microphone tends towards the response of a pure velocity operated type.

Pressure zone microphone (PZM)

In many recording applications, the recording microphone must, of necessity, be placed near a hard, sound-reflecting surface. Two classic examples of this are (a) a speaker standing at a lectern and (b) a piano which must be recorded with the lid raised but not removed. In these two situations (and indeed many others), the sound which arrives at the microphone arrives via two routes; directly from the sound source and reflected by way of the lectern or the piano lid. This results in a series of peaks and troughs in the frequency response (termed coloration) due to the second path length selectively either reinforcing or cancelling the direct sound, dependent upon wavelength. Figure 3.6 illustrates this phenomenon and Figure 3.7 shows the advantage of the PZM microphone

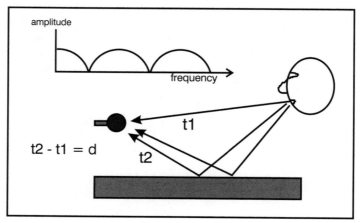

Figure 3.6 *Comb-filter coloration effect*

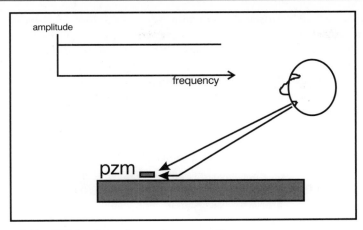

Figure 3.7 *PZM microphone in operation*

in such an application. PZM stands for Pressure Zone Microphone and such a microphone is designed so that a miniature capacitor microphone is so arranged that its diaphragm is parallel and very close to the sound-reflecting surface itself. This bestows a number of advantages. First and foremost that the direct sound and reflected sound no longer alternately reinforce and cancel but are instead effectively in-phase and therefore reinforce over the entire audio band. The PZM thereby ensures a smooth, natural frequency response in such an application. There is as well a 6 dB increase in sensitivity due to the coherent addition of direct and reflected sound. A commercial PZM microphone is illustrated in Figure 3.8.

Electret microphones

An important subclass of electrostatic microphone, named the electret type, makes use of a very light plastic diaphragm which is subjected to a

Figure 3.8 *Practical PZM microphone*

very strong electrostatic field during production and so remains permanently charged thereby obviating the requirement for a polarising voltage. Many modern incarnations of this type, despite their low price, are capable of a performance approaching genuine capacitor types. However, a pre-amplifier is still required and this must be energised, usually by means of a battery which is designed to sit within the microphone body.

Microphone usage – basic issues

The choice of microphone in any situation is governed by one very simple rule:

> Choose and position a microphone so that an adequate signal level is obtained for the wanted acoustic source while ensuring that the position and choice of microphone discriminates, as much as possible, against unwanted noises.

In every application where a microphone is required, this rule applies – whatever the sound source and unwanted sound may be. For example, suppose a narrator is to be recorded in a small uncarpeted room. In such a situation a very 'dry' sound is required; that is to say a sound quality lacking reverberation. In this case, the narrator's voice is desired, the 'sound' of the room is not. Moreover, due to lack of carpeting, the room probably has a hollow, ringing acoustic. In this case a cardioid microphone is essential, placed near enough to the narrator's lips that an adequate signal is returned to the mixing desk and so as to discriminate against the sound of the room. But not so near that plosive sounds are allowed to produce a blasting or popping sound from the microphone. (Plosive sounds are p's and b's and d's and t's which are accompanied by blasts of air as the lips interrupt the steady flow of air from the lungs. During the production of these sounds, known as vocal stops, the flow of air is abruptly resumed and it is this burst of air which, if allowed to reach the microphone diaphragm, produces a wideband 'popping' noise.)

If, on the other hand, the voice to be recorded is that of a singer in a quiet chapel, the choice of microphone will be quite different. In this case, the 'sound' of the acoustic surroundings is essential to the required sound quality; so a microphone is chosen which has either a figure-of-eight, or omnidirectional response. Although, once again, the practice of placing the microphone at a suitable working distance to avoid popping is valid. Indeed, the adoption of a reasonably remote microphone position (0.5 to 1 metre), aids greatly in reducing the dynamic range of a vocal signal. If working distances are much less than this, it will be found that the ratio of loud vocal tones to soft ones is very marked indeed – especially among actors who are trained to modulate the expression of their voices. It is worth noting that rock and pop vocalists tend to use smaller working distances in order to capture a more intimate, sensual vocal style. While clearly justifiable on artistic grounds, this practice has the incidental disadvantage that dynamic contrasts are often so marked that the signal cannot be adequately recorded and mixed with the instrumental tracks without the use of an electronic dynamic compressor which automatically reduces the dynamic range of the microphone signal. (Dynamic compressors are covered in Chapter 6.)

Proximity effect or bass tip-up

A further problem with small working distances (sometimes known as close-miking) is proximity effect or bass tip-up. This is a phenomenon only associated with directional (velocity or cardioid) microphones when used very close to a sound source whereby, as the distance between microphone and source decreases, so the proportion of low-frequency sound is exaggerated in comparison with middle and high frequencies. This is due to a situation in which the microphone is being used in the near-field (as described in the last chapter) where sound pressures may not be directly proportional to sound intensity and the particle velocity of the air molecules is not always aligned with the direction of propagation of the wave. In this region, the transduction of velocity in a directional microphone produces an anomalous, bass-heavy sound quality. This effect is sometimes exploited for artistic effect but, equally often, it is ameliorated by the use of low-frequency equalisation, either within the mixing console or (if the manufacturer has been thoughtful and provided a low-frequency roll-off switch) within the microphone itself.

Microphone technique

This section might justifiably occupy an entire book in its own right, because the craft of microphone choice and placement – while a science – is also an art. In other words, there exists no absolutely right way of miking anything. Some of the world's most arresting records have been made because a musician, an engineer or a producer had the courage to break someone else's arbitrary rules. Experimentation is the name of the game! However, there exist some sensible starting points and these are described below. The first question to ask when faced with recording a group of singers and/or instrumentalists is, 'should I use a few microphones or a lot of microphones?' In fact, usually the question is answered for us, because the different techniques underwrite entire musical genres. Nonetheless a good recording engineer should always keep an open mind.

Many microphones

In pop and rock music – as well as in most jazz recordings – each instrument is almost always recorded onto a separate track of multi-track tape and the result of the 'mix' of all the instruments combined together electrically inside the audio mixer and recorded onto a two track (stereo) master tape for production and archiving purposes. (The process and practice of multi-track recording is dealt with fully in Chapter 12.) Similarly in the case of sound reinforcement for rock and pop music and jazz concerts, each individual musical contributor is separately miked and the ensemble sound mixed electrically. In each case, the recording or balance engineer is required to make many aesthetic judgements in the process of a recording. Her/his main duty is to judge and adjust each channel gain fader and therefore each contributor's level within the mix.

A further duty, when performing a stereo mix, is the construction of a stereo picture or image by controlling the relative contribution each input channel makes to the two stereo mix amplifiers. In the cases of both multi-track mixing and multi-microphone mixing, the apparent position of each instrumentalist within the stereo picture (image) is controlled by a special stereophonic panoramic potentiometer or panpot for short. As we shall see in later chapters, stereophonic reproduction from loud-speakers requires that in each of these cases stereo information is carried by interchannel intensity differences alone – there being no requirement for interchannel delay differences. Consequently, the pan control progressively attenuates one channel while progressively strengthening the other as the knob is rotated, the input being shared equally between both channels when the knob is in its centre (12 o'clock) position.

A few microphones

Because rock musicians and producers hold great store in the creation of entirely artificial (synthetic) sound worlds, a multi-microphone technique is preferred. On the other hand, in a classical music recording, while it is possible in principle to 'mike up' every instrument within an orchestra and then – with a combination of multi-track and electronic panning – create a stereo picture of the orchestra, this is usually not done, partly because the technique is very costly and complicated and partly, because this 'multi-miked' technique has not found favour amongst classical music critics, all of whom agree that it fails to provide a faithful representation of the real orchestral experience. The exception to this is orchestral recordings for movie films which often do adopt this technique to generate a 'larger than life' sound from the orchestra. Obviously some logistical situations mitigate against the use of multi-microphone technique; that such a technique would not work for a recording of the dawn chorus should go without saying! As a result, most recordings of orchestras and choirs depend almost exclusively on the application of simple, or 'purist' microphone techniques where the majority of the signal that goes onto the master tape is derived from just two (or possibly three) microphones. Surprisingly there are no fixed rules as to how these main microphones should be arranged, although a number of popular deployments have evolved over the years.

To some extent, the way microphones are arranged achieves a certain character of sound. Often it betrays a 'house style'. For instance Deutsche Grammophon currently use two pressure zone microphones taped to huge sheets of perspex (an arrangement which is essentially the same as wide-spaced omnidirectional microphones much beloved by American recording institutions. This technique is illustrated in Figure 3.9. Each omnidirectional (pressure-sensitive) microphone simply transduces either the left-channel signal or the right-channel signal. Also notice that this technique captures the position of a particular instrumentalist by means of time differences between the signals arriving at the microphones.

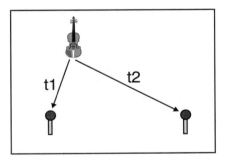

Figure 3.9 *Spaced-omni microphone technique*

British record companies have developed their own arrangements too, and hence their own house styles. Decca pioneered the eponymous Decca Tree while the BBC stuck almost exclusively to coincident crossed pairs (sometimes referred to as Blumlein stereo and illustrated in Figure 3.10) until relatively recently. Other companies, Philips for instance, use a combination of techniques. Notice that the coincident microphone technique encodes positional information by means of angled, directional microphones which capture intensity differences rather than time differences.

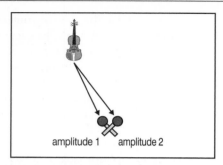

amplitude 1   amplitude 2

Figure 3.10   *Coincident microphone technique*

The problem with all these microphone arrangements is that each one fails (in different ways) to provide an absolutely satisfactory rendering of the sound field in which the microphones are immersed when the signals are replayed over loudspeakers. For example, the crossed pair technique probably produces the sharpest spatial picture but many recording engineers prefer the less accurate – but more natural – sound of spaced microphones. Unfortunately, until recently, a compromise seemed inevitable. In the process of producing a sharp and well defined stereo image, coincident microphones discard the time differences which do so much to convey to the listener the pleasant sensation of spaciousness that is apparent when spaced microphones are employed, even though spaced arrangements fail to produce a fully focused stereo image. However, recently new microphone techniques have been evolved which aim to combine the virtues of coincident and spaced microphones (Brice 1997), see Fact sheet #11 later in this text.

Reference
Brice, R. (1997) *Multimedia and Virtual Reality Engineering*. Newnes.

Fact Sheet #3: Signals
- The world of frequency
- Phase
- Signals in the time-domain and frequency-domain

The world of frequency

The phenomenon we came across in Chapter 2, that complex waveforms are composed of combinations of simple sine-waves, establishes a very important practical technique; that the response of an electronic circuit to very complex waveforms can be deduced from its behaviour when passing very simple signals. The only requirement being that a sufficient number of sine-waves at a sufficient number of frequencies are used to adequately determine the performance of the circuit.

The measurement technique which results from this line of reasoning is known as a frequency-response and it specifies the ability of a circuit to pass sine-wave signals at all frequencies. Of course, for practical applications, it's not necessary to measure (or specify) equipment at frequencies way beyond the frequency range of interest. For instance it's unnecessary to specify the response of an audio amplifier to radio frequency signals! However it's habitually the case that frequency responses are investigated with signals which extend beyond the serviceable frequency band for the equipment, this technique being utilised in order to establish an important parameter known as bandwidth.

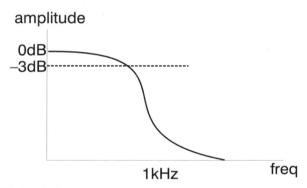

Figure F3.1   *A frequency response curve*

Take for instance the frequency response of the circuit shown in Figure F3.1. This was derived by testing the circuit in question with a signal of gradually increasing frequency and measuring the amplitude of the signals as they passed through the circuit. The results are expressed as a graph with the output shown relative to the input waveform (in a special ratio known as dB or decibels – see Chapter 2). As you can see, the circuit passes all low-frequency audio signals but prevents the passage of high-frequency signals. Note that the transition is not instantaneous; this is an effect seen in all real circuits. The band of frequencies passed by the circuit is termed the passband.

And the region where the frequencies are attenuated is known as the stop-band. The band in between being termed the transition region. The edge of the passband is usually designated as the point at which frequency-response has fallen by 30% or −3 dB.

From this data we can induce that complex signals, made from a combination of low and high frequency components, will be affected pro-rata; with the high-frequency signals 'stripped-off'. In subjective musical terms, we can therefore predict that such a circuit would sound 'woolly' with all the bass frequencies present and all the higher components (partials) missing. Such a circuit is known as a low-pass filter and the limit of the passband (−3 dB) is illustrated in the diagram too. Its alter-ego; a filter which passes only high frequencies and prevents the passage of low-frequency components is known as a high-pass filter. A combination of the two; sometimes formed by means of an LC circuit (see Fact Sheet 2) passes only a band of frequencies and is therefore termed a band-pass filter.

Phase

Actually, it's not possible to determine the entire performance of a circuit from its frequency response alone. For a complete description you have to know the phase response also. Phase response is another way of expressing delay vs. frequency; in fact phase response is sometimes defined in terms of change of phase in relation to change of frequency in which case it is called group-delay.

   i.e.,

Group delay = dP/dF

A non-distorting circuit is usually defined as having a linear phase response; which is another way of saying the group-delay at all frequencies is the same (because dP/dF is a constant).

If it were not, high-frequency signals arrive after low-frequency signals. In fact such a situation is a very common real-world occurrence and the results of this sort of phase distortion on a step waveform are illustrated in Figure F3.2. Note that in this particular example the relative amplitude of the low-frequency and high-frequency signals are not changed − only the time it takes for them to pass through the circuit.

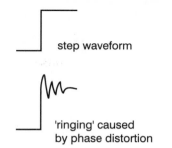

step waveform

'ringing' caused
by phase distortion

Figure F3.2 *'Ringing' on step waveform due to phase distortion*

For psycho-physiological reasons, the phase response of circuits is often not of prime importance in audio. A situation which has tended to exaggerate the importance of frequency response over phase-response (or group-delay) in this field. The same is not true in many other areas of electro-technology (analogue television and digital communications for example).

Signals in the time-domain and the frequency-domain

We have already met Fourier analysis in Chapter 2. The Fourier transform is the mathematical device which allows us to transform a time-based description of a signal into a frequency-based description of a signal (and back!). In this Fact Sheet we look at views of certain 'everyday' signals in these two different domains. All the graphs on the right represent a frequency-based representation of a signal; all those on the left, a time-based description of a signal.

In Figure F3.3a, a sine-wave is shown in familiar form on the left-hand side. As we already know a sine-wave is a pure sound of one particular frequency. This is illustrated on the right, where the same sound is represented as a single line in the entire frequency spectrum. Interestingly (and – if you're mathematically minded – have a look at the maths in Chapter 2 for the explanation), there's a symmetry in the relationship between time and frequency representation. So that, at the opposite extreme – look at Figure F3.3b – a single, very short pulse (known as a Dirac function). This translates into a continuous spectrum in the frequency domain. Somewhere in between (at c), the representations are the same in both domains; so that a train of pulses turns into a spectrum of infinitely many, spaced spectral components.

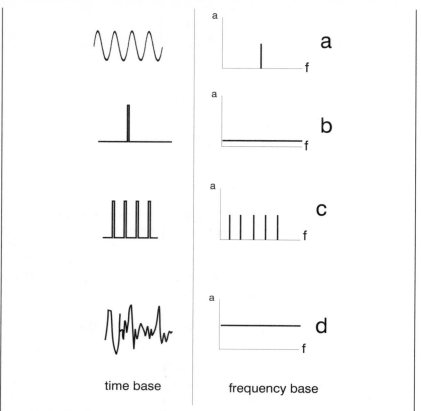

Figure F3.3   *Signals shown in the time and frequency base*

A further, very important waveform is illustrated at (d). This is known as white noise – a completely disorganised pattern in the time domain but a continuous spectrum (at least statistically) in the frequency domain. Other types of noise are discussed in Chapter 7.

4
Message in a Bottle – Valve technology

What is 'valve sound'?

Currently worldwide business in valves is an estimated several million dollars a year, most of which is in audio. But why have valves made a comeback? Many believe that digital music production and recording has a certain 'sterility' and the incorporation of valve circuitry in the recording chain assists in reducing this undesirable characteristic. Certainly, peculiarities of valve equipment and circuitry may influence the tone of the reproduced sound but it is necessary to observe the distinction, made in Chapter 2, concerning the differences between musical and recording/ reproduction electronic systems. Put bluntly, and despite widespread popular belief to the contrary, it is extremely unlikely that valves magically reveal details in the music obscured by mysterious and hitherto undiscovered distortion mechanisms inherent in solid-state electronics. However, what may be said with certainty, falling as it does squarely within the remit of electrotechnical science, is that valves do exhibit a number of characteristics which may perhaps (under certain auspicious circumstances) perform better than poorly designed semiconductor equipment or in a manner which beautifies or glamorises the reproduced sound.

Harmonic distortion

The first, and most obvious, peculiarity of valve circuitry is its inherent non-linearity. In applications such as amplifiers for electric instruments this almost certainly counts as a benefit; adding richness and brightness to the perceived tone. A simple triode amplifier has a curved transfer characteristic, like that illustrated in Figure 4.1.

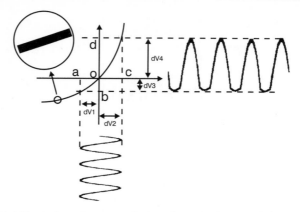

Figure 4.1 *Transfer characteristic of a triode amplifier*

A transfer characteristic is the term given to a graph which details the relationship of voltage input to voltage output. Typically the X axis represents the input and the Y axis represents the output. So, referring to Figure 4.1 a particular voltage *a* at the input results in another voltage *b* at the output of the electronic device or circuit. Similarly, the input voltage *c* results in the output voltage *d*. Crucially, observe that in the case of the triode stage, the ratio dV_1/dV_2, is not the same as the ratio dV_3/dV_4. This is the formal definition of amplitude related distortion in all electronic systems and the figure illustrates the effect the non-linearity has on a sine-wave input waveform.

The triode's transfer characteristic leads to the production of a set of even-numbered harmonics of rapidly diminishing strength; a fair degree of second harmonic, a small amount of fourth, a tiny amount of eighth and so on. Remember from Fourier's findings, it is the presence of these new harmonics which account for the change in the waveform as shown in Figure 2.2. Referring to Figure 2.7 in Chapter 2, even-numbered harmonics are all consonant. The effect of the triode is therefore 'benign' musically.

The pentode on the other hand has a transfer characteristic in which the predominantly linear characteristic terminates abruptly in a non-linear region. This difference does not account for the preference for triodes in high quality amplification (where they are chosen for their lower noise characteristics) but may account for guitarists' preference for triodes in pre-amplifier stages of guitar amplifiers where the triode's rather slower transition into overload enables a player to exploit a wider range of sonic and expressive possibilities in the timbral changes elicited from delicate to forceful plectrum technique.

While harmonic distortion may have certain special benefits in musical applications, it is unlikely that non-linearity accounts for the preference for valves in recording and reproduction (monitoring or hi-fi) applications. Non-linear distortion of complex waveforms results in the production of a plethora of inharmonic distortion components known as intermodulation distortion which, without exception, sounds unpleasant, adding 'muddle' to the overall sound.

Intermodulation distortion

Intermodulation distortion comes about due to the presence of non-linearities in an electronic system as well.[1] Take for instance the triode transfer characteristic illustrated in Figure 4.1. This characteristic is curved which, to the mathematically minded of you, will immediately imply a power law. In simple terms, a power law is a relationship which does not relate one set of values (the input voltage) to another (the output voltage) by a constant, which would yield a straight-line relationship – and incidentally a distortionless amplifier, but by a function which is made up of both constant and multiplication factor which is related to itself. Put a different way, not this:

$$E_o = k.E_i$$

where E_o is output voltage and E_i is input voltage
    But this,

$$E_o = k.E_i^p$$

Now the crucial concept here is multiplication. The power term p in the above expression illustrates that E_i is multiplied by a version of itself and, we know from Chapter 2, that when two sine-waves are multiplied together it results in sum and difference frequencies. The two frequencies are termed to *intermodulate* with one another. If we imagine connecting an input signal consisting of two sine-waves to an amplifier with such a characteristic, the result will betray the generation of extra tones at the sum frequency ($\sin A + \sin B$) and at the difference frequency ($\sin A - \sin B$). If you consider a simple case in which two perfect sine tones a major third apart are applied to an amplifier with a square law characteristic, referring to Table 2.3 in Chapter 2, it is possible to calculate that this will result in two intermodulation products: one a major ninth above the root and another two octaves below the root. This results in the chord in Figure 4.2. Now imagine this effect on every note within a musical sound and the interaction with every overtone with every tone and overtone within that sound and it is relatively obvious that distortion is an unlikely suspect in the hunt for the sonic blessings introduced by valve equipment.

Figure 4.2

Headroom

One benefit which valve equipment typically does offer in high-quality applications is enormous headroom. This may explain their resurgence in microphone pre-amplifier designs which must allow for possibly very high acoustic energies – for instance when close-miking a loud singer or saxophonist – despite the technical problems involved in so doing (see Figure 4.3 in which a quite beautiful modern microphone due to Sony incorpor-

Figure 4.3 *Sony's C-800G microphone*

ates valves for the signal path and semiconductor technology in the form of Peltier heat-pump technology to keep the unit cool and secure a low-noise performance). Technically valve microphone amplifiers have a great deal in common with the phono pre-amplifier illustrated in Figure 4.10 later in the chapter. (Indeed the justification for that valve design lay in its ability to handle the large peak signals.) But even in this instance the 'benefits' which valves bestow are a case of 'every cloud having a silver lining' because valves are only able to handle signals with very high peaks because they display significant non-linearity. Figure 4.1 illustrates this too, in the magnified portion of the curved transfer characteristic of a triode amplifier. The magnified section approximates to a straight line, the characteristic required for accurate and distortion-free operation. It illustrates that, in order to obtain a linear performance from a valve, it is necessary to operate it over a very small portion of its overall transfer characteristic, thereby providing (almost by default) an immense overload margin.

Interaction with loudspeakers

A characteristic of all valve power amplifiers (high quality and instrumental types) is the lower degree of negative feedback employed. Especially when compared with solid-state designs. Instrumental amplifiers, in particular, often opt for very little overall feedback or none at all. Because the signal in a valve amplifier issues from the high impedance anode (or plate) circuit, even the action of the power output-transformer[2] cannot prevent valve amplifiers from having a relatively high output impedance. This produces, via an interaction with the mechanism of the loudspeaker, a number of pleasant sonic characteristics that certainly profit the instrumental musician and may elicit a euphonious and pleasant (if far from accurate) response in a high quality situation.

It was noted in the last chapter that moving coil microphones and loudspeakers worked according to the same principles applied reciprocally. So that inside a microphone, a coil of wire produces a small electric current when caused to move within a magnetic field by the force of sound waves upon it. Whereas in a loudspeaker sound waves are generated by the excitation of a diaphragm connected to a coil of wire in a magnetic field when an audio signal passes through the coil. It should therefore come as no surprise to discover that a loudspeaker, excited into motion, generates a voltage across the driver coil. It is one of the functions of a perfect power amplifier that it absorb this voltage (known as a back-EMF). If the amplifier does this duty, it causes the current due to back-EMF to recirculate within the coil and this damps any subsequent movements the coil may try to make. A valve amplifier, on the other hand, especially one with a small degree of negative feedback, has a relatively high output impedance and therefore fails to damp the natural movement of a loudspeaker following a brief excitation pulse. This effect is illustrated in Figure 4.4. Because the natural movement of the loudspeaker is invariably at its resonant frequency – and this is always pretty low – the perceived effect of insufficient damping is of more bass, albeit of a rather

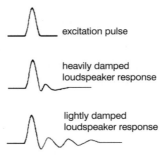

Figure 4.4 *Loudspeaker damping*

'tuneless' kind! In small rooms, which cannot support bass frequencies, and with small loudspeakers this effect can lead to an artificially inflated bass response which may suit some individuals.

Reduction in loudspeaker distortion

Another consequence of the high output impedance of a valve amplifier, and one that may be the nearest phenomenon yet to a mysterious semiconductor distortion mechanism, is the reduction in mid-range loudspeaker amplitude distortion. This is due to uncoupling the dependence of cone velocity on the changes in the magnetic circuit and lumped electrical impedance. In the relationship of cone velocity to amplifier output voltage, these terms (which are themselves determined by the position of the cone) are present. They are not in the relationship which relates amplifier output current to cone velocity. The relatively high output impedance of valve amplifiers thereby has some advantages in securing a reduction in harmonic and intermodulation distortion.

Valve theory

The Edison effect

I suppose electronic musicians and sound engineers might claim Thomas Alva Edison as their profane 'patron saint'. But that wouldn't be justified because Edison, who was probably the greatest inventor in history, changed the lives of ordinary people in so many ways that to claim him for our own would be to miss his other achievements. Among them electric light, the modern telephone, the typewriter, the motion picture as well as the father of modern industrial research! Nevertheless Edison did overlook patenting the electronic diode and thereby failed to reap the benefits of the first electronic, rather than electrical, invention. This fact is all the more amazing in that in the 1880s, he experimented with a two-terminal vacuum tube formed by putting an extra plate in the top of a light bulb; the filament acting as a cathode and the plate as an anode. (In America, the anode is still referred to as the plate.) Edison made the far-reaching observation that, in such an arrangement, conventional-current travelled in one direction only; towards the cathode. However, and perhaps because no mechanism had yet been advanced for why such an effect should occur (the electron was not discovered until 1899), Edison failed to see what application such an effect might have.

Invention of the diode

It is said that Edison never forgave himself for overlooking what, in 1904, J.A. Flemming had the vision to see; that the phenomenon which he had observed had commercial application in the detection of wireless waves. Braced with the discovery, by J.J. Thompson in 1899, of the electron, Flemming saw the potentialities (drawing on a hydraulic or pneumatic analogy) of an 'electronic valve' which allowed current to be passed in one direction while prohibiting its flow in the other. He realised that this 'valve' had applications in early wireless, where experimenters had been striving to discover something which would react to the minute alternating potentials produced in a receiver of wireless waves. Flemming's valve, by passing current in one direction only, produced a DC output from a radio frequency AC input, thereby acting as an 'indicator' of radio waves, what we today would call a detector.

Invention of the triode

In 1907, Dr Lee de Forest took out a patent in America for a valve like Flemming's, but with a third electrode, consisting of a mesh of wires interposed between the anode and the cathode. De Forest noticed that the potential of this grid exercised a remarkably strong control over the anode current and, moreover, saw that such a phenomenon increased enormously the potentialities of the valve because it could act as a detector and, most crucially, as an amplifier, and ultimately as a generator of oscillations – or an oscillator. So how exactly does the triode valve work?

Thermionic emission and the theory of electronic valves

The modern thermionic triode valve is a three-electrode amplifying device similar – in function at least – to the transistor (see Figure 4.5).

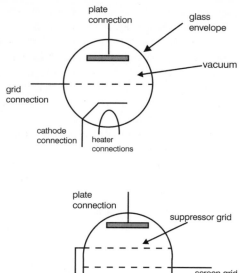

Figure 4.5   *Triode valve and pentode valve*

Externally the valve consists of a glass envelope in which the air has been evacuated. Each electrode is formed of a metal – each a good conductor of electricity. In an electrical conductor all the free electrons move about at random. When the valve is energised, one of the electrodes – the cathode – is heated (by means of a dedicated heater element which must have its own supply of electricity) to a point that this random motion of the electrons is sufficiently violent for some of them to leave the surface of the metal altogether. This phenomenon is known as thermionic emission. If there is no external field acting on these escaped electrons they fall back into the cathode. If another electrode within the valve is connected to a positive potential, the electrons on leaving the cathode will not fall back but will accelerate off towards the other electrode. This other electrode is called the anode. The passage of electrons from cathode to anode establishes a current which passes between these two electrodes. (Of course, convention has it that the current flows the other way, from anode to cathode.) The function of the third electrode, the grid (which is manufactured literally – just as it was in de Forest's prototype – as a wire grid, very close to the cathode), is to control the magnitude of the stream of electrons from cathode to anode. This is achieved by varying a control voltage on the grid, so that when its potential is appreciably more negative than the cathode, the electrons – leaving by thermionic emission – are repelled and are thwarted in their usual passage towards the anode and no current flows through the valve. And when, instead, the grid is made nearly the same potential as the cathode, the electrons pass unhindered through the grid and reach the anode, as if no grid electrode was interposed at all. At values in between the current varies proportionately. Because the grid is always held negative in relation to the cathode no current flows in this control circuit (unlike the transistor but more like the FET) so a very feeble control voltage can be made to control a much larger current – the essence of any amplifying stage.

Characteristic curves

If we plot the anode current through a triode valve at various anode voltages and grid voltages we obtain a family of curves like those illustrated in Figure 4.6. These are termed characteristic curves. Note that each curve relates to the relationship of anode current to anode volts for a particular grid voltage; in this case for a particular triode valve code numbered 12AU7. Notice also that each curve is relatively

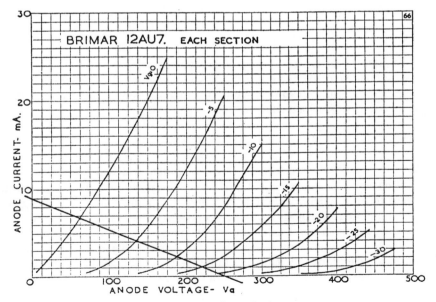

Figure 4.6 *Characteristic curves for triode*

straight except near the axis where it bends. Now look at the line drawn with an opposing slope to the valve curves. This is called a load-line and it represents the current/voltage relationship of a resistor R_a (the load) so arranged in circuit with the valve as shown in Figure 4.7. Observe that when the anode voltage is zero, all the current flows through R_a, and – if the anode voltage is equivalent to the rail volts – no current flows through R_a: hence the opposing slope. Because the supply rails for valve circuitry are very much higher than those associated with transistor circuitry, the rail supply is often referred to as an HT (high tension) supply.

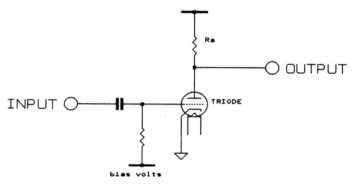

Figure 4.7 *Simple triode amplifier stage*

In fact, Figure 4.7 is a simple valve amplifier and we can now calculate its performance. By reading off the calculated anode voltages for grid voltages of –10 V (290 V at anode) and 0 V (60 V at anode) we can calculate that the stage gain will be:

(190 – 60)/10 = 13 times or 22 dB

This is typical of a single-stage triode amplifier and will no doubt seem pitifully small for those brought up on transistor circuitry. One important point to note when inspecting the curves of Figure 4.6 – try to imagine changing the slope of the load-line. You will see that it has very little effect on the stage gain; once again an unusual phenomenon if you are used to solid-state circuits. This is because the triode valve has a low anode impedance and the mechanism is more clearly understood by considering the equivalent circuit of a triode amplifier which is illustrated in Figure 4.8 and is, effectively, Figure 4.7 redrawn. (Remember that in signal terms, the load is returned to the cathode end of the valve via the very low PSU impedance.) As you can see, the anode impedance (or resistance r_a) is in parallel with the load resistor and so swamps the higher value load resistor. The stage gain is defined by g_m (mutual conductance which is a constant and is dependent upon the design of the particular valve) times the load (R_a) in parallel with the internal anode resistance (r_a). Note that the internal current generator has the dimension ($- g_m.e_g$) where e_g is the grid potential.

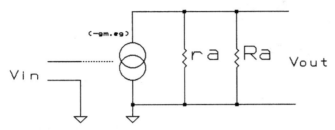

Figure 4.8 *Valve equivalent circuit*

The input impedance at the grid is effectively infinite at DC but is gradually dominated at very high frequencies by the physical capacitance between the anode and the grid. Not simply because the capacitance exists but because (as in solid-state circuitry) this capacitance (sometimes referred to as Miller capacitance) is multiplied by the stage gain. Why? Because the anode voltage is in opposite phase to the input signal (note the minus sign in the stage gain formula) so the capacitance is effectively 'bootstrapped' by the action of the valve itself.

Development of the pentode and beam tetrode

It was this troublesome Miller capacitance (especially in radio work, where high frequencies are employed) that led to the invention of the tetrode valve, in which was interposed another grid at a fixed positive potential between the grid and the anode. This extra (screen) grid being held at a potential at, or near to, the value of the supply rails thereby neutralising the Miller capacitance by screening the effect of the anode upon the control grid. Unfortunately, the simple tetrode, while successful in this respect (the second grid reduced the effect of the Miller capacitance to about one-hundredth of the value in a triode!) has the very great disadvantage that the electrons, when they reach the anode, dislodge extra electrons; an effect called secondary emission. In a triode these extra electrons are 'mopped up' by the anode so nobody is any the worse for it. But in a simple tetrode, especially when the anode volts fall below the screen-grid volts, it is the screen grid which attracts these secondary electrons and they are thus wasted. In practical tetrodes, steps must be taken to prevent this effect and this is achieved by forming the grids, and indeed the entire valve, so that the electrons flow in beams through both structures, the resulting valve being known as a beam-tetrode. All modern tetrodes are beam-tetrodes. The alternative solution is to install yet another grid between the screen grid and the anode called the suppressor grid, which is kept at the same potential as the cathode and which is often connected internally. This brings the total number of internal electrodes to five and the valve is therefore known as a pentode. The zero (or very low) positive potential of the suppressor grid has the effect of turning the secondary electrons back towards the anode so that they are not lost.

The addition of the extra grid, in the case of the tetrode (or grids, in the case of the pentode), although primarily to extend the frequency range of a valve way beyond the audio range, has a remarkable intensifying effect on the performance of a valve. It is this enhancement and not their superior RF performance which ensures their use in audio applications.

What is the nature of this enhancement? Well, the equivalent circuit shown in Figure 4.8 is equally valid for a pentode but the value of r_a, instead of the 5k to 10k in a small triode, is now several megohms. The stage gain is thereby radically improved. Typically a small pentode valve will have a stage gain of 100 times (40 dB), ten times better than a triode stage. Audio output valves are often either pentode or beam-tetrode types.

Valve coefficients

The maximum voltage amplification which a valve is capable of, given ideal conditions, is called the amplification factor; generally designated with the Greek letter μ. This is not truly constant under all conditions (except for a theoretically ideal valve) and varies slightly with grid bias and anode voltage.

Technically μ is defined as follows: expressed as a ratio, the incremental change in plate voltage to the incremental change in control-grid voltage in the opposite direction – under the conditions that the anode current remains unchanged (i.e. with a current source for an anode load) and all other electrode voltages are maintained constant. So:

$$\mu = -(dV_a/dV_g)$$

The two other principal valve coefficients, mutual conductance and anode resistance, we have met already. However, although we treated them earlier as constants, both coefficients are dependent to some extent on applied electrode voltages. Mutual conductance (g_m) is defined thus:

$$dI_a/dV_g$$

That is, the change in anode current, for a given change in grid voltage, all other voltages remaining the same. Anode resistance (r_a) is defined as:

$$dV_a/dI_a$$

There is an exact relationship between these three principal valve coefficients, provided they have all been measured at the same operating point:

$$\mu = g_m.r_a$$

Sometimes you will read, especially in older literature, references to reciprocals of these coefficients. For completeness these are defined below:

$1/\mu = D$, the Durchgriff factor or penetration factor

$1/r_a = g_a$, anode conductance

Practical valve circuits

The graphical technique explained in the last section is worthwhile in that it gives an intuitive feel for the design of valve circuits. It is, however, rather cumbersome and is not usually required for the majority of simple design tasks. Moreover it leaves several questions unanswered, such as the optimum bias conditions etc. For this, and virtually everything else the practical designer needs to know, all that is usually required is the manufacturer's data. Let's take as an example the 12AU7 valve (ECC82) which we met in the previous section. Using graphical techniques, we were able to ascertain that the stage gain of this particular triode valve circuit would be about 13 times, or 22 dB. We've already noted that the stage gain of a triode valve amplifier depends only very little on the value of anode resistor. This phenomenon, which we might term the 'device-dependent' nature of valve design, is certainly not consistent with most readers' experience of transistor stages and may take some getting used to for modern engineers. This consistency in performance makes it possible to tabulate recommended operating conditions for a valve in a manner that would be quite impossible for individual transistors. Usually, the manufacturers state a number of different recommended operating scenarios. Data for the 12AU7 (ECC82) valve consist of the following:

Operation as a resistance coupled amplifier

Anode supply voltage	100	250 volts
Anode load resistor	100	100 kΩ
Cathode bias resistor	4	3 kΩ
Peak output	17	50 volts
Stage gain	11	12

The circuit illustrated in Figure 4.9 is derived from the tabulated conditions. Note the addition of several extra components. First, the inclusion of a resistor in the cathode circuit. Second, the addition of the 1 meg grid bias resistor and input coupling capacitor C_{in}. Valves are virtually always biased as shown, with a positive potential being derived at the cathode (by means of a series resistor through which the anode current flows) and the grid

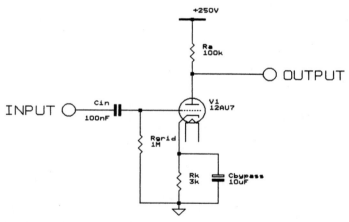

Figure 4.9   *Practical triode amplifier stage*

returned to ground via a very high value resistor. This latter resistor is absolutely necessary, for although a valve takes only a minute bias current via its grid circuit, it does take some and this resistor is included to provide such a current path. The advantage of this biasing scheme, known as cathode biasing, is inherent stability. If the current through the valve starts to rise, the potential at the cathode increases with respect to the grid, and so the grid bias moves more negative, thereby reducing the anode current. The only disadvantage of this scheme in a practical amplifier is the generation if a signal voltage in phase with the anode current (and thus out of phase with the grid signal voltage) due to the varying grid current in the cathode bias resistor. If the resistor is left unbypassed, it generates a negative feedback signal which raises the output impedance of the stage and lowers stage gain. Hence the third inclusion; the large value capacitor placed across the cathode resistor to bypass all audio signals to ground at the cathode circuit node.

The value of C_{in} is chosen so that the low-frequency break point is below the audio band. Curiously enough, this network and the impedance of the anode supply are related: one of the common problems encountered with valve amplification is low-frequency oscillation. Almost certainly a surprise to engineers weaned on transistor and op-amp circuitry, this results from the predominance of AC coupled stages in valve equipment. Just as high-frequency breakpoints can cause very high-frequency (supersonic) instability in transistor equipment, wherein phase lag due to several networks causes the phase shift through the amplifier to equal 180° when the gain is still greater than 1; phase lead – due to the high-pass filtering effects of several AC coupled stages – can cause instability at very low (subsonic) frequencies in valve circuits. The audible effects of this type of instability are alarming, producing a low-frequency burble or pulse. In fact a sound which is admirably described by its vernacular term – motor-boating! Low-frequency oscillation typically results from feedback signal propagation via the power supply which (as a result of the elevated voltages in valve equipment) may not be as low an impedance as is typical in solid-state equipment. In order to prevent this, valve equipment designers typically decouple each stage separately. Now, let's look at some practical valve amplifiers so as to apply our knowledge of valve circuitry.

A valve preamplifier

Figure 4.10 shows a design for a vinyl disc pre-amplifier that I designed and which ran for some years in my own hi-fi system (Brice 1985). It is a slightly unusual design in that it incorporates no overall negative feedback, passive RIAA equalisation and employs a cascode input stage.

In transistor equipment, the problem with passive equalisation is the risk of overloading the first (necessarily high-gain) stage due to high-level treble signals. With valves this does not present a problem because of the enormous headroom when using a power supply of several hundred volts. The usual choice for a high-gain valve stage is a pentode but these valves generate more shot noise than triodes because of the action of the cathode current as it splits between the anode and screen. Instead I used a cascode circuit. Like so many other valve circuits this has its origins in radio. Its characteristics are such that the total stage noise is substantially

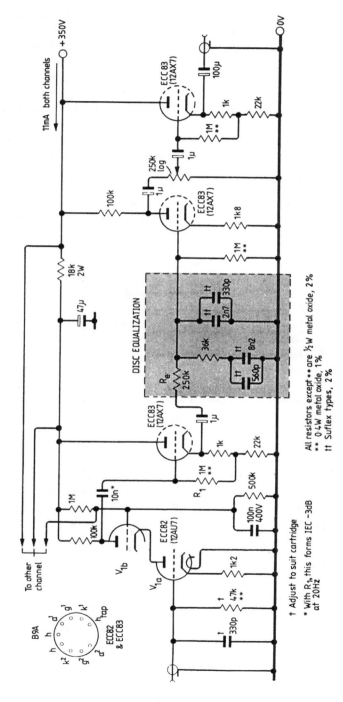

Figure 4.10 Phono pre-amplifier

that of triode V_{1a}. But the gain is roughly the product of the anode load of V_{1b} and the working mutual conductance of V_{1a}. In other words it works like a pentode but with lower noise!

The RIAA equalisation is shown in a dotted box. This is to dissociate myself from this section of the circuit. If you have your own ideas about RIAA equalisation then you can substitute your own solution for mine! On a general note, I disagree with those who say there is no place for valves in low-level circuitry. Well designed valve circuitry can give superlative results. Hum can sometimes be a problem. In this design I left nothing to chance and powered the valve heaters from a DC regulated power supply (Figure 4.11). The HT was also shunt stabilised using cold-cathode glow-discharge tubes. The power supplies were built on a separate chassis. The

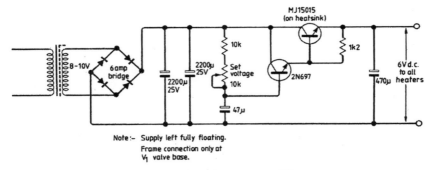

Figure 4.11   *Power supply for phono amplifier*

more common ECC83 valve would be suitable as the first stage cascode valve except that the ECC82 is more robust in construction and therefore less microphonic. I have found, from bitter experience, that there is no alternative but to select low-noise valves individually for the first stage valve.

Power amplifier

Figure 4.12 is a circuit diagram of a practical 20 watt Class-A beam-tetrode amplifier. This amplifier utilises a balanced (or push-pull) output stage employing two 6L6 beam tetrode valves. This is the most common form of audio amplifier configuration. Why? The answer is balance. As with so many other practical engineering solutions the lure of symmetry and equilibrium triumph in the human mind. In Figures 4.13 and 4.14 a single-ended and balanced (push-pull) output stage are drawn for comparison. The balanced version has a number of important advantages over its single-ended cousin.

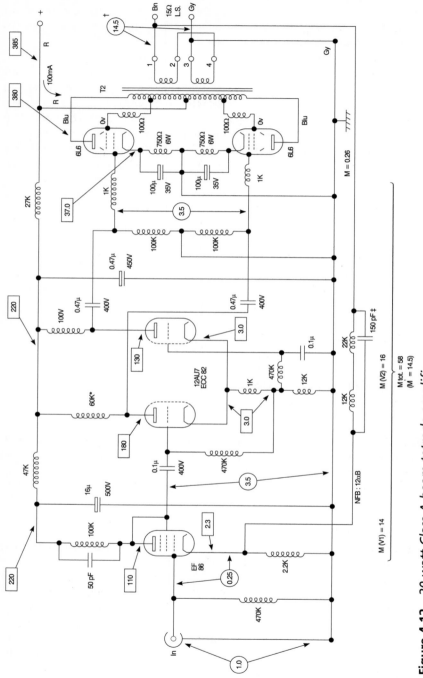

Figure 4.12 *20 watt Class-A beam-tetrode amplifier*

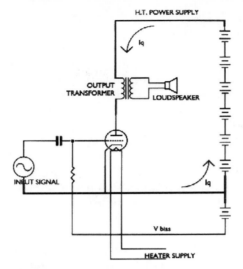

Figure 4.13 *Single-ended output stage*

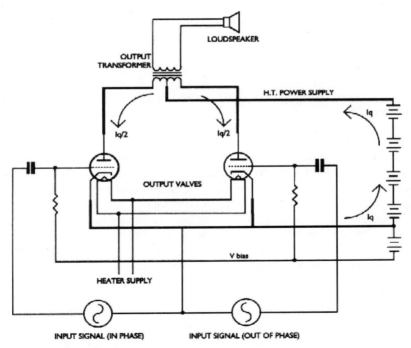

Figure 4.14 *Balanced output stage*

Magnetisation of the output transformer

All but a very few special valve amplifiers employ an output transformer. This component can be thought of as an electrical gearbox coupling the high impedance valve outputs to the low impedance loudspeaker. Transformers work using the principle of electromagnetism which you may remember investigating in school physics where you may have proved it was possible to create an electromagnet by winding wire around an iron (or steel) nail and passing current through the wire. Had it been possible in that school experiment to control the current through the magnet and measure the power of the magnet and how it related to current (perhaps by the number of paper-clips it could pick up) you would have found a linear relationship up to a certain point; the number of paper-clips would be directly proportional to current, up to a certain value of current. After that value of current had been reached, however, the magnet would not pick up any more paper-clips no matter how much more the current was increased. It would simply serve to warm up the magnet. This effect is known as magnetic saturation (and we will meet it again in relation to magnetic tape). It is due to all the magnetic domains within the nail eventually being used up. After that point the nail simply cannot become more magnetised. Exactly the same limitation exists with output transformers. The process (as within all transformers of converting electricity into magnetism and back again) is distortion free, so long as the transformer core does not become saturated. When that begins to happen, the process becomes non-linear and audible distortion will start to be produced.

If you compare Figures 4.13 and 4.14, you will notice, in the case of the single-ended output stage, that a continuous standing or quiescent current (I_q), flows from the power supply, through the transformer and valve and back to the power supply. Because audio signals are assumed (for design purposes at least) to be symmetrical, this standing current must equal half the maximum that the output valve is designed to carry. A typical amplifier with 10 watts output would require a standing anode current of about 70 mA. This much current would produce a magnetic core flux density of perhaps 5000 to 6000 gauss.

Now consider Figure 4.14. Here the quiescent current flows from the power supply into the centre of the output transformer. From here, it splits – half in one direction into one output valve and half in the other direction into the other output valve. The current, once shared between the valves, recombines in the common cathode circuit and flows back to the power supply. The great advantage of this configuration is that, because the current flows in at the middle of the transformer winding and away in opposite directions, the magnetic effects within the core of the transformer cancel out and there is thus no quiescent magnetisation of the core in a balanced stage.

Reduction in distortion products

It is a widely (and incorrectly) held belief that valves are inherently more linear than transistors. This is absolutely not the case. The transistor is a remarkably linear current amplifier. On the other hand, the valve is a relatively non-linear voltage amplifier! For example, a 6L6 output valve used in a single-ended configuration like that shown schematically in Figure 4.13 will produce something in the region of 12% total harmonic distortion at rated output. But the balanced circuit has an inherent ability to reduce distortion products by virtue of its reciprocity in a very elegant way. Essentially each valve produces an audio signal which is in opposite phase at each anode. (For this to happen they must be fed with a source of phase-opposing signals; the role of the phase-splitter stage which will be considered below.) But their distortion products will be in-phase, since these are dependent upon the valves themselves. So these distortion products, like the transformer magnetisation, will cancel in the output transformer! This is empirically borne out: Table 4.1 annotates my own measurements on an 807 beam-tetrode amplifier used in single and balanced configuration. The numbers speak for themselves.

Table 4.1

Single ended output stage

Anode V (mA)	Screen V	Power output (W)	Distortion
500 (50)	200	11.5	12%

Push-pull balanced output stage

Anode V (mA)	Screen V	Power output (W)	Distortion
500 (50 + 50)	300	32.5	2.7%

Returning to the complete amplifier pictured in Figure 4.12. Note that the first stage illustrates a practical pentode amplifier; the supply to the screen grid of V1 via R4 (2M2). This must be decoupled as shown. The suppressor grid is connected internally. This is followed by the phase-splitter stage. In the amplifier illustrated, the phase splitter is known as a cathode-coupled type and is relatively familiar to most modern engineers as a differential-amplifier (long-tailed pair) stage. The signal arriving at the grid of the left hand valve is equally amplified by the two valves and appears at the two anodes in phase opposition. These signals are then arranged to feed the grids of the output valves, via coupling capacitors.

Mark the slightly different anode loads on each half of the phase splitter which are so arranged to achieve a perfect balance between circuit halves.

Reference
Brice, R. (1985) Disc Preamplifier. *Electronics and Wireless World*, June.

Notes

1 In fact harmonic distortion is really only a special type of inter-modulation distortion.
2 The anode circuit of a valve is, in many ways, similar to the collector circuit of a transistor (i.e. very high impedance). You can think of a valve output transformer as an electrical gearbox coupling the high impedance valve outputs to the low impedance loudspeakers. See later in chapter.

Fact Sheet #4: Semiconductors

- Semiconductors
- Transistors
- Different flavours

Semiconductors

If Ohm's law was really a law, there would be an awful lot of electronic components in gaol! Because there's a whole range of important electronic components which do not obey the 'law'. These are neither insulators, resistors or conductors. Instead they're somewhere between all three and are hence termed semiconductors.

Semiconductors display important non-linear relationships between applied voltage and the current through them. Look at graph Figure F4.1: this is the graph of the current through a semiconductor diode at varying voltages. When a negative voltage is applied, no current flows at all: and when a positive voltage is applied, nothing much happens before the voltage reaches 0.7 V, after which the current rises very swiftly indeed.

The graph was obtained by plotting the relationship of I / V in a silicon semiconductor diode. The most important effect of the diode is its peculiarity that it allows current to flow one

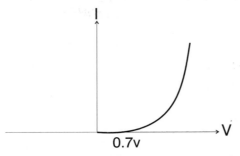

Figure F4.1 *Current/voltage curve for a silicon diode*

way only. It's therefore useful in obtaining DC current from AC current – a process which is required in most electronic equipment in order to turn the mains supply into a DC supply for the electronics.

The significance of the positive 0.7 volts 'twilight-zone' in the silicon diode is due to deep quantum physics. Effectively electrons passing through the diode have to achieve a sufficient energy level to get the diode to start to conduct and for this they require a certain voltage or electro-magnetic force. This 0.7 volts is known as the diode's offset voltage.

Transistors

Transistors spring from a common technological requirement that a small controlling input may be used to manipulate a much larger source of power; just as pushing the car's accelerator pedal (which requires very little effort) can cause you to hurtle down an autoroute at 200 k.p.h.! In the nineteen-twenties electronic valves were developed for this purpose, now these have been superseded (except in some audio equipment – see Chapter 4) by transistors which may be thought of as semiconductor valves!

Figure F4.2 illustrates a simple transistor amplifier. The input voltage being applied at the base of the transistor and the output developed across the resistor Rload. The graph illustrates the input voltage and the output voltage at the relevant ports. Note the three terminals of the transistor which are also labelled; base, emitter and collector respectively, the input being at the base and the output at the collector. This is the most common arrangement for a transistor amplifier.

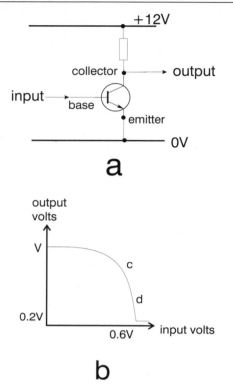

Figure F4.2   *A transistor used as a primitive amplifier*

From the graph it's pretty apparent that without qualification this doesn't represent a very useful amplifier! It's only over the region annotated c–d, that the transistor may be used as a linear device. For this to happen the input signal must be biased into the region close to 0.7 V (the offset voltage), and be prevented from swinging beyond the limits of this sector. Unfortunately this isn't as simple as it sounds due to the incredibly high gain of the amplifier in this region which would involve controlling the voltage of the base within thousandths of a volt (millivolts or mV). Worse still transistors are very susceptible to changes in temperature (offset voltage is actually related directly to heat). All of which would make for a very fragile amplifier without the adoption of several important circuit techniques illustrated in Figure F4.3; which represents a practical transistor amplifier stage. Note the important addition of Rb1 and Rb2 which are so arranged to provide a constant bias voltage at the base terminal. However,

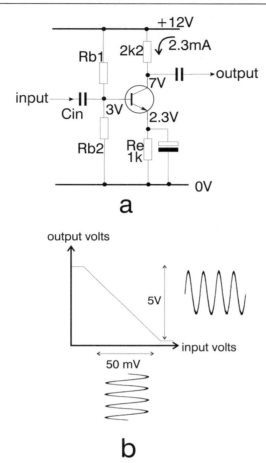

Figure F4.3 *A practical transistor amplifier stage*

you'll note that the bias voltage chosen is not close to 0.7 volts but is instead several volts. The emitter is 0.7 V less than the voltage on the base but the remainder of the bias voltage exists across resistor R_e. Notice that the signal is input to the stage via capacitor C_{in}. Notice also the capacitor which is placed across R_e. This is a large value capacitor and effectively ensures that at audio frequencies the impedance of the emitter circuit is close to zero. However at DC, the emitter circuit is dominated by R_e. This ensures that the AC gain of the amplifier stage (at all audio frequencies) is very much higher than the gain at DC. This is a very common technique in transistor equipment from simple transistor amplifiers to 1000 watt power amplifiers.

Calculating DC conditions

In the circuit illustrated, the collector current is set to 2.3 mA. We know this because 2.3 volts exists across R_e. We can also calculate the voltage at the collector because most (all but about 1%) of the current which flows in the emitter circuit, flows in the collector circuit. In other words, the current doesn't come from the base circuit. This is a very important point because remember the whole idea behind an amplifier is to control a much larger force by means of a small controlling force; if current flowed from the base circuit it would sap energy from the controlling circuit.

So, if 2.3 mA flows in the collector circuit through R_c, the voltage at the collector (using Ohm's law) will be about 0.0023 × 2200 = 5.06 volts less than the power supply,

$$12\,V - 5\,V = 7\,V$$

Stage gain

We can work out the gain at DC by imagining changing the bias voltage to 4 V (i.e. one volt more). The voltage between the base and the emitter will remain about the same (0.7 V) so in this case the emitter current will be 3.3 V/1k = 3.3 mA and the volt-drop across R_{load} will be 3.3 mA × 2.2 k = 7.3 V.

So a change of one volt at the base will cause 7.3 − 5 = 2.3 V change at the collector, in other words a gain of 2.3 times. Notice that this value is close to the value of R_{load}/R_e, in fact this may be used as an approximation for all practical calculations.

The stage gain at AC of the transistor stage in Figure F4.3 is given by,

$$Gain = gm. R_{load}$$

Where gm is a termed mutual conductance and is an expression which defines the degree of change in collector current for a given change in base voltage. This sounds simple except that in a transistor gm is related to collector current by the expression,

$$gm = 40\,mA/V \text{ per mA}$$

In other words, *gm* is not constant in a transistor but is dependent upon its operating condition. (Compare this with valves in Chapter 3.) Nevertheless, the simplification is usually made that the *gm* is calculated at the bias condition and this is used to predict the gain of the stage – and this works very well indeed due to the high gain of a transistor.

So, in Figure F4.2, the stage is 'sitting in' 2.3 mA of current (this is often termed quiescent current) so,

$$2.3 \times 40 = 92 \, mA/V$$

which implies that the collector current will change by 92 mA for every 1 V change in base voltage. This will cause a 92 mA × 2200 = 202 V swing at the collector output. Which is the same thing as saying the stage has a gain of about 200 times or 46 dB.

Of course the collector can never swing 202 V, instead it can probably swing between very close to the rail and to within about 0.2 V of its emitter volts (2.3 V), a swing of about 10 V. In this application, input voltage would have to be limited to about,

$$10 \, V/200 = 0.05 \, V \text{ or } 50 \text{ millivolts (mV)}$$

if the amplifier is not to distort the output waveform

Different flavours

The amplifier we looked at above was designed using a NPN transistor, which has a conventional current path from collector to emitter; this is indicated by the arrow which shows the current leaving by the emitter! But there's actually another type of transistor called a PNP type which works exactly the same way (from an external point of view – internally the physics are quite different) in which the current arrives by way of the emitter and exits by the collector. Not surprisingly, this is distinguished by an arrow which points in the opposite direction. Modern convention has it that a PNP transistor is also drawn the other way up to its NPN cousin; because it makes circuit diagrams easier to read if current always flows from the top of the page to the bottom.

The different types of transistors are very useful in building amplifier structures which would otherwise require AC coupling; as was current in valve equipment. Figure F4.4 illustrates a three stage amplifier using a PNP transistor in the second position. The circuit is said to be DC coupled and, not only does

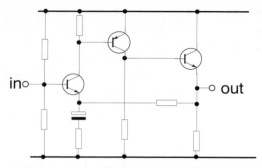

Figure F4.4   *Three-stage transistor amplifier using a combination of npn and pnp transistors*

the PNP transistor allow the designer to dispense with biasing arrangements for the second and third transistor, but ensures that there is gain (and critically no phase shift) at low frequencies. This enables the amplifier to be used in all sorts of roles which would be unsuitable for the AC coupled, valve amplifier.

In fact the little amplifier in Figure F4.4 is so good that it can be used for a wide variety of varying operations. It is therefore termed an operational amplifier or op-amp for short. Fact Sheet #7 explains some of the roles for an op-amp.

5
Roll Over Beethoven – Electric instruments

Introduction

Aside from the influence upon telecommunications, the invention of the microphone, the loudspeaker and the electronic valve amplifier, brought about a revolution in the art of music making. For several centuries, a firm distinction may be made between large-scale music making – orchestras, choirs, military bands and so on and music for intimate entertainment, or chamber music. The ability to amplify instruments and solo voices meant that for the first time chamber music could become a large-scale musical activity. The cultural revolution of rock-and-roll – and later rock music – is as much about how the music is made as it is about the musicological synthesis of blues and American country music. For the first time in history, and due solely to the progress in electronics, the world-view of a few young men (in those days it was just men) could capture the hearts and minds of hundreds, thousands – and with the intervention of radio, millions of young people. Little wonder then that the establishment has always had an uneasy relationship with rock music! Technologically a stone's throw from the early microphones is that icon of rock-and-roll rebellion – the electric guitar. From Scotty Moore's chiming lead guitar on the early Elvis records to Hendrix's angst-ridden, tortured performances, no other instrument characterises the octane-charged sound of rock-and-roll better than the electric guitar. So it is with this symbolic and seminal musical voice that we begin our look at electric instruments.

Electric guitars

A modern electric guitar is illustrated in Figures 5.1 and 5.2. In the diagram, the guitar is also labelled to illustrate the major components. The earliest electric guitars were created by attaching a contact microphone

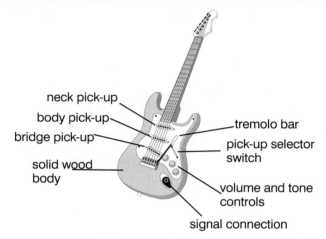

neck pick-up

body pick-up

bridge pick-up

solid wood
body

tremolo bar

pick-up selector
switch

volume and tone
controls

signal connection

Figure 5.1   *Diagram of electric guitar*

to the top sound-board of a conventional acoustic guitar, the resulting signal being fed to an external amplifier. However, the modern electric guitar was born with the invention of the electromagnetic pick-up and a typical arrangement is illustrated, diagramatically, in Figure 5.3. In principle, all electric guitar pick-ups are formed this way, with a coil wound on a permanent bar-magnet former. The magnet is arranged so that

Figure 5.2   *Fender Stratocaster guitar*

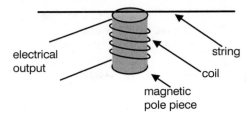

Figure 5.3   *Electromagnetic pick-up*

it points with one pole towards the string and the opposing pole away from the string. As the string is excited by the player, and moves in close proximity to the magnetic circuit, the flux in the circuit is disturbed and hence a small electric current is induced in the coil. Early pick-ups used a single magnet for all the strings but later models used separate magnets, or separate pole pieces at different heights relative to the strings, so as to compensate for the different sensitivity of the pick-up in relation to each of the guitar's six open strings. Guitar pick-up coils contain very many (often several thousand) turns of fine-gauge wire and are thus very sensitive to minute string movements. Unfortunately, this also renders them very sensitive to electromagnetic interference, and especially sensitive to induced hum due to magnetic fields emanating from the large transformers which find their way into the power supplies of guitar amplifiers! To counter this, Gibson introduced the humbucker pick-up. These comprise two magnets and two coils – wound electrically in series – but arranged in magnetic opposition, as shown in Figure 5.4. The vibrating string will, of course, create a similar signal in both these coils, and these will add due to the series connection. But an external field will induce a signal of opposite phase in either coil, and these will cancel due to the series connection.

Most guitars are fitted with a number of pick-ups and furnished with a selector switch to allow the player to choose their favoured sound. Pick-ups nearest the bridge tend to sound more 'trebly' and bright. Those nearest the

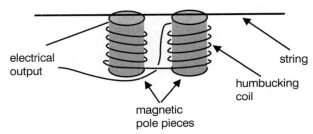

Figure 5.4   *Humbucker pick-up*

fingerboard have a more 'bassy' sound. Because players like to have a local control over amplification level and tone colour, all guitars provide volume and tone controls on the guitar itself. The pick-ups themselves have a relatively high output impedance, so it is necessary that they work into a very high impedance source. For this reason, most guitar volume potentiometers are very high value, perhaps 250k or 500k. Similarly, tone control circuits operate at very high impedance. As you may have already guessed, because of this, the action of the guitar cable itself – as well as the amplifier input impedance – all have a marked effect on the overall sound of an electric guitar set-up. This situation has helped fuel the enormous mythology which surrounds electric guitars, pick-ups and their associated amplifiers. The circuit schematic for the internal circuitry of the famous Fender Stratocaster guitar is drawn in Figure 5.5.

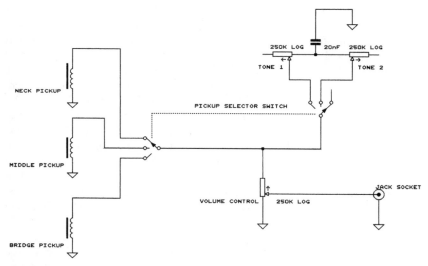

Figure 5.5 *Stratocaster circuit*

Electric organs

Despite their enormous technical complexity pipe organs have played a cardinal role in the development of Western music. Since the middle ages the great organ builders have exercised amazing ingenuity, coupled with an intuitive knowledge of harmonic analysis, to construct the grand instruments in churches and cathedrals. No doubt many builders strove to incorporate sounds (stops) in their creations which had never been heard before. The wonderful names of Open Diapason, Cornopean, Geigen Diapason bear witness to this. But they also strove to emulate existing instruments. Again, you only have to look at the names of the stops – Viole d'Orchestre, Clarinet and Trumpet – to know this. Organ builders often designed their acoustic instruments so that the keyboard action could be coupled together and pipes made to sound above the note actually depressed thereby adding harmonic colour.

Unfortunately large pipe organs not only complement great cathedrals, they also require cathedrals to house them! Many of the great organs in northern European cathedrals have bass pipes 32 feet long – which is a pity if you happen to enjoy playing the organ and would like to do so in your own home! Hence the invention of the first electric organs, for amateurs to share the experience of playing the organ at home. However, they are found increasingly as a substitute for pipe organs in churches, where acoustic organs are often beset by ageing problems because of their construction in wood, leather and in other degradable materials. All electric organ design starts with simple electronic oscillators.

Fundamental sound generation and synthesis

Chapter 2 introduced the idea that any waveform may be synthesised by the addition of a suitable number of sine-waves of appropriate amplitude and phase, a technique known as additive synthesis. Interestingly, however, some of the most simple waveforms when viewed in the time base are the most complicated when viewed in the frequency base and are thus the hardest to synthesise using a frequency domain synthesis approach. Fortunately some of these are straightforward to generate synthetically using a time based approach. But let us look first at a couple of practical circuits for the generation of that most basic of waveform synthesis building-blocks – the sine-wave.

Figure 5.6 illustrates two sine-wave oscillators. The *LC*, tuned-circuit oscillator is more suitable for the generation of radio-frequency waves (100 kHz and above). The circuit shown is known as the tuned-drain-tuned-gate type. Its action may be understood by considering the action of the internal Miller capacitance which couples the tuned circuit in the drain circuit (output port) to the tuned circuit in the gate circuit (input port). In the case of the circuit shown, the input port is tuned by a crystal. The feedback path, the forward gain of the FET stage, and the tuning effect of the *LC* circuit and crystal form the three necessary components of any electronic oscillator:

1 gain,
2 feedback
3 tuning.

This type of circuit is noted for its frequency stability because the crystal is in a condition of very low loading. The second circuit in Figure 5.6 is known as the Wein-bridge oscillator and is formed around a high gain op-

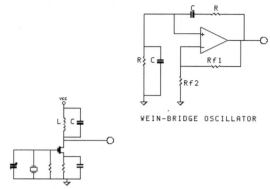

WEIN-BRIDGE OSCILLATOR

TUNED-DRAIN TUNED-GATE RF OSCILLATOR

Figure 5.6   *Sine-wave oscillators*

amp. Once again an op-amp with an FET input stage is ideal here due to its very high input impedance which has virtually no effect on the tuning components. To understand the Wein-bridge oscillator, consider the combination of the series *RC* circuit and parallel *RC* circuit which couples the op-amp's output port to its non-inverting input port. This circuit produces zero-phase shift (and thus maximum feedback) when the magnitude of the reactance of the capacitors equals the resistance of the resistors. At this frequency, the signal appearing at the non-inverting input of the op-amp is one-third the size of the signal at the amplifier's output. The amplification stage has thus to provide a gain of three in order to make up for this loss and maintain oscillation. So R_{f1} is organised to be twice the value of R_{f2}. Of course, if the gain increases beyond three, then very soon the amplifier will start to overload and destroy the purity of the waveform. In order to maintain sinusoidal purity, R_{f2} is sometimes formed by a small filament bulb which – should the output start to rise – will burn brighter, raise its resistance and thus back off the gain of the amplifier. The Wein-bridge oscillator is well suited for the generation of sine-waves at audio frequencies.

Now consider the ramp waveform illustrated as the output of the circuit in Figure 5.7. This waveform represents the simplest time function possible since, over the period between minus *PI* and plus *PI*, the waveform – expressed as a function of time – is time itself! Such a waveform is theoretically realisable by adding an infinite number of sine-waves like this:

$$f(t) = \sin t - \tfrac{1}{2}(\sin 2t) + \tfrac{1}{3}(\sin 3t) - \tfrac{1}{4}(\sin 4t) + \ldots \text{ etc.}$$

However, it is obviously much easier to obtain this waveform by constructing a mechanism which causes an electronic quantity to grow

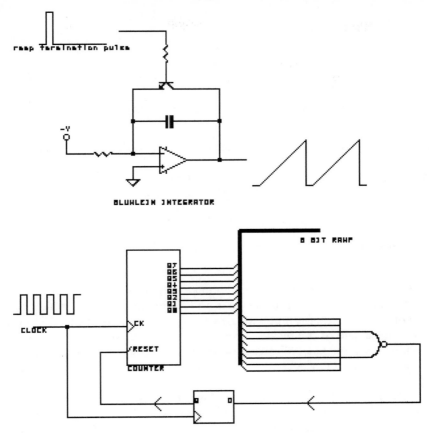

Figure 5.7 *Ramp (or sawtooth) waveform generation*

linearly with time – the voltage across a capacitor which is charging with a constant current, for instance. Just such a circuit is illustrated in the figure. All that is required is to return the value of the voltage to its minimum value at the fundamental periodicity of the waveform. This is achieved by means of a switched, high current discharge path formed by the transistor – the terminating periodic signal repeatedly energising its base emitter circuit. Also in this same figure is a practical circuit for a digital ramp generator which creates a binary integer value which grows linearly with time by stepping a counter circuit with a regular clock signal. This technique has very wide application as we shall see.

Circuits for the generation of parabolic waveforms are given in Figure 5.8. This is achieved by taking a ramp signal and squaring it. Though possible in the digital domain too, an alternative digital approach usually

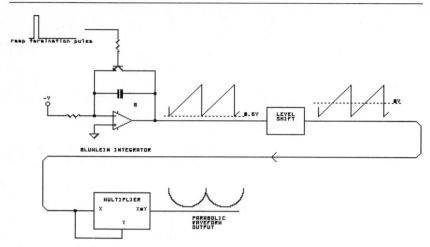

Figure 5.8 *Parabolic wave generation*

taken. In this case, the incrementing counter is used to address an EPROM. Stored inside this device is a look-up table of values which may define any waveform required. This technique demonstrates the power and advantage of the digital waveform generation. Note that the higher addresses of the EPROM can be used to switch between different pages of addresses. In this way a number of periodic waveforms may be selected at will.

Figure 5.9 illustrates techniques for the generation of square waves. Like the ramp, a square wave may be synthesised by an infinite number of sine-waves in the manner:

$$f(t) = \sin t + \tfrac{1}{3}(\sin 3t) + \tfrac{1}{5}(\sin 5t) \ldots \text{ etc.}$$

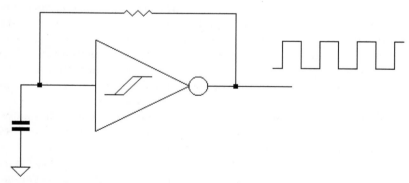

Figure 5.9 *Square-wave generator*

But it is much easier, since the waveform may be thought of in the time domain as switching between two states for equal lengths of time, to arrange a circuit to do just this, in the form known as an astable multivibrator.

Electric organs are required to emulate the various basic organ tone families (stops in organ parlance). There exist four basic categories of organ tone:

Flute tone
Diapason tone
String tone
Reed tone.

There are two basic types of electronic organ: divider organs and free-phase organs. The divider type uses a digital top-octave generator (one oscillator for each semitone) and chains of divide-by-two bistables to provide the lower octaves. This has the dual advantage of simplicity and consistent tuning across octaves. However, this approach has a number of disadvantages. First, because the generators are digital, all the fundamental notes are square-waves and therefore contain nothing but odd harmonics; ideal as the basis for reed tones but useless for the production of string and diapason tone. Second, because all the notes of every octave are locked in frequency and phase, there exists none of the richness of tone associated with a real pipe organ: a phenomenon that can be traced to the slight mis-tuning of similar notes in different octaves in a real organ, each producing a myriad of beats between fundamentals and overtones. (Indeed much of the expressiveness of orchestral string tones, or choral vocal tones, can be traced to the same phenomenon, to the slight differences in pitch between individual players or singers. We shall return to this counter-intuitive desirability of slight mistuning several times in our journey through musical electronics.) At any rate, the divider-type organ tends to produce a 'sterile' tone disliked by musicians.

The alternative is known as a free-phase electronic organ. Theoretically the free-phase organ has a different oscillator for each note of the keyboard. This is ideal but practical issues of complexity often force a compromise whereby a single oscillator is shared between adjacent semitones; the keyboard switch doing double duty as both a key to make the oscillator sound and to set its pitch between, for instance, G and G-sharp. The rationale behind this lies in the nature of Western tonal music which rarely requires adjacent semitones to sound simultaneously. Designers have shown considerable ingenuity in the implementation of such circuits. In the past organ builders have tended to use *LC* oscillators in practical designs but, more recently, there is a tendency towards Wein oscillators despite the considerable complications which arise in keying such an oscillator without producing an audible 'click' at the start and end of the note, and in the arrangement to have the oscillator sound at two different pitches.

In either case the predominantly sine-wave nature of the signal produced from the *LC* or Wein oscillator is ideal as the fundamental basis of the flute stop tone. Waveforms with higher harmonic content are required to simulate the many other pipe sounds. The simplest circuit in this respect is the diode clipper shown in Figure 5.10. This 'squares up'

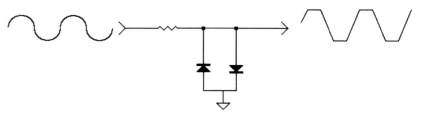

Figure 5.10   *Diode clipper*

the incoming sine-wave producing a primitive reed-stop tone. Note that this is the first example in this book where a non-linear circuit is used in a musical way. As mentioned in Chapter 1, this essential non-linearity is the essence of musical electronics and distinguishes it from recording and reproduction electronics which value above anything else linearity in frequency response and transfer characteristic.

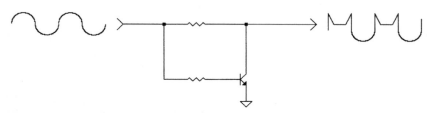

Figure 5.11   *Non-linear circuit for complex tone generation*

An interesting non-linear circuit is illustrated in Figure 5.11. This circuit is used to introduce second and higher even-numbered harmonics. The output of the circuit is illustrated also. This type of circuit is used to simulate the sound of the open diapason and, with modifications, can be employed to produce a wide variety of tone colour.

Hammond and Compton organs
At first, electronic organs sought only to emulate the tone of the acoustic organ, a feat which is now so well accomplished that only experts can tell if

the organ in a church is electronic or the acoustic original. But it wasn't long before the designers of these electronic instruments began experimenting with harmonic combinations which the increased flexibility of electronic – instead of mechanical – coupling allowed them. Just such an ambition led to the development of the classic Hammond B3 organ.

Figure 5.12 *Hammond B3 organ*

The designers of the Hammond B3 organ (Figure 5.12), predating as it did the development of solid-state electronics, wisely forewent the use of electronic oscillators to produce the fundamental sine tones and instead opted for an electromechanical scheme whereby rotating mechanical discs with shaped edges influenced the magnetic field of electromagnets wound near the edge of the disc. The principle, illustrated in Figure 5.13, is thus a variable reluctance electro-mechanical oscillator and is pretty

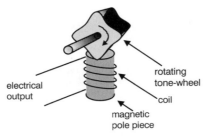

rotating
tone-wheel

electrical
output

coil

magnetic
pole piece

Figure 5.13 *Variable reluctance oscillator*

well unique. Other manufacturers displayed equal lateral thinking – Compton used rotary tone generators too but these operated by means of a variable capacitive technique. Identical electromechanical components were used for each note of the scale, the different pitches being achieved by the choice of pulley ratio used to drive the tone generators from a common mechanical drive. Hammond's ambitions went far beyond that of reproducing a pipe organ sound and instead aimed at recreating the sounds of other instruments. Their additive synthesis technique involved the analysis of real instrumental sounds (using a Fourier analyser) and the recreation of these by means of the suitable selection and addition of sine-waves generated from the continuous oscillator 'bank'. Fascinatingly, it is fair to say that Hammond almost totally failed to achieve what they set out to do with the Hammond organ, to simulate the sounds of other instruments! However, in so doing they created a 'classic' sound in its own right. This is another theme which we shall see over and over again in the following pages, that of 'successful failure' and the essential unpredictability of musical electronics.

Additive synthesis represents an entirely logical approach but it is, of course, beset with practical difficulties – the sheer number of oscillators for a start! The complexity of a two-manual (keyboard) electronic organ may be inferred from the above but consider that the tone colour of the pipe organ is relatively straightforward! We shall see later on that designers have had to find means other than additive synthesis to simulate the sounds of other musical instruments.

Theremin

One of the earliest electronic instruments, this monophonic (single tone) melodic instrument was originally developed in Russia about 1920 by Lev Sergeyevich Termin. When Termin moved to the USA he changed his name to Theremin and improved and patented his original invention. His instrument, the eponymous Theremin, was produced in sizeable numbers by RCA in the 1930s. Theremin's other inventions included the Terpsitone (a Theremin worked by the body movements of a dancer), the Rhythmicon (a sort of early sequencer/drum machine), an early attempt at colour TV, and a bugging device for KGB use! In 1939, in mysterious circumstances, Theremin returned to the Soviet Union, without his wife – whom he had met and married in America. There, he fell in and out of favour with the authorities, and then with various employers, so that he could only work on his musical instruments in his spare time. He was still working a 40-hour week (and living in a cramped room in a communal flat) when he was in his eighties. For decades, most people in the West assumed that Theremin was dead. When news that he was alive eventually filtered through, he was invited to various music festivals in Europe, and then, in 1991, to Stanford University's Centennial Celebration, where he was guest of honour at a concert attended by Bob Moog, Tom Oberheim, Don Buchla, Dave Smith, Roger Linn, and other electric music gurus. Theremin died in 1993 aged 97.

Theremin's own words (1924) describe the nature of the instrument admirably:

> In broad aspect, the means of the invention comprises an oscillating system capable of producing audible sound tones and adapted to be influenced or affected by an object or objects, such as the hands or fingers of an operator moved in proximitive relation to an element thereof, together with a sound re-producer operatively connected to said system . . . In order to generate clear sound or musical tones, and permit ready control thereof, a plurality of oscillators are employed, having a frequency above the audible range but interacting with each other to produce interference or beat-notes of audible frequency. The frequency of one or more of the oscillators is controllable by the operator to produce beat-notes of the desired pitch. The apparatus in preferred form also embodies means for controlling the volume and timbre of the music.

The circuit diagrams accompanying the 1924 Theremin patent are reproduced for interest as Figure 5.14, but these are rather difficult to read, so the operation of the Theremin is illustrated in schematic form in Figure 5.15. A circuit diagram of a version (attributed to Moog) is given in Figure 5.16. Notice that the instrument contains three radio-frequency generators (operating in the hundreds of kHz region). Radio-frequency oscillators 1 and 2 are pretuned to exactly the same frequency. Clearly, the resultant output from the non-linear circuit (the RF mixer) will be the sum and difference signal; the sum being subsequently filtered, leaving the difference signal alone to be passed on to the following amplifier stage. Oscillator 1 differs from oscillator 2 with the addition of the extra tuning capacitance, across the main resonant circuit, formed by the metal aerial and its interaction with ground. The player has only to bring their hand or body within a small distance of the aerial for there to be a change in oscillation frequency and a resultant audible tone issuing from the process of multiplication. The nearer the player approaches the plate, the more depressed the oscillation frequency of oscillator 1 and the higher the resultant pitch of the Theremin's audio-frequency output. In this most simple form, the Theremin is thus able to produce an endless range of frequencies from the subsonic to the inaudibly high in a long sustained glissando.

The expressive potential of such a system is inevitably limited, hence the addition of the third oscillator and its associated circuitry. This third

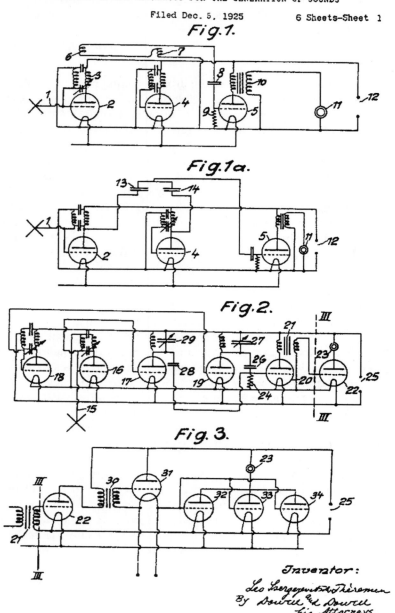

Figure 5.14   *Theremin patent illustrations*

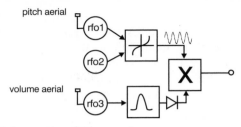

Figure 5.15 *Schematic of Theremin*

RF circuit produces a tuneable output, once again variable by means of the interaction of the player's anatomy in proximity to another metal aerial or wand. But this oscillator does not interact with another oscillator, instead its output is fed to a resonant circuit, tuned to the lower end of the variable oscillator's range. As the player approaches the aerial, the generated frequency drops and the output across the resonant filter rises. Suitably rectified, this signal becomes a control voltage which is employed to alter the gain of the final audio stage. The complete instrument thus has the ability to change pitch and volume and thereby produce articulate musical phrases. It is generally played with two hands, one to adjust the pitch, the other to adjust the intensity. In fact, the original RCA Theremin worked the other way about; with movement

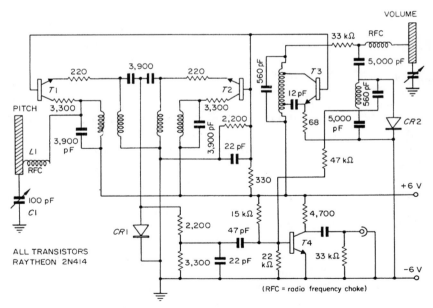

Figure 5.16 *Transistorised Theremin (attrib. Moog)*

away from the antennae causing the note to sound. This inconvenient and counterintuitive arrangement is reversed (or is switchable) on all modern Theremins. A modern Theremin, due to Big Briar, Inc. (554-C Riverside Dr., Asheville, NC 28801, USA) is illustrated in Figure 5.17. Big Briar produce beautiful, hand-made Theremins in three basic styles – two of Theremin's original designs, and one by a contemporary sculptor.

Figure 5.17   *Big Briar Theremin*

Despite being very difficult to play, the Theremin has achieved limited artistic success. It may be heard in several orchestral pieces including Andrei Paschenko's *Symphonic Mystery for Theremin & Orchestra* (1924) and Schillinger's *First Airphonic Suite* (1929). *Ecuatorial*, by Varese, originally called for two Theremins. The Russian composer Shnittke has written pieces for Theremin, as did Percy Grainger. Pop uses include the Beach Boys' *Good Vibrations* and the instrument has been used on many film and early TV soundtracks. The sound effects on *The Lost Weekend* and other great films were performed by one Samuel Hoffman, who also played Theremin on a 1950s album called 'Perfumes Set to Music'. Interestingly, the Theremin's real success is symbolic; it remains the emblem of experimental electronic music. A status that it perhaps enjoys because it is one of the very few instruments designed in historical memory to employ a truly novel playing technique.

The Univox

Designed in 1946 by Derek Underdown and Tom Jennings, and manufactured by JMI Company, the Univox (Figure 5.18) was a small, portable valve-based monophonic organ with a range of tone selections and vibrato effects. The sound of the Univox is familiar to everyone from the Tornadoes' 1962 single *Telstar*. Tom Jennings went on to form VOX.

Figure 5.18 *Univox*

The Hohner Clavinet

The Clavinet was, commercially and artistically, the most successful keyboard produced by German company Hohner, who designed it to replicate the sound of a clavichord (Figure 5.19). Various early models culminated in the Clavinet model 'C'. This, in turn, was refined into the D6 – a portable, amplifiable keyboard. This had a fully dynamic keyboard, see Figure 5.20, so the harder the player hit the keys, the louder and more vibrant the tone produced (Coates 1997). This design ensured the degree of tactile feedback that is so necessary for a truly 'musical' experience and, no doubt, accounted for its success with recording and performing musicians, among whom it was widely

Figure 5.19   *The Hohner Clavinet*

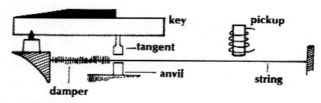

Figure 5.20   *Clavinet key action*

regarded it as a 'funky keyboard'. The Clavinet's most famous recordings include *Superstition* by Stevie Wonder and *Nut Rocker* by Keith Emerson of Emerson, Lake and Palmer.

Electric pianos

The most famous electric piano is, without doubt, the Fender Rhodes. This, and its many imitators, are actually more of an electronic glockenspiel (or vibraphone) than an electronic piano because the sound-producing mechanism is formed from struck metal bars, the hammers being actuated via a conventional keyboard mechanism. The Fender Rhodes piano dates from the early 1940s when Harold Rhodes, an American serviceman, built a 'baby piano' in which metal rods were struck directly by the wooden keys themselves. It was an immediate success with the servicemen, for whom it was built to entertain, and hundreds were constructed. Later on, the pitch range was increased and a damping mechanism devised. Finally, an adaptation of the electric guitar-type pick-up was added so that the piano could be amplified. It was this unit that attracted the attention of guitar maker Leo Fender and thus the Fender Rhodes, as we know it today, was born.

The operation of the Rhodes is simple. The wooden key activates a hammer via a cam. When the key is depressed, the dampers are lifted above the sounding bars which are struck by the hammer. This bar (known as a tine) vibrates and disturbs the magnetic circuit formed by the permanent magnet within the pick-up. This movement is thereby transduced into an electric current. Figure 5.21 is an illustration of the Fender Rhodes action (ibid.) and Figure 5.22 details the transduction mechanism. Compare this illustration with that of the electric guitar pick-up and the waveform generation mechanism of the Hammond organ. The

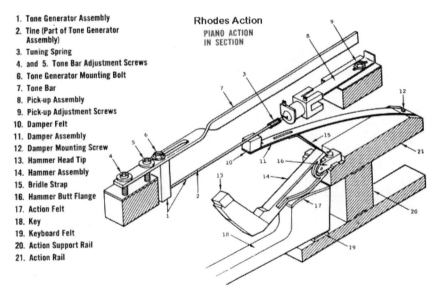

1. Tone Generator Assembly
2. Tine (Part of Tone Generator Assembly)
3. Tuning Spring
4. and 5. Tone Bar Adjustment Screws
6. Tone Generator Mounting Bolt
7. Tone Bar
8. Pick-up Assembly
9. Pick-up Adjustment Screws
10. Damper Felt
11. Damper Assembly
12. Damper Mounting Screw
13. Hammer Head Tip
14. Hammer Assembly
15. Bridle Strap
16. Hammer Butt Flange
17. Action Felt
18. Key
19. Keyboard Felt
20. Action Support Rail
21. Action Rail

Rhodes Action

PIANO ACTION IN SECTION

Figure 5.21 *Fender Rhodes key action*

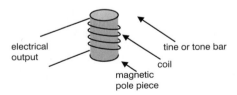

electrical output

tine or tone bar

coil

magnetic pole piece

Figure 5.22 *Fender Rhodes transduction mechanism*

Fender Rhodes was made in two types: a Stage model which was entirely passive – just like a guitar – and a Suitcase model which required mains to power the integral amplifier system (like that illustrated in Figure 5.23). Due to the physical nature of the mechanism – which permitted a large variation in expressive tone by means of the force used to strike a key – coupled with the keyboard's naturally unlimited polyphony, ensured the

Figure 5.23 *Rhodes Suitcase model*

Rhodes was, and continues to be, a widely used instrument. Indeed, so ubiquitous is the Fender Rhodes that it is simply impossible to draw attention to a few particular performances.

Electronic pianos

A good acoustic piano is a very desirable addition to a musical home. However, even the smallest grand piano may be difficult to accommodate in today's smaller houses, especially in advanced industrial regions where space is at a premium. A good upright piano is one solution but (as anyone who has tried moving a piano knows) weight and 'deliverability' are an issue for people living in flats or who are forced to move on a regular basis. These and other considerations have ensured a market for electronic pianos which aim to simulate the real 'piano experience', both in terms of sound and in terms of the physical, tactile feel of the keyboard in a package which is smaller and lighter (and cheaper) than a real piano. An example of an electronic piano due to Roland is illustrated in Figure 5.24. This model

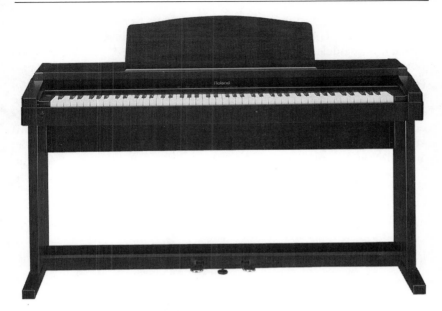

Figure 5.24 *Roland electronic piano*

has a full 88 note piano keyboard with each key individually weighted via an oil-damped mechanism to simulate the feel and responsiveness of a good grand piano. This particular model uses Roland's proprietary Structured Adaptive Synthesis (SAS) which is an eclectic blend of techniques, honed to give the most realistic sound possible. Other manufacturers have adopted sampling (see Chapter 7) as the basis of the sound generation for their electronic pianos. The techniques used by modern manufacturers in the design and implementation of these instruments are so good that electronic pianos represent a real alternative to pianos used at home and for education. They offer very significant advantages too for the sound engineer, because any piano is a notoriously difficult instrument to record – and a poor piano is virtually impossible to record well. Not only do electronic pianos offer headphone outputs which may be used for private practice but all offer line level (stereo) outputs which may be taken via direct injection (DI) into the sound mixer.

Martenot

The Ondes Martenot (literally, Martenot Waves) was invented by Maurice Martenot, professor at the Ecole Normale de Musique in Paris. The first model was patented on 2 April 1928 under the name Perfectionnements aux instruments de musique électriques (improvements to electronic music instruments). The first versions bore little resemblance to the later production models as illustrated in Figure 5.25. Indeed, the earliest units bore a closer resemblance to the Theremin, consisting, as they did, of two table-mounted units controlled by a performer who manipulated a string attached to a finger ring, using the body's capacitance to control the sound characteristics in a manner very similar to the Theremin. This device was later incorporated as a fingerboard strip above a standard keyboard. The Ondes Martenot was first demonstrated in Paris in 1928 and it won first prize at the 1937 International Exhibition of Art and Technics. Many of the first composers to hear and take up the instrument

Figure 5.25   *The Ondes Martenot*

were fascinated by the sounds it could produce, as it combined great responsiveness to touch with its eerie and ethereal electronic tones. The instrument became popular among members of Les Six in France (particularly Milhaud and Honegger). One of the early virtuosi of the Ondes was Martenot's sister, Ginette Martenot. Later instruments also had a bank of expression keys that allowed the player to change the timbre and character of the sounds. One version even featured micro-tonal tuning.

Martenot's aim, to produce a versatile electronic instrument that was immediately familiar to orchestral musicians, paid off, because the Ondes Martenot is probably the most widely accepted of all electronic musical instruments in the classical oeuvre. The Ondes Martenot therefore has a surprisingly wide repertoire, far wider than that of the Theremin. Works were written for the instrument by distinguished composers – Edgard Varèse and Olivier Messian among others, the latter orchestrating the *Turangalâla Symphonie* and *Trois Petites Liturgies de la Presence Divine*. Other composers include Maurice Jarre, Jolivet and Koechlin. The Martenot often figures either as a solo instrument (as in works such as Marcel Landowski's *Jean de la Peur*) or as an orchestral instrument, employed from time to time within a score for certain special effects. The birdlike calls and trills distinctive of the work of Olivier Messaien are a good example of this usage. Other composers wrote for ensembles of Ondes, sometimes as many as eight at a time!

Mellotron

Despite Beatles' producer George Martin's reservations that this instrument came about 'as if a Neanderthal piano had impregnated a primitive electronic keyboard, and they'd named their deformed, dwarfish offspring 'Mellotron' (Martin 1994), this (albeit primitive) analogue sampler had a profound effect on the tonal palette of popular music of the 1960s. It operated by means of a length of tape upon which were recorded recordings of real instruments. When a key was pressed, the length of tape was drawn over a playback head until it was exhausted. In Martin's words, 'whereupon a strong spring snapped it back to the beginning again. This meant that if you held down a note longer than a couple of seconds, the machine would give a loud hiccup and stop while it rewound and reset itself.'

The Mellotron was the conception of a Californian, Harry Chamberlin, in the late 1940s. True to its pedigree as the world's first sampler, the original model had 14 loops of drum patterns and was aimed at the home organ market. For the next ten years, Chamberlin designed and manufactured a series of keyboards culminating in a two 35-note console machine; the first console being devoted to the 'sampled' instrumental sound, the second to rhythm tapes and sound effects. In the 1960s, Chamberlin hired a salesman who, frustrated by the inventor's inability to resolve various technical problems, took the idea to Bradmatic Ltd. in England, who supplied tape heads for the Mellotron. He suggested they engineer a new model of the Mellotron and this they duly did. (Unfortunately the salesman failed to tell Bradley that the concept wasn't his and, similarly, omitted to inform Chamberlin about the new 'arrangement'!)

After much acrimony, in 1966 Chamberlin agreed to sell the technology to the Bradleys who renamed their company Streetly Electronics and commenced production of the mature Mellotron keyboard. (Chamberlin continued on a parallel development path with a series of instruments known simply as the Chamberlin.) It was the Bradleys' new Mellotron keyboard that attracted the attention of British bands who were searching for new additions to their tonal palette. Among them; the Beatles, the Rolling Stones and the Kinks. In 1966, John Lennon composed a small phrase which McCartney played on the Mellotron; it was the beginning of *Strawberry Fields Forever.* This four-bar phrase alone, forming as it does the opening of one of the most innovative records of all time, guarantees the Mellotron a place in the annals of sonic history. The exterior and interior of a 1960s Mellotron are illustrated in Figures 5.26 and 5.27 respectively.

Figure 5.26 *The Mellotron*

Tape-bow violin

Somewhat akin to the Mellotron is the tape-bow violin, the invention of Laurie Anderson who was born in 1948 in Chicago, Illinois. Anderson studied sculpture at Columbia University and engaged in various performance artworks while at college. After qualifying, she remained in New York where she met Phillip Glass. During work with a number of electronic musicians, Anderson designed the tape-bow violin, an instrument with magnetic tape instead of a bow, and a playback head instead of

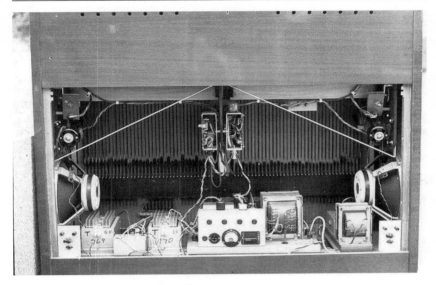

Figure 5.27   *Interior of Mellotron*

strings. The musical 'sample' recorded on the bow could be made to play by drawing the bow across the tape head as in conventional violin technique. The invention's power lies in the fact that variations in bowing can bring about very flexible sample manipulation. Anderson is most famous for her hit single *O Superman* which was a small part of her huge United States I–IV series. Anderson's latest work involves collaboration with Brian Eno and Lou Reed.

References

Coates, B. (1997) Melbourne Music Centre Web Pages.

Martin, G. and Pearson, W. (1994) *Summer of Love - The Making of Sgt. Pepper.* Macmillan.

Theremin, L.S. (1925) US Patent: Method of and Apparatus for the Generation of Sound. Serial No. 73,529.

Fact Sheet #5: Digital logic

● Logic gates and functions
● Logic families
● Circuits with memory

Logic gates and functions

It's a common observation that a large problem is better broken down into smaller, more manageable problems. Whilst this is an intuitive sentiment, its validity has been proved by logicians and mathematicians: it really is the case that large mathematical and logical problems are the sum of many, much simpler, operations. For instance, multiplication is a complex operation: but it's really just repeated addition. That's to say,

$$(6 * 6) = (6 + 6 + 6 + 6 + 6 + 6)$$

Digital circuits, inside DSP chips and computers, employ binary arithmetic in which only two number types exist, 0 and 1. In physical terms these are represented as two, distinct voltage ranges around 0 V (known as low-level or LO) and a higher voltage range, about 3.3 V or 5 V (known as high-level or HI). In logical terms these states are also referred to as FALSE and TRUE. In other words, we can think interchangeably about the two states within each circuit of a digital circuit as,

Number		Voltage		Logical state
0	=	LO	=	False
1	=	HI	=	True

The third nomenclature ('logical state') explains why, however complicated – or even intelligent – the external behaviour of a digital computer or DSP might appear, at a microscopic level, all the processing is derived from very simple electronic circuits which perform very simple, logical functions. These simple circuit elements are therefore known as logic gates.

The first type of logic gate is the simplest of all; a logical 'inverter'. If it receives a logical FALSE (a LO, or a numerical 0), it outputs a logical TRUE (HI or 1). It is formed from one transistor (or metal-oxide enhancement FET) and it is illustrated in Figure F5.1, in both its physical implementation and schematic representation. It is the latter which you will see in circuit

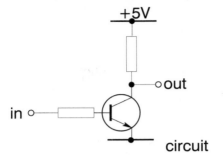

+5V

out

in o

circuit

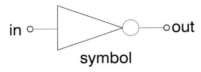

in o ———▷◯——— o out

symbol

Figure F5.1   *Inverter circuit and schematic symbol*

diagrams. Let's just represent the function of this circuit in a table.

Input	Output
0	1
1	0

This contrary circuit can be thought of in terms of a logical NOT. The output is 'NOT' the input.

Now things get a bit more interesting with a gate which makes a logical decision. Figure F5.2 illustrates the physical implementation of such a gate. Notice it is made up of two transistors (see Fact Sheet 4) and crucially has two inputs. The gate's job therefore, is to 'make a choice' based on the state of these two inputs. The two active elements share the same collector load; which is also the output of the gate. With LO voltages on their bases, both transistors will remain cut-off and the voltage on their collector will be HI. But, if either of them receives a HI voltage on their base, that transistor will saturate and the voltage at the common collector terminal will fall to

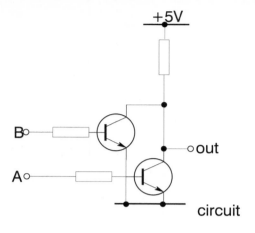

circuit

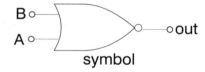

symbol

Figure F5.2   *NOR gate circuit and schematic symbol*

the saturation voltage, a LO. The same condition will result if both transistors receive HI inputs. In tabular form,

```
           B
         0   1
       ┌────────
     0 │ 1   0
   A   │
     1 │ 0   0
```

Note that the two possible conditions of each of the two inputs are shown at the left and at the top of the table. The logical conditions of the possible output states are shown in the centre part of the table. Because this table characterises the logical states resulting from the logical states of the combination of its inputs, it is known as a 'truth table' and this type of logical function is known as 'combinational logic'.

The truth table for this circuit reveals that the output is low (or FALSE) when either A input is true OR when the B input is true. The inherently inverting form of this circuit slightly complicates the issue here because the gate is therefore not known as an OR-gate, but as a NOR gate; meaning the combination of a NOT and an OR. The subsequent addition of an inverter after this gate would construct a logical OR gate. The NOR gate is illustrated as a circuit element is illustrated too in Figure F5.2.

This little two-transistor circuit is the basis for all decision making in digital circuits. No matter how powerful the computer or how smart the behaviour, at a circuit level all decisions are broken down into small NOR type decisions. Now, imagine putting two NOT gates before the NOR gate: the output would only be HI when both the inputs were HI. The truth table would look like this,

	B	
	0	1
A 0	0	0
A 1	0	1

We can say that the output is TRUE only when input A AND B are TRUE. The circuit is therefore an AND gate. Figure F5.3 illustrates its circuit schematic and the schematic element for its brother, the NAND gate; formed by the downstream addition of an inverting NOT gate.

Another, important combination of inverters and NOR gates is known as the exclusive-OR, or EX-OR gate. In the EX-OR gate, the output is TRUE only when either A OR B are TRUE, but not when both are true. The truth table is therefore,

	B	
	0	1
A 0	0	1
A 1	1	0

The schematic element of this, and the EX-NOT (or NOT EX-OR) is illustrated in Figure F5.4.

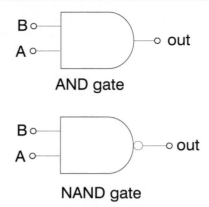

AND gate

NAND gate

Figure F5.3　*AND gate and NAND gate*

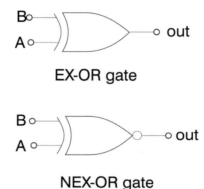

EX-OR gate

NEX-OR gate

Figure F5.4　*EX-OR gate and NEX-OR gate*

Logic families

You may hear the term 'logic family' from time to time. Each family actually refers to a different technology of logic gate. The logical functions never differ but the voltages which represent high (1) and low (0) are different. Also some families have input and output resistances which mean that gate outputs can effectively drive a virtually unlimited number of gate inputs (of the same family). These families are said to have high fan-out and high fan-in. Low speed CMOS is a logic family which posseses this advantage. Other families limit the number of gate inputs, like the TTL family and its derivatives.

Circuits with memory – sequential logic

So far, we have looked at, so called, combinational logic. But digital electronics' power rests particularly on its ability to store and to recall, not only data, but also program instructions. It is this capacity of digital circuits which differentiates it from – and accounts for its superiority to – its analogue antecedents.

The basic memory device is formed from two simple gates as illustrated in Figure F5.5. Assume that inputs A and B are both TRUE (HI), look at the top gate and imagine that the other input (equivalent to output Y) is also HI. Because the gate is a NAND, the output X will be LO. The bottom gate will therefore be in the following condition, input B will be HI, its other input (effectively X) will be LO and its output (Y) will therefore be HI; which is what we imagined. In other words, this circuit will sit happily in this condition; it is said to be a 'stable' state.

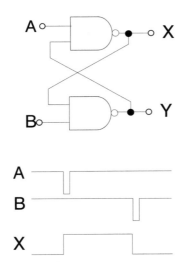

Figure F5.5 *Simple bistable or 'flip-flop' circuit*

Because the circuit is symmetric, it should be pretty obvious that the circuit will just as happily sit in the alternative condition where X is HI and Y is LO. (If it's not, work through the circuit with the NAND truth-table.) This arrangement of gates is stable in two distinct states and is known as a bi-stable. Another – more colourful name is flip-flop. This circuit element is the basis for all digital, electronic memory.

To set the state of the memory it's necessary momentarily to pull either the A or B input LO. The action of these two inputs is illustrated in Figure F5.5 too. Note that pulling A LO causes the flip-flop to go into one of its possible states and input B causes it to go to the other. For this reason inputs A and B are usually termed the RESET/SET inputs and this type of bi-stable is called a RESET-SET (or R-S) flip-flop.

The addition of further gates enhances the bi-stable so that it may be forced one or other state by the action of a single data (D) input, the logical state (0 or 1) of this input being gated by a further clock pulse. This type of flip-flop is called a D-type bi-stable, or more often, a 'latch' because data is 'latched into' the memory element. D-type flip-flops are probably the most common circuit in modern digital electronics as program and data memory is formed from bi-stable elements.

6
Wild Thing – Electronic effects

Echo and reverberation

A true echo is only heard when a reflected sound arrives a twentieth of a second or more after the direct sound first reaches our ears. Compare that sound with the sound that accompanies the voice of a priest or of a choir as their effusions stir the roar of reverberation in the atmosphere of a vast, medieval cathedral. This reverberation is made up of echoes too, but by a mass of echoes following more swiftly than those of a discrete echo. In fact, reverberation has several, distinct phases as illustrated in Figure 6.1. The delay between the original sound and the first reflection is known as the pre-delay, there follows several distinct reflections which gradually fuse into a more-or-less continuous reverb 'trail'. These die away until they are –60 dB below the intensity of the original sound, by which time the

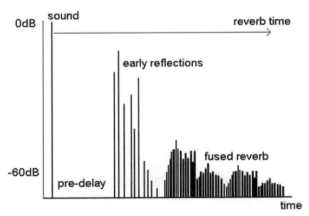

Figure 6.1 *Anatomy of reverberation*

reverberation is deemed to have stopped; the time from the instant of the original sound to the −60 dB point being termed the 'reverberation time'.

Clearly most recording studios are not large enough for an echo to be a natural consequence of their design. Neither are most cavernous enough to possess the acoustics of a cathedral. And a good thing too for it is far easier to add artificial reverberation and echo than it is to eliminate the natural form. This then is the philosophy behind most modern approaches to smaller studio design – aim for a dry natural acoustic and augment this with artificial reverberation when required.

Artificial echo was originally accomplished by means of a tape delay device as illustrated in Figure 6.2, the signal being fed to the record head

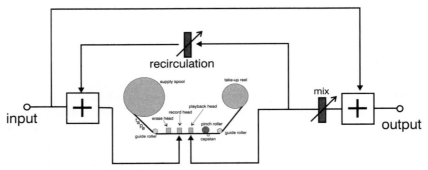

Figure 6.2   *Tape-based echo unit*

and the 'echo' signal picked off the replay head which was situated separately and 'downstream' of the record head. The distance between the two heads and the tape speed, determined the delay. On commercial units, the tape speed was usually made continuously variable so as to realise different delay times. This arrangement, obviously, only produced a single echo. In order to overcome this limitation, to this simple device a circuit was added which allowed a proportion of the output of the replay head to be fed back and re-recorded. By this means was an infinitely decaying echo effect performed (approximating fused reverb). By altering the tape-speed and the degree of feedback – known in this context as re-circulation – differing reverberant trails could be achieved. Just as the early microphones had, in their time, fathered the vocal style of crooning (because they performed best when capturing slight sounds very close to the diaphragm), so the tape-based echo unit spawned an entire vocal technique too.

Modern digital delay devices have shunned tape techniques but accomplish the same effect by delaying suitable digitised audio signals

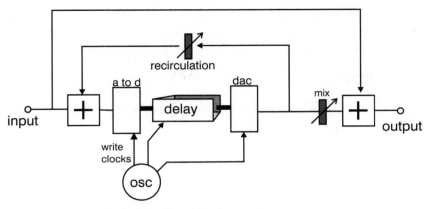

Figure 6.3   *Digital store-based echo unit*

(see Chapter 10) written into, and read out of, a RAM store; see Figure 6.3. Alternatively hybrid digital/analogue techniques are utilised which exploit 'bucket-brigade' delay lines. Both these techniques have all the obvious advantages of a purely electronic system over its electro-mechanical precursor. With one exception. And that is that oftentimes the rather poor quality of the tape transport system in the early devices introduced a degree of randomness (in the form of wow and flutter – see Chapter 9) into the replay system which help ameliorate a 'mechanical' quality which the resulting reverberant sound otherwise has. Digital devices exhibit this quality quite distinctly; particularly at short delay times when the tail takes on a characteristic 'ring'. This unwanted outcome manifests itself more clearly still when the initial delay shortens as in the case of synthesised reverberation of smaller spaces.

When a simple delay and re-circulation technique is employed to synthesise a reverberant acoustic it can take on a very unnatural quality indeed. Better results are obtained when a number of unequally spaced delay points (taps) are used and these separate signals fed back in differing proportions (weightings) for re-circulation. Top quality delay and artificial reverberation units go so far as to introduce several discrete delays in parallel and quasi-random elements into the choice of delay taps and weightings so as to break up any patterns which may introduce an unnatural timbre to the artificial acoustic. Fortunately digital techniques have come so far that reasonable results are obtainable at very low cost. Figure 6.4 illustrates a screen from the program Cool Edit Pro which illustrates an interesting 'number crunching' solution in which reverb is calculated (although not in real time – yet) and the program permits the parameter entry defining the shape and reflectivity of the modelled reverberant space. Another, state-of-the-art solution, dubbed 'convolu-

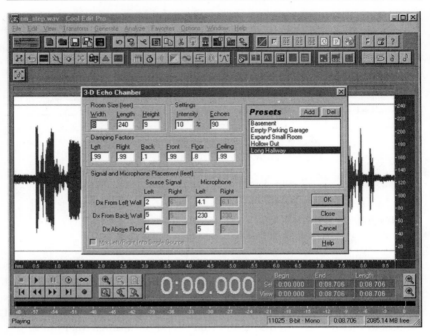

Figure 6.4 *Software-based echo and reverberation*

tional reverb' is presented in Fact sheet #13. Artificial delay and reverberation are almost always incorporated in the audio system via the audio console effect send and return (see Chapter 12).

Tremelo

Rarely used nowadays except as a period gimmick, one of the earliest guitar effects was called Tremelo. This pumping, amplitude modulation effect was originally accomplished within the guitar amplifier and was implemented with a valve-based low frequency oscillator (LFO) which modulated a light-source and thereby a light dependant resistor. This latter component typically formed the lower leg in a potential divider circuit within the combo pre-amplifier. LFO frequency and degree of attenuation were usually controllable by means of potentiometers labelled 'speed' and 'intensity'. The effect was typically switchable by means of a foot switch which enabled and disabled the LFO.

Fuzz

Usually an effect to be guarded against in both design and operation of any audio circuit is the severe amplitude distortion known as clipping. Usually this is caused because a signal is applied to a circuit at a sufficient amplitude that it drives the circuit beyond its available voltage swing. The waveform is thus 'lopped off' on either positive or negative excursions or both. For guitarists this effect is amongst their stock-in-trade. (Grunge has re-established this sound in recent years.) Known variously as fuzz, overdrive or plain distortion, the manner in which the circuit overloads becomes an integral part of the sound timbre. So much so, that for guitarists, a whole mythology surrounds this subject! The first commercially available unit intended solely for the purpose of generating severe waveform distortion (another non-linear, musical function) was The Gibson Maestro Fuzztone (1962). It was this unit that was used on the Rolling Stones record *Satisfaction* in 1965 although the effect had been known for many years and was simply achieved by over-driving amplifiers or loudspeakers or both!

Inside digital sound processors, distortion can be carefully controlled by passing the linear PCM signal through a look-up table stored in ROM

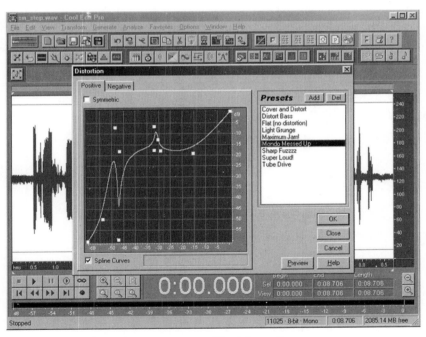

Figure 6.5 *Distortion curve specified in software program*

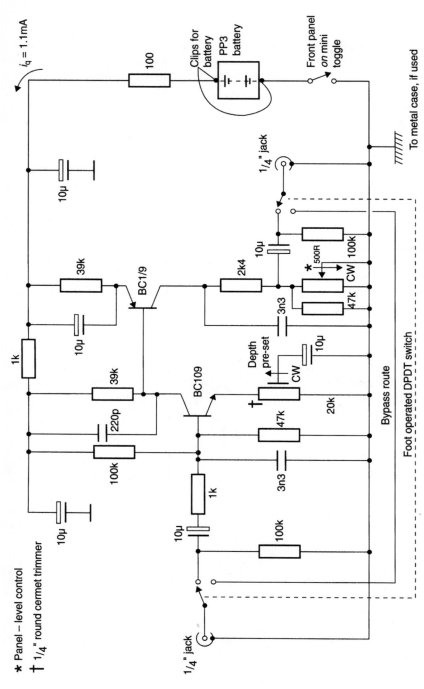

Figure 6.6 *A practical 'fuzz' circuit*

with any desired transfer-function. Similarly a personal computer (PC or Mac) can incorporate a programmable transfer function as shown in Figure 6.5. Nevertheless analogue alternatives are often preferred (see Fact sheet #12 for why!) and may be extremely simple. A design which has been used for some years, and which has appeared on many professional recordings, is illustrated in Figure 6.6. Effectively the transistor pair create a high gain amplifier – enough to drive the output signal well beyond the supply rails. The collector load on the second stage is split to reduce the overall gain back to around unity and to provide an adequately low output impedance. Control of the AC emitter load of the first transistor alters the gain of the amplifier and therefore the depth and character of the distorted output.

Wah-wah

Wah-wah is a dramatic effect derived from passing the signal from the electric guitar's pick-up through a high-Q, low-pass filter, the turnover frequency of which is adjustable usually by means of the position of a foot-pedal as illustrated in Figure 6.7. Most wah-wah effects seem to be

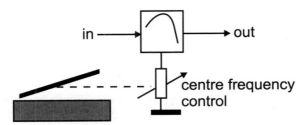

Figure 6.7   *Wah-wah pedal*

based on a similar circuit topology, although circuit values differ considerably from model to model. A typical circuit is illustrated in Figure 6.8. The circuit is a feedback amplifier with a frequency selective circuit (10 nF and 660 mH) in the feedback loop. The position of the pedal determines the degree to which the 10 nF capacitor is bootstrapped and thereby the degree to which its value is multiplied by the forward gain of the first transistor of the pair. In this way, the resonant frequency of the LC circuit is varied, influencing the filter frequency response. The circuit response is typically a low-pass response with a small (+3 dB) peak at about 500 Hz with the pedal in the up position (small amount of feedback): to a very peaked (+15 dB at 1 kHz) low-pass response with the

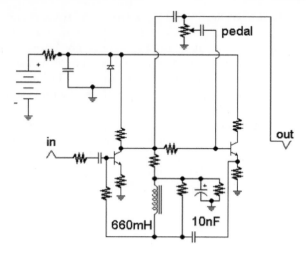

Figure 6.8 *Classic wah-wah circuit*

pedal in the down position. Typically, due to the very small value of output coupling capacitor (10 nF), there is considerable loss of LF response as well; making the overall effect substantially band-pass.

As already mentioned, various versions of this circuit are in current circulation and there exists a great variation in the values; specifically of the feedback capacitor and the output coupling capacitor. If this latter component is increased to 100 nF from its more usual 10 nF, the output is a very markedly peaked response (+20 dB) which sweeps between about 400 Hz and 1 kHz. This gives the 'same' circuit a very different performance and explains why wah-wah pedals are widely 'tweaked' by their owners to garner a different sound.

In use, the guitar player exploits a combination of standard guitar techniques together with associated pedal movements to produce a number of instrumental colours from an almost percussive strumming technique, to a lead guitar style (usually in combination with fuzz effect) in which the guitar, filtered with pseudo-vocal formants, almost 'cries' in a human-like voice.

Pitch shifting

Pitch shifting is used for a number of aesthetic reasons, the most common being the creation of 'instant' harmony. Simple pitch shifters create a constant interval above or below the input signal like this harmony at a major third:

Figure 6.9

As you can see, only three of the harmony notes exist in the key of C major (the scale on which the original part is based) all the rest are dissonant and are unsuitable when harmonised by virtually any chord within the key of C major. You might think that such a limitation was pretty devastating, however, various automatic transpositions produce less disastrous results. For instance, a harmony at a perfect-fifth produces the following scale:

Figure 6.10

which is usable except for the F-sharp. Harmony at a perfect fourth is even better since it,

Figure 6.11

has only one note which is not present in the key of C major, like the harmony at the perfect fifth. But the note is B-flat which is a prominent 'blue' (i.e. blues-scale) note in C major. It is therefore often acceptable in the context of rock music. For this reason the instant transpositions of perfect-fourth up (or its lower octave equivalent, perfect-fifth down) are the most common transpositions employed in simple pitch shifters (with the exception of octave transpositions). Guitarists in particular most often employ a pitch shifter in one or other of these two roles. Intelligent Pitch Shifters can be programmed to produce a harmony which is related to a selectable musical key, so that a musical harmony can be created. Like this harmony at a third:

Figure 6.12

Technically, pitch shifting is achieved by converting the input signal to a PCM digital signal, writing audio data into a short term store, and reading it back out at a different sample rate. Thereafter, the resulting PCM signal is converted back to an analogue signal. Because the short-term store is used over and over again it is referred to as a circular buffer as illustrated in Figure 6.13. Pitch shifting by various musical intervals is achieved by adjusting the ratios of the input and output clocks. Natural ratios are preferred for (as Pythagoras noticed two-and-a-half thousand years ago) these are related by simple numerical ratios (see Chapter 2). Note that most pitch shifters allow for a bypass route so that the original sound can be mixed with the harmony in a desired proportion before the signal leaves the unit.

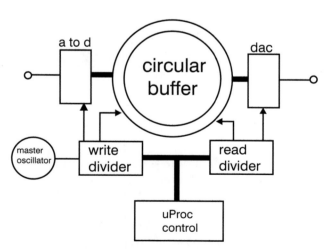

Figure 6.13 *Pitch shifting using circular buffer*

Flanging, phasing and chorus

Another application of the pitch shifting technique described in the last section is in creating audio effects which, to the uninitiated, sound nothing like pitch shifting at all! Instead effects such as chorus, flanging and so on, create a swirling, thickening texture to a sound. Originally these techniques were created using analogue techniques but digital implementations have entirely superseded analogue methods.

Remember, the problem the PZM microphone set out to alleviate? By eliminating multiple path lengths between sound-source and microphone element, the use of this special microphone prevents a 'comb-filter' effect whereby successive bands of frequency are reinforced and cancelled as was illustrated in Figure 3.6. Actually, although such an eventuality is extremely undesirable in the context of recording speech sounds, the phenomenon produces an interesting acoustic effect; a kind of hollow ring. Even more interesting is the effect as the microphone is moved in relation to sound source and reflecting body. This causes the frequency bands of reinforced and cancelled output to change. Imparting on the captured sound a strange, liquidity – a kind of 'swooshing, swirling' ring. Of course, such an effect is not practically obtainable using moving microphones[1], instead it relies on utilising an electronic (or electro-mechanical) delay-medium to recreate an acoustic delay. This effect has come to be known as flanging. The Beatles' ground-breaking producer, George Martin claims the invention of flanging is due to Ken Townsend, the Abbey Road engineer at the time of *Sgt. Pepper's Lonely Hearts Club Band* (Martin 1994). It came about due to the slight lack of synchronisation between two 'locked' tape recorders; a technique Townsend called ADT (see Chapter 9). Perhaps precisely because the sound quality was so unusual (recording engineers having striven to avoid its acoustic equivalent), John Lennon even employed this effect on vocals to depict the dreamy experiences of childhood.

A modern flanger dispenses with a tape mechanism to create the delay and, instead, a digital lag circuit is almost always employed. The flange-rubbing hand being replaced by a low-frequency oscillator (LFO). The amplitude of the LFO signal controls the depth of the flange. This is equivalent to the amount of 'de-synchronisation', and it is controlled as shown in Figure 6.14. The speed of the flange, controls the frequency of the LFO and the degree of the effect is controlled in a mixing stage as shown. Attempts at non tape-based analogue flange techniques involved the use of adjustable, cascaded all-pass filters providing the necessary delay elements. These circuits only produce a very small amount of delay per circuit and – even with a relatively large number of delays cascaded together – the delay was small in comparison to that required for a full flange effect. These devices thereby produced a particular, gentle effect,

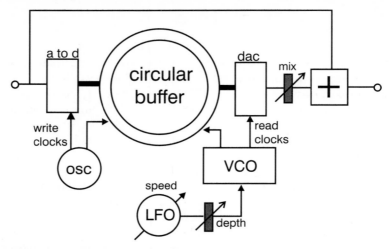

Figure 6.14   *Modern digital flanger*

sonically apart and worthy of its own name – phasing; a term based on the fact the circuits produce phase-shift, rather than full delay. In a modern digital processor, the terms phasing and flanging really describe the same effect; the term phasing being used to describe very light degrees of flanging with delays up to about 1 ms. Flanging uses delay variations in the region 1 ms to 7 ms. Chorus is the next term in this continuum, in a chorus effect, the feedback fraction and the minimum delay-time are limited so as the ensure the depth of the comb-filter effect is much less pronounced than in the flanger or phaser. In a chorus effect the delay typically varies between 20 to 40 ms. Phasing and flanging and chorus find their metier in the hands of guitarists. Or should I say feet! Because, more often, this effect is incorporated in a pedal. A refinement which facilitates switching the effect in and out without the necessity for the guitarist's hands to leave the instrument. A collection of guitarist effects pedals are illustrated in Figure 6.15. (As an exception, look at Figure 6.16 which illustrates the chorus dialogue-screen as part of Cool Edit Pro.)

Ring modulation

De rigueur in early electronic music studios was an electronic processing tool known as a ring modulator (Figure 6.17). A ring modulator is essentially an audio signal multiplier; one signal acting as the multiplicand, the other as the multiplier. All sound signals (no matter how apparently complicated) are made up of combinations of sine-waves of

Figure 6.15   *A panoply of effect pedals*

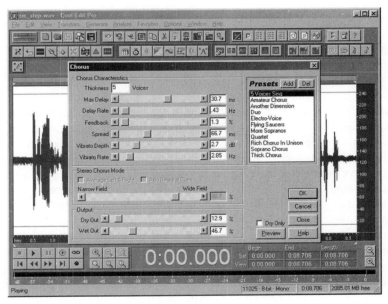

Figure 6.16   *Chrous effect generated in software*

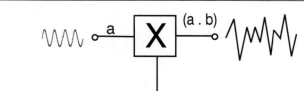

Figure 6.17 *Ring modulator*

various phases and frequencies. So the action of the ring modulator is best understood by considering the simplest process possible; the process of multiplying two sine functions together, which (from Chapter 2) may be discovered from the mathematical identity,

$$\sin A \cdot \sin B = \tfrac{1}{2}(\cos A - B) - \tfrac{1}{2}(\cos A + B)$$

which illustrates that the result of such a multiplication comprises a cosine function of the sum and difference of A and B. Evidently, the output of a ring modulator circuit is, like the output of the fuzz circuit, very rich in harmonics. However, the remarkable feature of the ring modulator circuit is that the output contains only harmonics (and sub-harmonics); all the fundamental tones disappear. Therein lies the unique property of this instrumental effect.

Dynamics processors

Every electronic system (digital or analogue) has a finite dynamic range. Suppose we want to send or store a signal, but its dynamic range exceeds that of the electronic system. Either the quietest parts of the audio signal will be lost in noise, or the loudest sections will distort. This situation is far from rare. So for engineering purposes, it is often desirable to shrink the dynamic range of a signal so as to 'squeeze' or compress it into the available channel capacity. The studio device for accomplishing such a feat is called a compressor. Ideally, if the compression law is known, a reciprocal process may be performed at the receiving or playback end of the chain and the signal may be 'expanded' to its original dynamic range using a circuit known as an expander. This complementary companson (compression followed by expansion) approach is the one taken in all forms of digital compression and analogue, complementary noise reduction.

The principle adopted in all forms of compression and expansion is the control of a variable gain element (a multiplier) whose inputs are (a) the audio signal itself, and (b) a multiplier based on the overall level of a short section of the signal at (a).

Because in both cases a compressor and expander operate to manipulate the dynamic range of a signal they are sometimes grouped together and termed dynamics processors. The effect of various types of dynamics processors are illustrated in Figure 6.18. Each curve in the

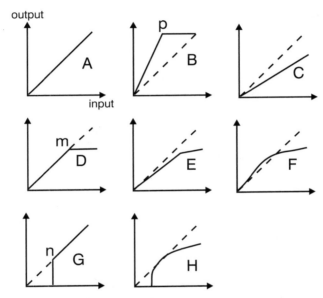

Figure 6.18 *Dynamics processing*

figure relates to the input-versus-output characteristic of the particular processor concerned. Do not confuse these curves with the curves for instantaneous transfer characteristic shown elsewhere. Remember that these curves delineate the input/output relationship over a short period of time (from ms to several seconds). Curve (A) illustrates a one-to-one relationship; as found in an amplifier. Curve (B) illustrates the effect of expansion, whereby the output signal amplitude increases more quickly than the input signal. Of course expansion cannot be used on its own without causing distortion and this too is illustrated in curve (B), by the flat-top region of the curve after the overload point p. Expansion can, as has already been said, be used after a reciprocal compression circuit, and the effect of such a circuit is illustrated by curve (C). Note that curves (B) and (C) represent a linear compression and expansion process. Such a process is often useful for engineering purposes for reducing overall dynamic range to that suitable for a particular transmission channel. This is an excellent and widely adopted technique (usually confusingly referred to as noise-reduction) but it suffers from the necessity for carefully controlled complementary circuits and conditions at either end of the channel. Certainly in the analogue world, this calls for precision equipment and skilled line-up practices.

For this reason, for the majority of music and speech broadcasting, complementary noise-reduction is not used. Instead, linear circuits are used – and a combination of skilled balance engineering and controlled compression applied to extract the maximum possible dynamic performance from a given dynamic range. Signal compressors are often used creatively as well, in order to add both perceived volume and consistency to a vocal or instrumental performance. The generic form of a compressor/limiter circuit is depicted in Figure 6.19. The term limiter is

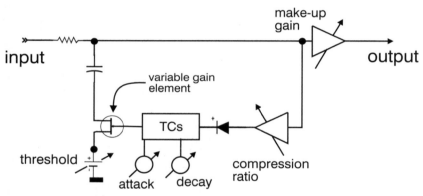

Figure 6.19 *Generic form of feedback compressor-limiter*

used to refer to a device which is employed to impart a very high degree of compression upon a signal at a particular level; usually a few dB below the absolute-maximum system modulation level (see Figure 6.18D). This type of device is engaged, for example, just prior to a transmitter to prevent illegal (and possibly dangerous) over modulation. The function of a compressor and limiter are depicted in Figure 6.18(E) which also portrays their fundamental similarity. In a limiter, the knee of the compression curve (the threshold) is set way up the transfer function, after which a very high degree of compression is imparted on the input signal. In a compressor, a much lower degree of compression (compression ratio) is imparted at a lower threshold. Most compressors provide controls to adjust both compression threshold and ratio so it is logical to speak of these units as compressor/limiters because they can function as either. Often, favoured compressors – and this is especially true of valve types – combine these functions in one overall smooth characteristic as shown in Figure 6.18(F).

When using a compressor, the peak signal levels are reduced in the manner illustrated in the curves of Figure 6.18. Obviously this would be of little use if the signal (now with compressed dynamic range) was not amplified to ensure the reduced peak values fully exercised the available 'swing' of the following circuits. For this reason, a variable gain amplifier stage is placed after the compression circuit to restore the peak signal values to the system's nominal maximum level (as shown in Figure 6.19). Notice that the perceptible effect of the compressor, when adjusted as described, is not so much apparently to reduce the level of the peak signal as to boost the level of the low-level signals. Unfortunately, this brings with it the attendant disadvantage that low-level noise – both electrical and acoustic – is boosted along with the wanted signal. The solution is a primitive expansion circuit known as a noise-gate, the effect of which is to suppress all signals below a given threshold and only 'open' in the presence of wanted modulation. The effect of this circuit is illustrated in curve (G) of Figure 6.18. Notice that all signals below the threshold value n, are 'cut-off' by the action of the gate. Because low-level noise is a real problem in a compressed signal, a good commercial compressor limiter will often include a noise gate circuit as well; the overall circuit transfer characteristic being combined as shown in Figure 6.18(H).

The majority of commercial compressors are still analogue, hence the choice of an analogue compression element (FET) for the circuit illustration. However, the FET device is exhibited as forming the lower leg of a potential divider across the signal path. It is therefore a form of multiplier; indeed many analogue compressors use analogue multiplier circuits as the compression element. Analogue multiplication techniques involve the use of current 'steering' via two alternative circuits to achieve such a multiplication. The circuit in Figure 6.20 demonstrates the general principle (although a practical circuit of this type wouldn't work very well). Essentially the audio signal at the base of the lower transistor is turned into a current in the collector circuit of the same transistor by the transistor's transconductance mechanism. Notice that a resistor is included in the emitter circuit to linearise this current. This collector current divides into two circuits, through T1 and T2; the ratio of the current being dependent on the voltage on the base of T1: Higher than the signal on T2 base (V_k) and the current will flow predominantly through T1 and appear as a voltage signal on R_r. Lower than the signal on T2 base and the current will flow predominantly in T2. By altering the

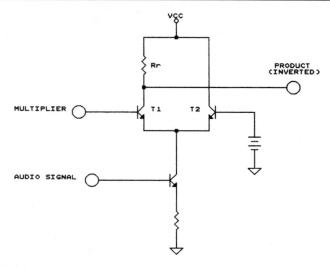

Figure 6.20 *Analogue multiplier circuit*

value of the voltage on the base of T1, a variable proportion of the original signal voltage (suitably inverted) can be recovered across R_r.

Digital multiplication can be achieved many ways but a good technique involves the use of a look-up-table (LUT). Essentially multiplier and multiplicand are used, together, to address one unique reference in a read-only-memory (ROM) where the result can be looked up, as in Figure 6.21. In a digital audio compressor, the circuit element could be a discrete

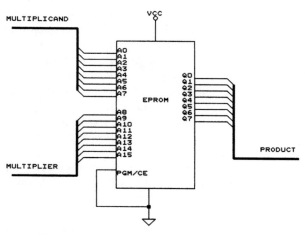

Figure 6.21 *Digital multiplier circuit*

multiplier but, much more likely, the circuit's operation is be performed as an algorithm within a DSP IC (see Fact sheet #12).

Feed-forward and feedback compressors

In the circuit illustrated, the control side chain is fed from the output end of the signal path. It is thus known as a feedback side-chain compressor. It is also possible in any type of dynamics processor to derive its control signal upstream of the compression circuit itself and is known as a feed-forward compressor. Both types have their advantages and disadvantages.

Two practical compressors

Just to illustrate the wide variety of compressor implementations the musician-engineer will encounter, let's look at two practical compressors; one, a software only implementation is pretty self-explanatory and is illustrated in Figure 6.22. The second is a valve-based circuit, illustrated in Figure 6.23.

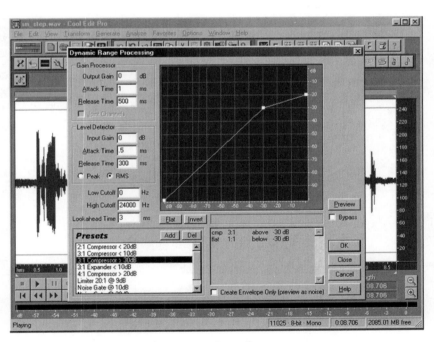

Figure 6.22 *Compression as set in software program*

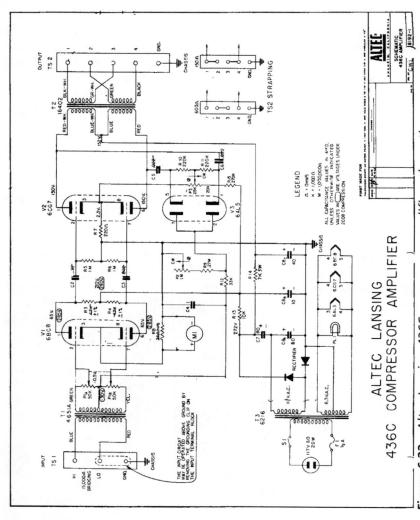

Figure 6.23 Altec Lansing 436C compressor amplifier schematic

The Altec Lansing 436C compressor amplifier (Figure 6.23) dates from 1963, but Jim Dowler (of ADT) tells me that an original 436 was in use at Olympic studios when many of the early punk classics were being recorded. The reason why the AL436C is interesting here is its consummate 'valve-ness'. The circuit relies on the inherent non-linearity of valve equipment to provide the compression characteristic. In this design, the input valve has a particularly curved $V_{g-k/Ia}$ characteristic. Provided input signal excursion is limited, a valve of this type can therefore operate at various regions of the curve, at various gains, or various mu; hence 'vari-mu'.

The circuit works like this: large negative signal peaks on the anodes of the push-pull output stage V2 cause double-diode V3 to conduct and drag down the bias volts to the input stage V1. This action is damped by the time-constant formed by R9, P2 and C4 which allows for some adjustment of the compressor's release time. The level threshold, at which the compression action is required, is adjustable too; by means of the simple expedient of biasing the cathodes of the double-diode V3 with a constant bias derived from the slider of P3. The compression action is obtained because, as signal peaks depress the bias volts on V1, the stage 'slides' further down the input valve's curved $V_{g-k/Ia}$ characteristic: the required voltage change for a given change in anode current is reduced and the stage-gain is lowered. The compression control-signal itself remains inaudible because it is applied equally to both input valves: whereas the music signal is applied differentially – in other words, the common-mode control-signal is cancelled out in the differential amplifier.

Noise reduction

The circuit of a typical analogue dynamics processor is given in Figure 6.24. This is a single-ended noise reduction system. This type of noise reducer is very common in recording studios and may even be incorporated within the mixer electronics. Its employment is made necessary where a noisy signal arrives at the mixer and must be expanded (to separate the wanted signal from its noise-floor) without having first enjoyed the benefit of a complementary compression process. Hence the term single-ended. The engineering challenge, presented by a dynamic process of this type, is to provide a beneficial level of noise reduction without introducing the artefacts which so often plagues practical noise reduction. The particular 'noise signature' of an expander system is a result of the relatively constant noise on the input signal which, when modulated by the changing gain of the expander, takes on a varying quality which has been described as sounding like 'breathing' or 'pumping'. All forms of noise reduction using signal dependant amplifica-

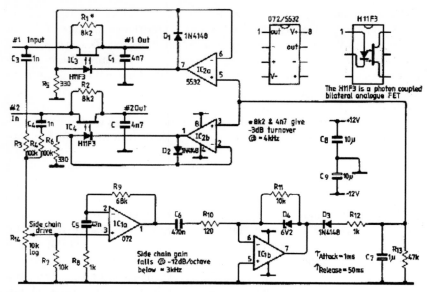

Figure 6.24   *Practical noise-reducer circuit*

tion suffer to a greater or lesser extent from this phenomenon and it places an upper limit on how large an expansion ratio may be used for acceptable results.

The noise reducer utilises the expansion technique whereby low level audio signals are amplified less than high level audio signals. A threshold is set so that the residual system noise has insufficient energy to cause the variable gain amplifier to change to its higher gain regime. It is an important feature of any design that it must ensure that the presence of useful audio signal raises the gain of the amplifier sufficiently quickly that the transient start of the audio signal is not destroyed and that the period of time the amplifier remains in the high gain regime once a high level signal has ceased is appropriate. Too long and the amplifier will be unable to return to the low gain state in the 'gaps' between wanted signal. Too short and the expander will mutilate the reverberant part of the audio signal. Some noise reduction systems rely simply on a switch which remains open when signal level is below a defined (pre-set) level and closes, as fast as possible, after the signal level exceeds this threshold. Such a type is known as a noise gate and it is this type which is usually integrated within the mixing desk. The noise-reducer in Figure 6.24 is more subtle and has some features in common with the digital audio compression schemes examined in Chapter 6; hence its consideration here. In this implementation, advantage is taken of the psychoacoustic phenomenon of masking.

We saw in Chapter 2 how the presence of a single pure tone at 60 dB above the threshold of perception, causes the desensitisation of the ear by as much as 20 dB in the octave and a fifth above the original tone with the 'threshold shift' being as much as 40 dB near the tone frequency. Music has very many pure-tones present simultaneously but, for the majority of the time, its masking effect only operates at low to middle frequencies. This is because system noise, whether generated by thermal agitation or from quantisation errors, has a very flat energy versus frequency characteristic. Music, on the other hand, has an average energy versus frequency characteristic which falls with frequency. Obviously, a signal which has an uneven energy versus frequency characteristic like music, will sometimes fail to mask one which has – like noise. This is especially significant, since the music signal fails to mask the noise in the high frequency portion of the audible range where our ears are most sensitive to detecting the noise part of the signal.

The circuit shown in Figure 6.24 operates by automatically controlling the frequency response of the forward signal-path of the noise reducer using controlling information derived from the high-frequency content of the programme. The principle, which is used in some commercial noise reduction systems (see further reading), works rather as if the 'treble' control of a pre-amp is constantly and quickly varied, so that in the absence of music signal the control is left fully anticlockwise, thus attenuating the system noise in the HF part of the audio spectrum where it is most subjectively annoying and only moved clockwise when wanted signal, containing high frequencies, occurs. This is achieved thus:

The gain controlling element in the noise-reducer is a photon coupled bilateral analogue FET in which a gallium arsenide infrared emitting diode (IED) is integrated with a symmetrical bilateral silicon photo-detector. The detector is electrically isolated from the control input and performs like an ideal isolated FET. The part is designed for control of low-level AC and DC analogue signals and has a response time of less than 15 microseconds. Within certain operating limits, it's possible to think of the device as a variable resistance element, its resistance controlled by the current flowing through the integral IED. The turnover frequency, in the absence of signal, is defined by R1 and C1. When no current flows in the IED, the FET has a very high resistance indeed. It is the action of this low-pass filter which secures the noise reduction. When signal rises above a pre-set level and current starts to flow in the IED the corner frequency of this low pass filter is made progressively higher and higher.

As a general rule, only one side chain is necessary for stereo dynamics processing. In fact, unless the control is common to both channels the stereo image can appear to 'wander' as the gain of the two channels varies independently of one another. The side chain comprises the mixing resistors R3 and R4 which take an equal proportion of the left and right signals to the sensitivity control. R7 pads VR1's control law to be more manageable. IC1A is straightforward HF gain stage. IC1B gives a very large gain until the output swings minus 6.2 V when the Z1 goes into zener conduction and the gain reduces very quickly. The output can only swing positive by about 0.6 V when the zener diode goes into forward conduction. C6 and R10 roll-off the voltage gain below 2.8 kHz. The IC1B stage virtually provides the rectification necessary to derive a control signal, D3 ensures no positive signals are applied to the control-voltage storing C7. The two time constants formed by R12 and R13 with C7 determine the attack and release times of the noise reducer. (Strictly speaking, the side-chain should full-wave rectify the audio signal since the magnitude of negative and positive peak signals can differ by up to 8 dB.) The control voltage stored on C7 drives the dual op-amp IC2. The infrared emitting diodes are connected within the feedback loop of the op-amps, so the current through these is controlled linearly by the voltage on the non-inverting inputs of IC2 and the value of R5 and R6. D1 and D2 are included so that IC2A and B do not saturate in the unlikely event of their receiving a positive voltage input. The IED's offset voltage is continually compensated for by the action of the feedback loop of IC2A and IC2B and this guarantees the dynamic response of the circuit is dominated by the time constants formed by R12, R13 and C7.

The circuit of Figure 6.24 could easily be configured to provide high frequency compression rather than expansion. The circuit could then be used to compress HF information above the system noise floor and expand it afterward. Similar to digital compression schemes, such an implementation has much in common with commercial two-ended noise reduction schemes such as Dolby Lab's Dolby B.

Audio enhancers

In recent years the professional audio market has seen a flurry of processing units intended to add 'power' and 'punch' to a mix. Most of these processors use a combination of dynamic spectral enhancement (a combination of tone-control and dynamics processors) and a technique, originally pioneered by Aphex, termed 'excitation'.

Dynamic tone-controls come in a various forms but essentially the process is one of compression; except that the compressor is made to act over a limited range of frequencies: and not across the whole spectrum, as in a classic compressor. Excitation is used to generate more HF energy but is considerably more clever than simple tone-control. In an exciter, the input signal is high-pass filtered and is used to feed two inputs of a multiplier circuit. This circuit will produce sum and difference frequencies, as we have seen elsewhere. The output is itself high-pass filtered (to remove difference frequencies) and this signal is mixed with the unprocessed signal; the degree of added processed signal being proportional to the amount of 'excitation' added. The resulting spectrum is modified such that there are frequencies present which simply did not exist in the original sound. Some enhancers offer the ability to rectify one of the signals prior to multiplication. This has the effect of adding even – rather than odd – harmonics which alters the characteristic of the 'excited' sound. Whilst a very powerful signal processing tool, audio enhancement effects can easily by overdone.

De-essers

One type of dynamic tone control which is very widely used in recording and mastering is the de-esser. The need for a de-esser results from different recording medium's differing capacity to record and reproduce HF energy; particularly the sibilant energy associated with the sung 's'. A closely miked recording of a singer will often contain very substantial HF content which might record and reproduce well on all-digital studio equipment, but once transferred to – for example – an analogue cassette or to video tape results in a nasty mushy 'shhh' sound where the original recording has an 'sss' sound. In the case of analogue tape this effect is due to HF tape saturation. Such a phenomenon presents the mastering engineer with a very difficult problem because, whilst she could reduce overall HF levels by means of tone-control, this would detract from the incisiveness of the rest of the recording. What is needed is a progressive, dynamic tone control which can be made to act so as to 'duck' HF frequencies only when the treble energy reaches the particular, troublesome threshold. Such a device is known as a de-esser. Professional equipment varies in its complexity, with the most flexible units offering adjustable filters for both compression and side-chain drive.

Vocoder

Not only was the human voice almost certainly the earliest musical instrument, it is still arguably the most expressive. And it has the unique quality that, because singing involves language as well as pitch, it can express ideas. The Vocoder is a device which allows the unique expression of the human voice to modulate an instrumental sound which may be monophonic or, more often, polyphonic. In order to understand the Vocoder, it's worthwhile taking a few minutes to understand the production of vocal sounds.

The fundamental sound source involved in vocal production is a rather low frequency complex tone produced when air from the lungs travels up the windpipe and excites the vocal folds in the larynx. The source of sound energy is known as the glottal source because the space between the vocal folds, and between which the air from the lungs is forced, is known as the glottis. The spectrum of this glottal source is subsequently modified by the vocal tract which comprises the pharynx (throat), the nose and nasal cavities and the mouth. The shape of the vocal tract can be varied extensively by moving the tongue, the lips and the jaw. In so doing, the spectrum of the glottal source is modified as it is filtered by the various resonances formed in the discrete parts of the vocal tract. Each of these resonances is known as a formant and each is numbered; the lowest frequency formant being termed the first formant, the next – the second and so on.

The principal features of the Vocoder are its two inputs; one for an instrument and another for a microphone. The block diagram for a simple instrument is given in Figure 6.25. Vocoder operation relies on the

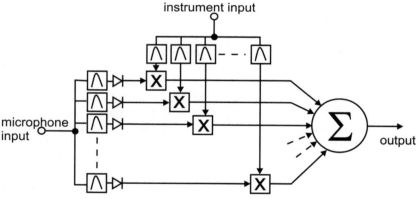

Figure 6.25 *The Vocoder*

amplitude envelope of the vocal formants modulating the instrumental inputs via audio signal multipliers (VCAs in an analogue Vocoder). In circuitry terms this involves splitting the vocal signal and the instrumental signal into a number of frequency bands by means of band-pass filters. The greater the number of bands, the better the performance of the Vocoder function. (In a digital Vocoder, the frequency spectrum can be split into a great many bands by means of a wave filter – like that discussed in relation to audio signal compression in Chapter 10.) Following the band-dividing filters, the vocal signal path passes to a number amplitude envelope-

Figure 6.26 *Digitech Vocalist Workstation*

detector circuits (peak rectifiers in an analogue circuit). These envelope signals are then utilised as the variables applied to each of the multipliers following every band-dividing filter in the instrumental signal path. In this way, the frequency spectrum of the speech is 'imprinted' on the instrumental sound. You can draw a physiological parallel by saying it is as if the lungs and vocal folds were replaced with the instrumental sound whilst the function of larynx, mouth and nasal cavities remain the same.

Not only is the Vocoder capable of some exotic colouristic effects (listen to *O Superman* by Laurie Anderson) but the physiological analogy may also have suggested to you an application whereby, the instrumental input can be a synthesised tone; similar to that produced by the lungs and vocal folds. If that synthesised tone – or tones – is under MIDI control, the Vocoder can be used as an artificially enhanced voice – always in tune and able to sing in perfect harmony with itself! Digitech produce several products of just this type in which they combine this function with intelligent pitch shifting to produce a powerful, versatile vocal processing device. The Digitech Vocalist Workstation is illustrated in Figure 6.26.

Talk-box guitar effect

Lying somewhere between the wah-wah pedal and the Vocoder is the talk-box guitar effect. This exploits the unique and expressive acoustic filter formed by the various resonances of the vocal tract and mouth to modify the sound of an electric guitar. This is done by driving a small loudspeaker with the amplified guitar signal, feeding it through a horn and into a plastic tube. The tube is then clipped or gaffer-taped up the microphone-stand into a position so that it can be fed into the mouth of the guitar player. The resulting sound is recorded via the microphone feed. Talk boxes feature in the recordings of Aerosmith, Frampton and Joe Walsh amongst others.

Reference

Martin, G and Pearson, W. (1994) *Summer of Love – The Making of Sgt. Pepper.* Macmillan

Note

1 Although it is practical to generate such an effect by moving loudspeakers in relation to a fixed listening (or microphone) position. The Leslie loudspeaker works in just such a fashion, as explained in Chapter 14.

Fact Sheet #6: Music foundations

- Note names
- Musical notation
- Scales
- Harmony
- Modulation
- Serialism

Note names

Every note in each octave of the keyboard illustrated in Figure F6.1 has a name – as illustrated. In the English speaking world all the white (natural) notes are named after the first seven letters of the alphabet and the terms sharps and flats are applied to the nearest black notes. (The reason why there are two names for these notes is explained in Chapter 2.) Despite the common observation that music is an international language, before you start to jam with your German pen-pal be aware that not all countries use this system. In France for example the system known as 'solfeje' is used and in Germany a system which is a bit like – but crucially unlike – our own is practised.

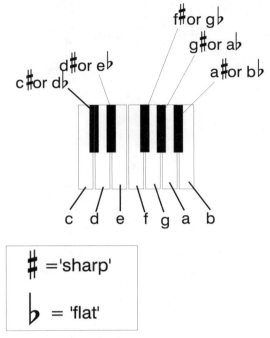

Figure F6.1 *Musical note names*

Musical notation

It may come as a bit of a surprise to someone who is not a musician to know that musical notation only presents a very rough guide as to a musical performance. Musical notation does nothing to capture the nuances of a musical performance that play such an important role in communicating an emotionally moving experience. Instead it concentrates on the 'nuts and bolts' of pitch and rhythm. As an interesting comparison, take a look at Figure F6.2, it illustrates a player-pianola roll: each hole represents a key press; its horizontal position determines the pitch of the note (i.e. the position of the note on the keyboard) and the length of the hole represents the length of time the key should be depressed. Of course a pianola does not have a brain to interpret this information, instead the roll is drawn at a constant rate across a mechanical sensor which deciphers the holes into key presses. Which is why a pianola sounds like a machine playing a piano and not like a human being!

Music notation is very similar to the pianola roll. Essentially it shows the pitch of a note by a blob; its vertical position

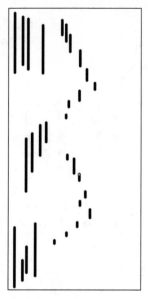

Figure F6.2   *A pianola roll*

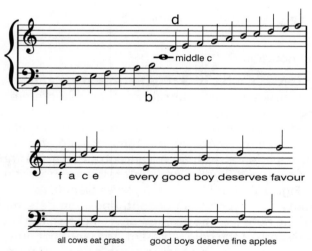

Figure F6.3   *Musical note notation*

representing pitch. The upper part of Figure F6.3 represents
the scale of C major over more than two octaves. (The scale of
C major is just all the white notes on a piano.) Notice how the
two bands of five lines break up the vertical scale like graph

paper in order that it's easier to see-at-a-glance the vertical position of the note. Each band of five lines is known as a stave. Have a look too at the funny squiggles at the extreme left hand position of the staves. These are called clefs and they identify the musical range over which the five stave lines apply. The lower squiggle is called a bass stave and this indicates that the bottom stave is concerned with bass notes. The upper squiggle is called the treble stave and indicates notes in the treble (high) register. Notice that one note (called by a special name – middle C) is in between these two staves so that the bass and treble staves could effectively be squashed-up and joined together as a stave of eleven lines. In fact, when this system was originally invented, that's just exactly what was done; the visually forbidding result being referred to by the equally momentous name – the Great Stave! The modern system is much easier to read, but it does result in one complication that the bass stave represents notes in a different position to the treble stave.

Most people find it easier to remember notes on the stave by means of their position, on the lines or in the spaces between the lines. The note on the lines in the bass clef are; G, B, D, F and A. And the notes in the spaces are; A, C, E, G as is also shown in Figure F6.3. Generations of music students remember these by the mnemonics; Good Boys Deserve Fine Apples and All Cows Eat Grass respectively! In the treble stave, the lines represent the notes; E, G, B, D, F and the spaces; F, A, C and E. These are remembered by Every Good Boy Deserves Favour and FACE, respectively as shown.

The principle of dividing pitch range by means of lines is so sound that it may be extended to rhythm as well. In this case vertical lines are drawn through the stave to divide the music into discrete time intervals called bars. Rhythm represents the way we choose to divide the inexorable passage of time. To all of us, the most familiar way of doing this is with the tick-tock rhythm of a clock. In music this is referred to as a duple rhythm; with two beats in a bar. But there are many others like the waltz which has three beats in the bar and most pop and rock music which has four beats.

The duration of notes is annotated a little differently to the system we saw for the pianola. Take a look at the top of Figure F6.4. It shows a note lasting a bar. This 'whole note' (or semi-breve) may be divided into smaller divisions as shown in the rest of Figure F6.4. Firstly into two half-notes (or minims), into four quarter-notes (crotchets) into eight eighth-notes (quavers) and so on.

Figure F6.4 *Note durations*

Figure F6.5 is a piece of music which puts all of this together. Notice in addition to the staves, clefs, time-signature, bars and notes there are written instructions on how the music should be played; how loud (or soft) and at what speed. In the example they are written in English but often these instructions are written in Italian (another peculiar musical tradition). Notice also that tradition has it that the words to a song are written below the notes.

Scales

Chapter 2 contains a description of the development of the scales current in traditional and popular western music. But lest you gain the impression that scales are somehow God-given, it's important to remember; melody came first. Scales are simply an intellectual attempt to rationalise the typical gamut of notes composers have chosen naturally in order to express themselves. Artistic theorising is not artistic creation. In art theory you start with the painting and deduce the different coloured paints that were on the original easel. That said, it is true that very many thousands of traditional (and indeed modern) songs are based on the use of five notes said to comprise the Pentatonic scale (Figure F6.6a).

Figure F6.5   *Song score*

From the ancient Chinese and the ancient Celts to Gershwin and heavy metal, inventive men and woman have naturally invented melodies that encompass these notes. Other important 'natural' scales consist of the major and minor scales (whose development is traced in Chapter 2) and the blues scale which is developed from a codification of traditional African melodies introduced into the west, originally by enslaved peoples of African origin (see Figure 6.6b). The poignant character of melodies constructed on these scales convey the kind of triumphant melancholy which their composers sought, no doubt, to communicate.

a) pentatonic scale

b) 'blues' scale

✳ = 'blue' notes

Figure F6.6   *(a) Pentatonic and (b) blues scales*

Harmony

Harmony is the sounding of two notes together. By contrast with melody, which is very ancient indeed, harmony is a modern invention. In fact there is very little evidence that harmony (except doubling of a melody at an octave – which happens naturally when a man and woman sing the same tune) existed at all much before the Renaissance. The breakthrough in harmony came when composers realised melody could be doubled at an interval other than at the octave; the perfect fifth being the favourite choice. This type of harmony is called Organum and is typical of early Church music. Here we have an interesting irony, it was the classification of scales, intended to codify melodic phenomenon which permitted the rationalisation of a harmonic system. Indeed, especially today, the value of a knowledge of scales is useful, not for melodic purposes (except perhaps as the basis for improvisation) but as the backbone of harmonic theory. Take for example the major scale of Figure F6.3. If each note is harmonised by the note a fifth (two lines of stave) above we have the pattern shown in Figure F6.7a. Why does this sound good? Well, if you look at Table F6.1, which annotates the harmonic series, you'll see that the interval of the fifth plus an octave (a major twelfth) is the second overtone of a note. So, when a fifth is sung above an

Table F6.1

Harmonic	Musical note	Comment
Fundamental	C	
2nd (1st overtone)	c	octave
3rd	g	twelfth (octave + fifth)
4th	c'	fifteenth (two octaves)
5th	e'	seventeenth (two octaves + major third)
6th	g'	nineteenth (two octaves + perfect fifth)
7th	b-flat' (nearest note)	dissonant; not in natural scale
8th	c"	three octaves
9th	d"	major 23rd (three octaves + second)
10th	e"	major 24th (three octaves + third)
11th	f" (nearest note)	dissonant; not in natural scale
12th	g"	major 26th (three octaves + fifth)
13th	a"	dissonant; not in natural scale
14th	b-flat"	dissonant; not in natural scale
15th	b"	major 28th
16th	c"'	four octaves
17th	C#"'	dissonant; not in natural scale
18th	d"'	major 30th
19th	d#"'	dissonant; not in natural scale
20th	e"'	major 31st

original melody it simply serves to strengthen a note which is already there in the original voice!

By induction, what might be the next most obvious harmony note? The fourth overtone, two-octaves and a major third above the fundamental. This produces the pattern shown in Figure F6.7b which actually presents the virtually complete harmonic theory for thousands of pieces of western music! These patterns of three notes (termed a triad) on each degree of the major scale represent the seven basic chords of music constructed in a major key. They are usually annotated by Roman numerals as shown. Of the seven, three are especially important; the first (I), the fourth (IV) and the fifth (V). Thousands of wonderful songs have been written with just these three major chords. Next most important are the minor chords on the 6th and 3rd degrees of the scale (VI and II respectively). The minor chord on the third degree is less often used and the chord based on the last note of the scale

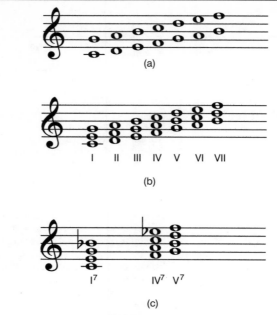

Figure F6.7 *Harmonic theory*

(the so-called leading note) is called a diminished chord and is the least often used. All the chords in other keys are constructed in similar ways; including all the chords in minor keys which are built up on the harmonic minor scale.

The next step

It shouldn't be especially amazing, that the next most common sets of chords (comprising four notes) are constructed by emphasising the next note in the harmonic series (ignoring octaves and octaves and fifths). This is 6th overtone. This operation produces minor seventh chords on the second, third and sixth degrees of the scale and produces a major seventh chord on the fifth degree of the scale, the so called dominant-seventh chord which has an especially important role in western classical music. But interestingly, this procedure, doesn't always produce a note which is in the original scale, notably on the first and the fourth degree of the scale. In the case – for example – of C major, the chord on the first degree (the tonic) acquires a Bb. And the chord on the fourth (the sub-dominant) acquires an Eb (see Figure F6.7c) For this reason these chords are avoided in much classical western art music.

It is however of great cultural significance that the extra notes present in the chords on the first and fourth degree of the scale, were present in the melodic scales used by dispossessed people of African origin (the 'Blues' scale). One can easily imagine that as composers of African origin sought to harmonise their music (perhaps with the chapel harmonium – inspired by the hymns they were obliged to sing on a Sunday) they reached for the four note chords on the first, fourth and fifth degree of the scale thereby creating the hallmark of blues music which is largely constructed on a 12-bar pattern of these chords.

Modulation

Musical modulation is the process of changing key. This is usually done for dramatic or expressive purposes. A wily composer can sometimes modulate several times in the space of a short melody, without you hardly noticing. (Until you come to try to whistle the tune and realise it's harder than you thought!) Mozart was particularly adept at this. When a modulation happens seamlessly it's said to be 'well-prepared'. Other composers treat modulation as an architectural device devising single-key episodes into a grand campaign of key changing; often deliberately managing the key change abruptly in order to cause a jarring effect. The music of Beethoven particularly characterises this approach. All music written with a sense of key (and especially the movements between keys) is designated, tonal music.

In modern art music, the concept of key has largely broken down (see below) but in jazz, pop, rock and dance, tonal music is still very much alive. Pop, rock and jazz very much in the tradition of Mozart, reserving well-prepared modulation for brief expressive highlights within the melodic framework. Dance, due to its inclination to 'paint much larger canvasses' has tended to adopt a more structural (and deliberately unprepared) approach to modulation.

Serialism

Serialism takes as its starting point a pre-determined (pre-composed) sequence of the twelve possible chromatic tones on a piano (all the white and black notes in one octave). This basic compositional cell may be used backwards or in various transpositions and inversions, but it always determines the choice of the notes in use at any one time; sometimes melodically, sometimes harmonically. Much serial music aims, not only to break harmonic norms but rhythmic ones as well, as illustrated in the extract in Figure F6.8.

Figure F6.8 *Score using serial techniques*

7
Pet Sounds – Electronic synthesis

Introduction

The theory of electronic oscillators was covered in Chapter 5. There we met several practical examples of LC-based oscillators and Wein-bridge types for the generation of sine-waves, astable multivibrators for the generation of square waves as well as ramp generators for the generation of sawtooth waveforms. We noted that each waveform had a different harmonic structure and, consequently, a different timbre when reproduced over loudspeakers. These circuits form the basis of analogue sound generation and synthesis. In many ways, Chapter 5 pre-empted some of the material which might have waited until here. For what is an electric organ if it is not a synthesised pipe organ? Later in the chapter, we will return to the subject of the emulation of real instruments, the oft misunderstood term for music synthesis. In the first half, I want to concentrate on the development of the analogue synthesiser which owes its genesis to Robert Moog, the inventor of the first commercial synthesiser and the artistic inspiration of the composers working in the early post-Second World War electronic music studios; composers of the stature of Stockhausen, Eimert and Berio. Analogue synthesisers contain certain cardinal circuit blocks each, originally, more at home in a laboratory than in a music studio! Each of these is now considered in turn.

Voltage controlled oscillator (VCO)

A voltage controlled oscillator or VCO is a circuit element that adjusts its oscillation frequency in response to an input voltage. The simplest (and most common control input) is a DC potential derived from a special musical keyboard which acts rather like a complicated switched potentiometer, with each key closing a switch on a continuous resistive

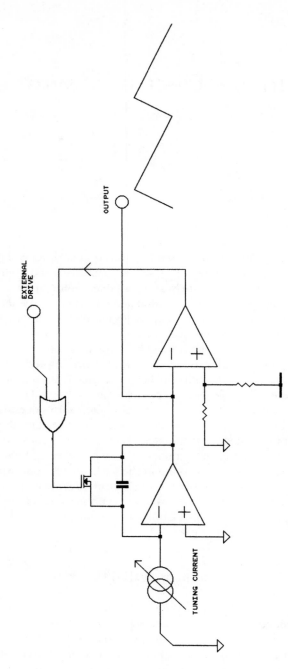

Figure 7.1 *Voltage controlled oscillator*

element. Design of such a circuit is not altogether straightforward because the oscillator must be made to swing over the entire audible range, a frequency range of some 11 octaves. Most often the oscillator is a sawtooth generator type like that illustrated in Figure 7.1. Notice that the rate at which the integration capacitor charges in the feedback loop of the op-amp is variable by means of the adjustable current source. It is interesting to compare this rather complicated circuit with the ramp generator shown in Figure 5.7 in Chapter 5. Notice that, in the circuit shown here, the circuit must itself generate the ramp termination pulse shown as being supplied externally in the circuit in Chapter 5. The self-generation of the termination pulse occurs due to the action of the comparator circuit which has a preset negative voltage on its positive input terminal. Once the ramp circuit output voltage (which is shown supplied to the comparator's negative input) has reached this threshold, the comparator changes state and closes the electronic switch shown connected across the integration capacitor. The charge-integrating capacitor is thereby shorted and the ramp terminates – allowing the whole process to start once again. It is worth pointing out that there is nothing to stop an external pulse being sent to this oscillator in the manner shown in Figure 7.1. This is often done in commercial synthesisers where the technique is called synching the oscillators. By setting the natural oscillation of one oscillator to a different frequency from that of its externally supplied synching pulse, some very complex waveforms are obtainable.

One major complication with voltage control for synthesisers is due to the nature of the relationship between control voltage and frequency. From a technical point of view the easiest control-law to generate is linear, or $V/F = k$, where k is a constant. But, in fact, from a musical standpoint, far better is a law which relates a constant change in pitch (frequency) to a constant change in control voltage. This is a logarithmic law and considerable complication exists within most analogue synthesisers to alter the control law of the VCO to that suitable for musical applications.

Voltage controlled filter (VCF)

A voltage controlled filter is a frequency selective circuit which may be made to alter its cut-off frequency under the control of an externally applied voltage. The most usual type in synthesiser applications is the voltage controlled low-pass filter, which has the most useful musical applications. A simplified schematic of a VFO is given in Figure 7.2. This unusual circuit operates like this: the cut-off frequency is programmable by means of the current sink 'tail' which may be made to vary its sink

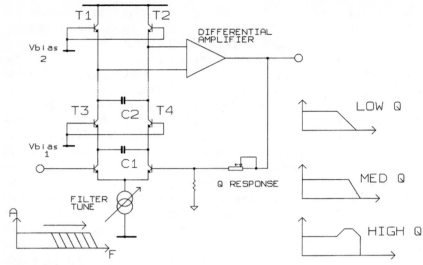

Figure 7.2   *Voltage controlled filter*

current as in the manner of a normal current mirror. This current divides between the two cascode pairs and into the collector loads of T3 and T4, themselves another cascode pair. At very low sink currents, the value of the collector loads (T1 and T2) will be relatively high because the output impedance from an emitter follower – which is what these loads are – is inversely proportional to emitter current. Similarly, the transconductance of the differential pair will be low too. The gain of the stage will therefore be the product of the lowish transconductance of the pair, multiplied by the relatively high impedance of T1 and T2 collector loads. At high tail current, these conditions alter so that the transconductance of the differential cascode pair will be high, but the impedance of the collector loads – from which the signal is taken differentially – will be low. The overall low-frequency gain of the circuit will thereby remain constant, irrespective of changes in tail current. What will alter, however, will be the available slew-rate (bandwidth) of the amplifier which will be severely limited by the ability to charge and discharge C1 and C2 at low standing currents. A situation which will improve – thereby increasing the bandwidth of the circuit – at high standing currents. Sometimes practical circuits repeat the cascode structure of this circuit many times to increase the number of poles in the filter, often earning the circuit the name ladder filter. An important further function of the circuit is the ability to feedback a proportion of the signal to the other side of the differential pair. Note that this is not negative feedback but positive feedback. This has the effect of increasing the Q of the circuit, especially near the turnover frequency.

It is therefore possible to produce a range of responses like those shown in Figure 7.2 which offer a gamut of musically expressive possibilities by permitting the possibility of imprinting high-Q formants on the fundamental wave. Sometimes this Q control (as shown in Figure 7.2) allows the possibility to produce instability at extremes of the control's range, thus turning the VCF into another, somewhat unpredictable, VCO.

Envelope generation

However complex and rich the tones produced by the foregoing means, each note of the synthesiser would still become tedious were it not for the addition of envelope generation circuits. What are these – and why have them?

Every musical sound has, in addition to pitch and timbre, a dynamic envelope. As a matter of fact, dynamic envelope is perhaps the single most salient 'character building' component of a particular sound! Consider asking a child to represent the sound by means of a drawing. Ask her to draw a drum sound and she is likely to ignore the depiction of pitch, even of timbre. Instead she might draw something like that depicted at the top of Figure 7.3. Similarly, ask her to draw the sound of a gently bowed violin note and she might draw something like that shown below the drawing of the drum sound. What has she drawn? Well, in the first case she has represented a loud sound that started quickly, was maintained for a short period of time and subsequently died away quickly. In the second case she depicted a sound which started from nothing, slowly built up to a moderate level, was maintained at that intensity for a relatively long period of time and then died away slowly. Clearly this dynamic character is vital to our appreciation of different types of musical sound. The lower third of Figure 7.3 formalises the child's drawing into the three formal epochs of any synthesised sound's brief stay on earth: attack, sustain and release (or decay).

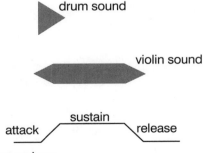

Figure 7.3   *ASR envelope*

Real musical sounds have a dynamic envelope and the modification of a primitive synthesised musical waveform into a usable musical sound involves the manipulation which imprints an attack, a sustain and a decay onto the fundamental sound source. This manipulation is twofold, it being a controlled multiplication function. The attack being the speed with which a signal is multiplied from zero (silence) to a constant (the sustain level) and the rate at which the sustain level decreases back to zero (releases or decays) once the keyboard key is released. The multiplication function is performed by the voltage controlled amplifier or VCA which is a signal multiplication device as described in the last chapter. (The circuit in Figure 6.14 demonstrated the general principle.) The controlling signal is derived from an envelope generation circuit sometimes known as an attack-sustain-release (ASR) generator.

Attack–sustain–release (ASR) generator

An attack-sustain-release generator is illustrated in Figure 7.4. When the key closes the control voltage – which is fed to the VCA – rises at a rate predetermined by the setting of VR1 and its interaction with C1. The

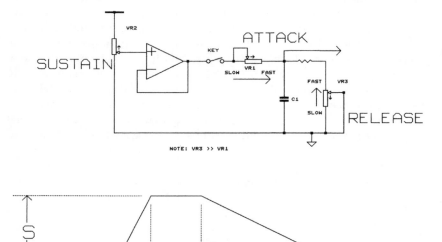

Figure 7.4 *ASR generator circuit*

control voltage will rise ultimately to the value set on VR2 which determines the sustain level. Finally, once the key is released, the control voltage will fall as determined by the setting of VR3 and C1.

Low-frequency oscillator (LFO)

The low-frequency oscillator is often a ramp generator or sometimes Wein-bridge type. External voltage control is seldom provided; instead the function of this oscillator is to control either VCA or VCO – in order to provide tremolo or vibrato respectively. Alternatively the LFO is often used to control a VCF, in order to obtain a protean musical timbre, one of the musical hallmarks of analogue synthesis.

Analogue noise generators

The thermal noise generated in a resistor is given by the expression:

$$E^2 = 4kTBR$$

where E = the RMS value of the noise EMF in circuit of resistance R at temperature T (Kelvin) with an effective bandwidth B. This simplifies (assuming normal room temperature and a bandwidth of 20 kHz) to:

$$E = (1.6 \times 10\text{--}20.R) - 0.5$$

Looked at in this way, every electronic component is an analogue noise generator! Most engineering jobs strive to minimise noise, not to generate it for its own sake. For that reason, analogue noise generators are relatively simple, often as inelaborate as a resistor followed by a high gain amplifier. Another source of noise is a reversed-biased diode or zener diode, followed by a high gain amplifier. This is a good generator of noise because the noise in a diode is (relatively) high due to the avalanche effect as high velocity electrons, entering the n-region from the p-region, liberate others in the valence bands of the n-region, a process which is inherently noisy. Furthermore it side-steps a problem with a resistor circuit in which (in order to get a high enough noise voltage) a very high value of resistor must be utilised and this invites the possibility of other signals being electrostatically, or electromagnetically, coupled into the circuit, hum being especially problematic in this respect. The diode, on the other hand, generates a comparatively high noise voltage across a low impedance. A circuit for an analogue noise generator is given in Figure 7.5.

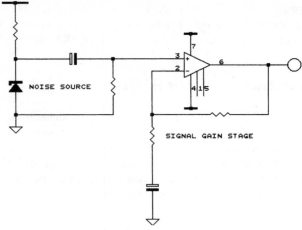

NOISE SOURCE

SIGNAL GAIN STAGE

Figure 7.5   *Analogue noise generator*

Colours of noise

Note that the circuit in Figure 7.5 is of a white-noise generator because all frequencies are present (at least stochastically). There is therefore an analogy with white light. However, due to the perception of sound, pure noise – which has all frequencies equally present – is perceived as particularly 'hissy'; that is with high-frequency energy predominating over low-frequency energy. While such a sound source may be artistically useful in some circumstances, often composers need a sound which is modified in some way. Examples of this include variations of low-passed filtered noise; so-called pink or red noise because again of an analogy with light (pink or red noise sounds more explosive or like the sound of the sea) and band-passed filtered noise which, if swept with a narrow band-pass fitter, can sound like the rolling waves or perhaps wind through trees. Digital noise generators are discussed in Chapter 10.

Analogue synthesisers

Analogue synthesisers incorporate all the foregoing components – VCOs, VCFs and noise generators as well as the voltage controlled amplifiers and ASR generators. But (to borrow a theory from gestalt psychology) the whole is so much more than the sum of the parts. Its power lies in its ability to cause each of these to interact in amazingly complex ways. Fundamental to the whole concept is the voltage controlled oscillator.

This may be controlled by a switched ladder of resistances, perhaps by means of a conventional musical keyboard, or by means of a constantly variable voltage – thereby providing a sound source with endless portamento like the Ondes Martenot and the Theremin. Alternatively it may be controlled by the output of another oscillator, the resultant being a waveform source frequency modulated by means of another. And perhaps this resultant waveform might be made to modulate a further source! By this means, the generation of very rich waveforms is possible and herein lies the essential concept behind analogue synthesisers.

Patching

In order to produce a practical, usable musical sound from an analogue synthesiser each of the circuit blocks mentioned above must inter-connect. A simple patch (the name used for a particular interconnection scheme) is illustrated in Figure 7.6. Notice that the keyboard controls two VCOs – to generate a pitch; the VCO is followed by the VCF to impart a character onto the basic sound source – and this is driven by the output of the LFO to create a changing formant, and the output of these modules is passed to the VCA block where the trigger signal from the keyboard (see next chapter) controls the ASR generator. However, Figure 7.6 illustrates but one possible patch. Remember, the power of the analogue synthesiser lies in its ability to cause the various sound and noise sources housed within it to interact. Commercial synthesisers differ greatly in the flexibility they present to the user, in terms of being able to route the various signals and thereby have control over the pattern of possible interactions. Some synthesisers offer very limited repatching, others permit virtually unlimited flexibility, so that practically any signal may be used as a parameter to control another process. Electrically this switching function may be provided by hardware switches, by electronic switches or even by a traditional telephone-style jack-field!

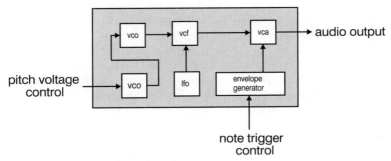

Figure 7.6 *Analogue synth patch*

Moog MINIMOOG

The Moog MINIMOOG is one of the most popular synthesisers of all time and is still used today on thousands of dance recordings worldwide. It contains three oscillators, the third of which can double as an LFO for modulation. This circuitry is followed by tone filters. The three oscillators can be routed and modulated with each other and herein lies the MINIMOOG's flexibility. Moreover, the open layout of the controller knobs and switches (illustrated in Figure 7.7) makes it easy for the composer/performer to experiment and thereby discover sounds quickly.

Figure 7.7   *Moog MINIMOOG*

The MINIMOOG was actually a second generation instrument, inspired by a series of instruments that Moog designed for experimental musicians; where each separate module could do one thing and one thing only towards shaping or generating a sound. These early modular instruments looked more like telephone switchboards than musical instruments because they incorporated patch chords that connected the parts together whereas the MINIMOOG was prewired. It was simple enough and quick enough to use so that it could be used on stage.

The inspiration for Moog's first modular instruments came from the experiments electronic music composers had been making since the end of the Second World War. Composers like Karlheinz Stockhausen and Herbert Eimert, and in France, at the Paris radio station, Pierre Schaeffer and Pierre Henry. Moog responded to the needs of these composers, who had previously had to make do with what they could find from other scientific and technical fields (using war surplus equipment and laboratory equipment – anything at all that could generate or modify a sound electronically). What Moog did, was to take everything that the musicians had found useful at that time and build it all in a neat form so that all the various components interfaced appropriately. Although many of the first customers were experimental musicians, it wasn't long before advertising agencies latched onto the MINIMOOG because the synthesiser turned out to be the perfect way to bridge the gap between sound effects and music. The MINIMOOG was in production for ten years which is a very long time for an electronic musical instrument, especially by today's standards.

FM sound synthesis

One of the lessons learnt from analogue synthesis – and especially from controlling one oscillator from another – was that extremely complex timbres may be generated by relatively simple means. In 1967 Dr John Chowning – a musician who had studied with Nadia Boulanger in Paris and subsequently set up a music composition programme at Stanford University – realised the possibilities of using frequency modulation to imitate complex musical sounds. This may seem like a complicated approach, but consider the alternative. Additive synthesis, as described in Chapter 5, is theoretically interesting but practically the number of separate oscillators required to produce complex musical tones – and especially transients – is prohibitively large. On the other hand, the complex nature of FM sidebands seemed to suggest that here might be a technique where complex musical tone structures could be built up using one or two oscillators. Chowning was right, but he had to wait for the advent of digital techniques to guarantee the necessary stability, predictability and repeatability required for this synthesis technique.

FM theory
Originally, it was believed that an amplitude modulated radio carrier wave would remain a pure single frequency transmission. It was, after all, only the amplitude of the wave that was being altered, so how could extra sideband frequencies come to exist? Many radio workers continued to believe this even despite the mathematicians who pointed out that if one sine-wave was modulated by another, sum and difference frequencies would be produced. In the case of frequency modulation, or FM, it is more obviously the case that the transmission will occupy a given bandwidth related to the deviation (the amount the carrier frequency is 'wobbled' by the modulating frequency). In fact, as is now well known, the situation is more complicated. The sideband frequencies produced around a frequency modulated carrier are related not only to the deviation as a proportion of the modulating wave frequency (the so-called modulation index) but to all the harmonics of the modulating frequency

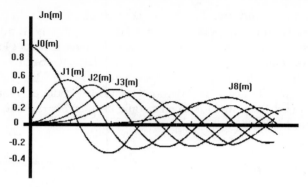

Figure 7.8 *First-order Bessel functions*

as well. The structure of the resulting sidebands issuing from all these variables are determinable using mathematical relationships known as Bessel functions. A collection of first order Bessel functions are illustrated graphically in Figure 7.8. These illustrate the harmonic content of the FM process, the abscissa value represents the modulation index (which is equal to the frequency deviation divided by the modulation frequency) and the ordinate represents the amplitude value of the carrier J0 – and the first eight harmonics, J1 to J8 respectively. Intuitively, you can see that when the modulation index is zero (i.e. no modulation) all the energy in the resulting output is concentrated in the carrier, that is, J0 is unity and all the other functions J1 through J7 are zero. As the modulation index increases slightly, J0 declines and J1 climbs. Note that J2, J3, J4 etc. climb more slowly. This illustrates why low modulation index FM has a spectrum similar to that of AM, with only first-order sidebands. And that a signal modulated with a very high modulation index will have a very rich spectrum indeed.

Frequency modulation produces complex harmonically related structures around a carrier, but doesn't that mean that the carrier would have to be different for each note of the keyboard? And what happens to the lower sidebands, these don't normally exist in the case of a musical sound? One answer is to use a carrier of 0 Hz! It is the modulating signal which determines the fundamental, and the depth of the modulation (or modulation index) which may be manipulated to produce the harmonic structure above the fundamental. Look at the figure of Bessel functions and imagine that a modulating frequency is employed which relates to the note middle C, 261 Hz. Suppose a deviation is chosen of 261 Hz, that is with a modulation index of 1. From the curves in Figure 7.8, it is clear that at $m = 1$, J1 equals about 0.45 and that J2 equals about 0.1, all the other harmonics still remaining very small at this low index. The resulting sound – suitably synthesised, amplified and transduced – would be a note at middle C with a second harmonic content of about 22%. (The 0 Hz carrier would, of course, effect no audible contribution.) This might make a suitable starting point for a flute sound. Now imagine a much greater modulation index is employed: $m = 3$, for example. Reading off from the curves, it shows that in this case $J1 = 0.34$, $J2 = 0.49$, $J3 = 0.31$ and $J4 = 0.13$. This would obviously create a much richer musical timbre.

If the index of modulating signal is changed over time, it is possible to create musical sounds with extremely rich musical transients, full of spectral energy which then segue into the relatively simple ongoing motion of the resulting note. In other words, synthetic sounds would be created just like real musical sounds, where a 'splash' of harmonics is created as the hammer hits the piano string, or the nail plucks the string of a guitar, or as the first breathy blast of air which excites the air inside a flute, all of which decay quite rapidly into a relatively simple ongoing motion. All these effects may be synthesised by generating a modulating signal which initiates the carrier modulating with a large deviation – thus creating a rich transient part to the sound – which then decays to a level where it causes only relatively small deviation, thus generating a relatively pure ongoing sound.

A complication arises as a result of the choice of the zero frequency carrier. Just because the carrier is 0 Hz, it doesn't stop there being sidebands in the negative frequency region. These 'fold back' into the positive frequency region and destructively interfere or constructively reinforce the sidebands present there already. These too have to be taken account of in generating FM synthesis algorithms. But there are other more interesting uses for the lower sidebands and these embrace the use of a very low-frequency carrier instead of one of zero frequency. When a low-frequency carrier is used, the negative frequency sidebands 'fold back' into the positive region to be interleaved with the positive frequency sidebands. In this manner, even more complicated timbral structures may be built up.

Perceived fundamental

Another approach is employed in simple FM (that is, FM with only one modulator and one carrier) which takes advantage of a psychacoustic phenoemon sometimes called, 'hidden bass' because it is used to enhance apparent bass response. If a loudspeaker fails to reproduce a fundamental bass frequency but, in so doing, it creates a relatively high degree of harmonic distortion, the fundamental is nonetheless heard; the ear 'infering' the bass from the ensuing harmonic structure. The same effect

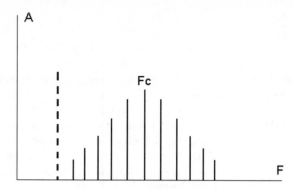

Figure 7.9 *The fundamental pitch doesn't have to be present to be heard!*

holds for higher frequencies too. So, if the carrier is modulated so as to produce the symmetrical sidebands shown in Figure 7.9, the perceived pitch of the note will actually be at the frequency shown by the dotted line; where no frequency actually exists. In this case we have the fascinating phenomena that the synthesiser produces only the harmonic structure and leaves the human ear to 'fill in' the pitch!

Complex FM and feedback

So far we have only considered simple FM, in which one carrier (usually termed an 'operator' in FM synthesis parlance) is modulated by another simple sine-wave operator. In practical FM, more than two operators are often used to produce complex musical sounds. Finally, feedback may be employed wherein the output of an FM modulation routine is 'recirculated' and used to drive the process. This technique can be used to generate very complex waveforms.

The FM synthesis technique was first used commercially in the Yamaha DX7 keyboard and, at the time, it caused a sensation in synthetic music. So successful was the method and Yamaha's implementation so excellent, that the 'sound' of the FM synthesis DX7 dominated the popular music of the 1980s. It is largely due to this fact that FM has today come to imply a rather passé algorithm. FM is remarkable in that it represents the high point of pure, electronic sound generation. It still remains the sound technique employed in cheaper PC sound cards although most better cards nowadays allow for a plug-in upgrade to include wavetable synthesis, a technique which utilises recorded, or sampled, sounds.

Sampling

Digital sampling systems rely on storing high quality, digital recordings of real sounds and replaying these on demand. The main problem sampling incurs is the sheer amount of memory it requires. Sampling is well suited to repetitive sounds (like drums and other percussion instruments) because the sample is mostly made up of a transient followed by a relatively short on-going (sustain) period. As such, it may be used over and over again so that an entire drum track could be built from as few as half a dozen samples. Problems arise when long, sustained notes are required, like the sounds generated by the orchestral strings. The memory required to store long sustained notes would be impossibly large. (And impossible to know how large it should be too; how long is the longest note someone might want to play?) Pure sampled-synthesis systems rely on 'looping' to overcome the limitation of a non-infinite memory availability. A loop is constructed as illustrated in Figure 7.10, the steady-state part of the note forming the loop.

An important part of sampling technique involves the use of one acoustic sample over a group of notes, the replay of the sample at the appropriate pitch being achieved by the applicable modification of the read-clock frequency in exactly the same way we saw with pitch shifting in Chapter 6. In theory, one sample may be used at every pitch. However, due to distinctive formants imprinted on the sound by the original voice or instrument, this can result in an unnatural sound if the transposition is

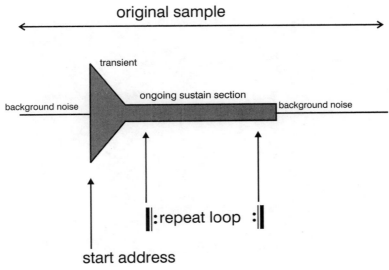

Figure 7.10

Figure 7.11   *Akai sampler*

taken too far, an effect known as Munchkinisation (named after the under-speed recorded singing Munchkins in *The Wizard of Oz*). This effect is ameliorated by recording several samples at different pitches and assigning these to various ranges of transposition; the term multi-sampling is used to describe this process. Furthermore, samples are usually recorded at various different dynamic levels from very quiet (pianissimo or pp) to very loud (fortissimo or ff). The sampler uses these different samples and a mixture of samples at different dynamics points, to achieve touch-sensitive dynamics from the controlling keyboard (see next chapter). Good sampling technique therefore involves judicious choice of looping-point, transposition assignments, and dynamics assignments. This is no mean task and successful sampling programmers are very skilled people. A modern sampler is illustrated in Figure 7.11. Fortunately, those less skilled have access to literally thousands of samples on CD-ROM (Figure 7.12). The ROM drive usually interfaces with the sampler via an SCSI interface (see Chapter 10).

Figure 7.12   *Sample CD-ROM*

Wavetable synthesis and other proprietary synthesis algorithms

The technique known as wavetable synthesis sits somewhere between classic synthesis using basic waveforms (sine, square, ramp, triangle) and sampling; where complete recordings of 'real' sounds are used as the basic grist for synthesis. 'Wavetables' are nothing more than small samples stored in EPROM (look-up tables of waves) which are used in combination with a number of techniques which include; carefully edited looping points, pitch shifting, interpolation and digital filtering to reduce the prohibitive memory requirement which a pure sampled instrument would require. Examples of this technique include LS Sound Synthesis from Roland (as used in the sound module illustrated in Figure 7.13).

Figure 7.13   *Roland MT–100*

Other interesting hybrid techniques include dynamic vector synthesis used in the Yamaha TG33 sound module (Figure 7.14) which was designed by the same team who worked on the Sequential Prophet VS and the Korg Wavestation. The TG33 creates sounds using up to four independent voice sources. Two waves are samples and the other two are created by digital FM synthesis: various wave combinations are available. The joystick can be swept in real time to adjust the contribution of each of the four independent signal sources and these 'vector' sweeps can be stored as part of the overall patch design. This technique uses what might be termed 'empirical' sound synthesis techniques, utilising a mixture of sampling and tone generation (by additive of FM synthesis) to arrive at

Figure 7.14   *Yamaha TG33*

their final result. At 'the cutting edge' sound synthesis tends to be like this; a non-purist subject based on subjective 'feel' rather than mathematical precision.

Modern trends in synthesiser design

The method of synthesis employed in the 'classic' 3-VCO and filter synthesiser is sometimes referred to as subtractive synthesis. In fact this term is a little misleading, but it relates to the situation whereby the fundamental tones produced by the VCOs are of a complex harmonic structure in the first place and these are subsequently filtered and modified in the filter and envelope circuits; thereby simplifying the harmonic structure by 'subtracting' harmonics. In exactly the same way, the voice may be considered a subtractive synthesiser because the vocal chords produce a crude but harmonically complex 'buzzing' fundamental tone which is capable of being modulated in pitch alone and on which is imparted the filtering formants of the mouth and nasal cavities. Nature, always a parsimonious architect, adopted such a system because it is efficient. However, while subtractive synthesis is a very potent arrangement, and FM techniques and sampling all offer their particular advantages, there is another approach which offers the ultimate in synthesis technology. We have met this before, it is known as additive synthesis.

Additive synthesis
We looked at additive synthesis before, when describing the classic tone-wheel Hammond organ. Hammond's implementation was Neanderthal by today's synthesis standards. However, despite the crude nature of their synthesis algorithm – and that simply means the limited number of

harmonic partials they had available – additive synthesis represents the 'ultimate' synthesis algorithm because it is Fourier analysis in reverse! Of course, Hammond's engineers didn't adopt a limited number of partials out of choice, they were thwarted by the technology of their period. Very Large Scale Integration (VLSI) integrated circuits offer the possibility of the, literally, huge number of oscillators necessary to produce convincing synthetic sounds by this means.

The 'analogue' interface

Other recent trends in synthesiser design involve revisiting some of the characteristics of the analogue synthesisers of the past. This has been driven by the fashion for analogue synthesiser sounds, which continue to remain in vogue in dance music. In some instances, manufacturers have responded by looking again at the manner in which sounds are created in analogue equipment. In fact, some modern synthesiser models have implemented digitally the 'classic' Moog 3 oscillator and LFO. However this has not been the most important analogue 'rediscovery'. As so often in analogue instruments, the sonically interesting effects of a particular analogue process are discovered in the open-ended and relative contrariness inherent in analogue circuits; the interesting marginal stability, the unforeseen interaction. Just the sort of phenomena, in fact, that are so hard to replicate in digital circuits. Good sampling technology is often more fruitful.

The more interesting long-term effect of the renaissance in analogue synthesisers has been in looking again at the user interface. Digital synthesisers have mostly always included programming but often this has required considerable patience, partly because the nature of the synthesis techniques often precludes straightforward interaction. More often, it is simply due to a lack of usable parameter control. Often a natural musician is thwarted in a search for a new and interesting sound by the irksome necessity of typing numbers into a spreadsheet-type user interface! Even in a MIDI controlled, SMPTE locked, sampled modern studio, music is a

Figure 7.15   *Roland E–500*

physical, tactile, sensual experience. Moreover the 'sculpting' of new sounds requires vivid and immediate feedback. Analogue synthesisers, unlike their digital cousins, exemplify these requirements. Parameters are usually controlled by rows of potentiometers – it might look complicated but everything is just a 'finger's-touch' away. This, of course, is relatively easy to implement in an instrument – irrespective of its internal technology pedigree. Just one such instrument is illustrated in Figure 7.15. From Roland, this digital keyboard features user control over various 'analogue-type' parameters on an array of rotary 'pots' and sliders. One can't help looking at this renaissance without wondering whether the elimination of the array of rotary controls on modern recording consoles won't be similarly missed and 'rediscovered' in a few years' time!

Physical modelling

Instead of concentrating on the sound events themselves, some workers in synthetic sound have chosen a different route; to model mathematically the results of a physical system. Using this technique, a guitar sound is modelled as a stretched string, with a certain compliance, mass and tension coupled to a resonant system and excited into sound by means of an excitation function (itself modelled to represent the action of plucking a string). That this technique is computationally intensive is something of an understatement, however it has already proved to be a fruitful synthesis approach and it also offers the possibility of modelling new sounds of impossible instruments which is especially interesting for composers searching for new sounds of ostencibly 'physical' or 'acoustic' origin.

Functional physical modelling

Writers of advertising copy (rarely the best source of technical accuracy!) have adopted a less strict definition of the term 'physical modelling', which has permitted a rash of so-called physical modelling (or PM) synthesis products. The nomenclature PM in these instances rarely describes the classic approach described above in which an entire instrumental sound is modelled from basic physical principles. Instead it describes a variety of techniques under the general heading of 'Functional PM'. These techniques include; re-circulating wavetable synthesis and waveguide filtering.

Functional PM takes as its starting point a rather more macro-model of a typical musical instrument. Rather than model an instrument from the ground up, in terms of fundamental moving masses and compliances, it commences the modelling at a higher, conceptual level. In Functional PM, an instrument is modelled as a source and resonator. The source is typically, relatively wideband and 'noisy' (i.e. many frequency components present): the resonator is a filter structure. This arrangement is typical of a real acoustic instrument. Indeed it is typical of vocal production too, as we have seen in the section on the Vocoder. Many examples justify this approach. Brass instruments are a resonating tube (a filter) excited by 'blowing a raspberry' into the mouth piece (noisy source). The clarinet, oboe and bassoon are all resonating structures in wood (a filter) excited by a reed structure which produces a rattling, squawking and relatively tuneless noise. Of course, this relaxation of the term PM allows all manner of synthesis techniques to be assumed under the title; even classic VCO + VCF subtractive synthesisers like the MINIMOOG may be regarded as PM – a copy writers dream! Nevertheless, there exist two approaches which are new and different and these are described below.

Re-circulating wavetable approach

This approach uses a fixed length look-up table (wavetable), initially filled with random samples (noise). These are re-circulated using a feedback network which incorporates a filter. Because of the iterative nature of the filtering – even if the filter characteristic is not particularly steep – the noisy start to the signal gradually becomes transmuted into an ongoing, stable sustain section which eventually decays away. Whether such a system actually models physical behaviour is doubtful. However, the resulting output from the system certainly resembles the physical characteristics of real sounds; whereby a highly organised, stable sound emerges from an initial attack – the 'splash' of the highly disorganised transient.

Waveguide filtering

Waveguide filtering may, more justifiably, be termed a PM technique since its aim is to model, not only the classic source and resonator parts of the instrument, but the highly important reverse effect of the resonator's effect on the source. Anyone who plays a wind or brass instrument will know that blowing into the mouthpiece when disconnected from the body of the instrument does not produce the same physical sensation. The excited air inside the instrument – once the ongoing sound has commenced – produces a complex back-pressure. This, in turn, alters the raw sound produced by the mouthpiece and lips. In waveguide filtering this 'reflection' of the resonator upon the source is modelled as delayed, frequency selective feedback upon the source function. The feedback is delayed because the effect upon the source is not present before the resonator is fully excited.

Granular synthesis

The idea of analysing sound events in terms of large number of tiny sound 'grains' is usually attributed to Gabor. The idea was siezed by Xennakis in his search for a theoretical framework for music which was outside both the notes, beats and bars of Western music theory and the tones, timbre and duration of the 'classic' electronic music studio of the nineteen-fifties and sixties.

Xennakis

Composer, theoretician Xennakis used Gabor's idea of sound-grains as a form of analysis and transmuted the idea into a method of composition whereby short sound events – often, but by no means always, realised as pizzicato – are conceived within a frequency-amplitude two-space. These 'frames' of sound events can be imagined a bit like frames of a cinema film. This mathematical conception of an evolutionary perceptual 'space' has evident heuristic power to a composer with a scientific frame of mind. Xennakis has used the technique, for instance, to orchestrate the physical process of diffusion whereby sound-grains act like molecules of a gas diffusing within another. Such a mechanism is not entirely determinsistic. So – whilst it's possible to conceive on a macroscopic level the overall process of diffusion – it's not possible to predict each molecule's individual time-space trajectory. The mathematics which deals with the description of large number of small 'things', each appearently behaving in a manner close to randomness, but actually behaving as part of a bigger overall event (like molecules in a diffusing gas) is known as Stochastic theory and is another of Xennakis' great passions.

Xenakis uses stochastic mathematical functions to inform the position of individual sound 'grains' within each sound 'frame' and the 'movement' of these grains as they change position 'frame-by-frame'. Whilst individual instrumental players would find it hard, if not impossible, to infer the overall process from their individual musical part, the listener – aware of the aggregate effect of the evolving musical grain 'cloud' – perceives the overall enterprise, this organisational principle replacing that which was played by song form or sonata form in pop, rock and classical music.

Xennakis' approach can be regarded as 'pure' or formal granular synthesis and, whilst many of his most celebrated works use orchestra, he has also worked extensively with electronics. In many ways the 'controllability' of electronics suits the formal, controlled nature of his compositional technique. Strictly speaking, grains ought to be of the simplest form of sonic material in which a sine-wave is envelope filtered by a window function so as to avoid on and off 'clicks' due to rapid

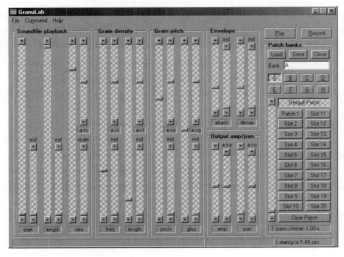

Figure 7.16 *GranuLab software*

waveform discontinuities. However, in a more modern interpretation of granular synthesis, grains are often themselves formed of small pieces of digital audio files in a process resembling a kind of microscopic sampling! Like the sine-wave 'primitive' these grains must too be filtered by a window function to avoid 'clicks'. But, once done, these tiny samples can be subjected to manipulation ('brassage') and aglommeration into larger sound events. Figure 7.16 illustrates the control screen of a software package known as GranuLab written by Rasmus Ekman. GranuLab takes .wav files as input from which grains can be selected looped, spread over pitch ranges and subjected to envelope modulation. Available for download at: http://hem.passagen.se/rasmuse/Granny.htm, GranuLab is an easy way to discover the world of granular synthesis.

The important aspect of this looser form of granular synthesis is the rediscovery, originally made by the post-war musique-concrete artists, that – if real world sounds are used as the grist for later electronic manipulation – the resulting sounds are the richer due to the complex nature of the raw material.

Waveshape distortion synthesis

Waveshape distortion synthesis (or simply waveshape synthesis) is really an extended form of waveform modification we saw in relation to musical effects; as in the distortion pedal or fuzz-box. If a sine-wave is multiplied by a non-linear function, it is transformed into a different waveform with a higher spectral richness. In the fuzz box, the non-linearity function is pretty crude. But there exist a family of mathematical functions (known as Chebychev polynomials of the first kind) which can transform a single sine-wave input into an output of any required harmonic spectrum. Whilst not possible to implement in an analogue synthesiser, these functions can easily be pre-loaded inside a computer or DSP and used as a look-up table, the address within the table being defined by the input function.

Let's look at an example of multiplication by some low-order Chebychev polynomial functions of the first kind. Incidentally, the term 'polynomial' will no doubt horify some readers but it's actually just a maths jargon word for a mathematical expression consisting of the sum of terms, each of which is a product of a constant and a variable raised to an integral power; that's to say, raised to the power of 0, 1, 2, 3, etc. Chebychev polynomials are no exception, here's the first four polynomials:

$$T0 = 1$$

$$T1 = a$$

$$T2 = 2a^2 - 1$$

$$T3 = 4a^3 - 3a$$

The function here is a. In a synthesiser it's usually a sine-wave input. Applying a sine-wave to a zero-order Chebychev function will produce the same sine-wave; because the signal is simply multiplied by one. However, applying a sine-wave to a first-order function will produce the output,

$\sin x. \sin x$

which, as we have seen many times already in this book, produces a DC (difference) term and a twice-frequency (sum) term. Now let's apply a sine-wave to the second-order function. Here the process can be written,

$\sin x [2(\sin x. \sin x) - 1]$

which we can write as,

$2 (\sin x. \sin x. \sin x) - \sin x$

or,

$2 (\sin^3 x) - \sin x$

Now, just a squaring produced a twice-frequency term, so cubing produces a three-times-frequency term. And the next polynomial (in which sine is raised to the power of four) will produce a $(\sin 4x)$ term and

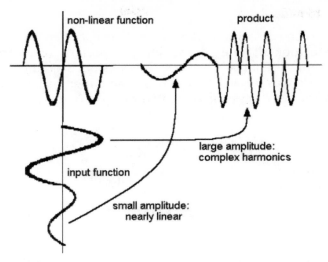

Figure 7.17 *Waveshape distortion*

so on. By constructing an overall transfer function derived from weighted sums of Chebychev polynomials, an output of any degree and blend of harmonics may be obtained.

One attraction of this form of synthesis approach is that the amplitude of the carrier often defines the resulting richnes of the waveform output. Look at Figure 7.17 wherein a sine-wave is being applied to a non-linear (sinusoidal) function. Because the central part of the function is relatively linear, the output is relatively pure. Only when the sine-wave value is greater will the output distort and become perceptibly 'richer': in the case of the figure, very much richer! Rather as we saw with FM synthesis, the advantage here is that waveform distortion synthesis results in musical sounds with relationships which mirror musical experience and sensitivity. For instance, the relationship between effort and spectral richness is particularly intuitive and satisfying. Waveform distortion synthesis is musically appealing because larger amplitudes, easily produced – for instance – from higher MIDI velocity values, can readily be made to produce more complex harmonic spectra.

Fact Sheet #7: Negative feedback and op-amps

- Negative feedback
- Effect on frequency response and distortion
- Op-amps
- Op-amp cookery

Negative feedback

Before considering operational amplifiers, it's first necessary to take a look at one of the most important discoveries in electronics; negative feedback. This was discovered by Harold Black. He was searching for a better amplifier system, but had used the very best components he could find. Where to go

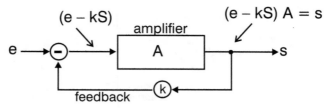

Figure F7.1   *Negative feedback*

next? He reasoned that if he took an amplifier and subtracted a proportion of the output signal from the input signal and fed this modified signal to the amplifier, this would result in a dramatic improvement in performance. To understand this it's necessary to use a bit of simple maths.

Black's original amplifier (Figure F7.1) has two imporant networks added to it: a network for deriving a proportion of the output signal (often as simple as a resistive potential divider in practice) and a subtractor network for subtracting the proportion of the output signal from the input signal. The diagram is labelled to show the signals which appear at the various circuit nodes. They are

e = input signal

s = output signal

k = feedback fraction

A = amplification.

With the modification the signal which is actually input to the amplifier is $(e - k.s)$. So the output will be,

$s = A. (e - ks)$

$\quad = A.e - A.k.s$

therefore

$s + A.k.s = A.e$

and

$s.(1 + A.k) = A.e$

or

$s = (e.A)/(1 + k.A)$

In itself this equation isn't very exciting. But if we substitute a few imaginary values, you can see the power of Black's idea. Firstly let's look at how negative feedback can be used to stabilise the gain of the original amplifier. For instance suppose the gain was 100 times and the feedback factor was 1/50th. With feedback the gain would be modified to be,

$s = (e. 100)/(1 + 0.02.100)$

$s = (e. 100)/3 = 33$ times.

Now imagine that, due to ageing or temperature change, the gain of the amplifier changed to only 50 times. Re-doing the equations we arrive at a result of,

$s = (e. 50)/2 = 25$ times.

In other words a fairly disastrous 50% reduction in gain has been reduced to a tolerable 16% change in gain! The only disadvantage is that overall sensitivity is lost (from 100 times to 33 times). But actually this isn't important because gain per-se is much easier to achieve than stability.

Effect on frequency response and distortion

By concentrating on change in gain, we can imagine the effect of negative feedback on other circuit effects. For instance, what might you think the effect would be of negative feedback on frequency response? Since frequency response is just another way of saying 'changing gain with frequency', negative feedback has the same calming effect on frequency response as on ageing and general stability.

Even more amazingly negative feedback has an effect on harmonic distortion. Once again (see Chapter 4) distortion mechanisms are really due to varying gain throughout the operating window of the amplifier and negative feedback reduces harmonic distortion in the same proportion as it reduces gain. So, for example, if the amplifier in Figure F7.1 without feedback, had a distortion of 1% THD for a given output swing, the distortion of the amplifier, with feedback, would be reduced to 0.3%.

Op-amps

An important consequence of the feedback equation (1), is that; as the gain term (*A*) increases, the resulting gain of the amplifier is dominated more and more by the feedback fraction (*k*).

For example, if you work back through the examples for amplifier stability. But this time, substitute a starting gain of 1000, falling to 500; you will find the gain (with feedback) changes from 48 times to 45 times. A 7% change. With a starting gain of 10,000 falling to 5,000, the gain-with-feedback (known as closed-loop) gain starts at 49.75 times, and falls to 49.50 times. A 1/2% change. Notice too that, as the gain of the amplifier (also called open-loop gain) rises, so the closed-loop amplification is simply the reciprocal of the feedback fraction.

The 'science' of op-amp design relies on this phenomenon. To wit, if an amplifier has a high enough gain (and various other attributes that we need not consider here), it's real world application becomes dominated by the choice of feedback network. With the advent of integrated circuits, it became obvious that provided silicon manufacturer's provided good, cheap complete amplifier ICs or 'chips', designers could simply regard an amplifier as a basic building block in the way that they would have previously used a transistor; as a circuit element. Such a circuit element being termed an operational amplifier or op-amp for short. Whilst there are still some corners of electro-technology where op-amps are not applicable, it's fair to say that the technology of analogue audio electronics is absolutely dominated by op-amp applications. Very few examples of discrete transistor design now exist, except in power amplification and special high-power op-amps are starting to make inroads here as well. In fact, it is true to say that it is now very difficult to better the design of op-amp circuitry, this being a testimony to the quality of the op-amps themselves – especially the most modern types.

Op-amp cookery

Op-amps permit the construction of powerful circuits with the minimum of fuss. They may be used to construct amplification stages (for instance for microphones or guitars as shown in Chapter 12), they may be used to undertake various mathematical operations; the most important in audio being the addition (or mixing) of two or more audio signals. They also provide the basic 'gain-engine' for active filters such as those used in tone-controls, equalisers and active crossovers and they are frequently used as the gain element in oscillator circuits; such as those described in Chapters 5 and 7.

Figure F7.2 illustrates just some of the circuits obtainable with op-amps and the relevant design equations.

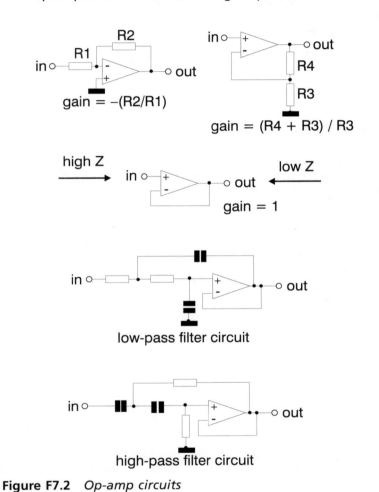

Figure F7.2 *Op-amp circuits*

8
Silver Machine – Sequencers and MIDI

Analogue sequencers

Digital synthesisers and their associated digital control system (MIDI) are now so ubiquitous that it might be possible to think the control of analogue synthesisers was a subject hardly worth covering. And yet, as modern musicians increasingly renounce modern digital instruments and turn instead to analogue synthesis in search of new inspiration and new expressive possibilities, the wise (and truly competent) recording engineer must be aware of older instrument control technologies. These are covered briefly here.

CV and gate control
What does the following really mean?

Figure 8.1

Well it indicates a particular note (the A above middle C) should be played for a single beat, the beat's duration being defined by the metronome marking above the stave. It also tells you how loud the note should be played; actually, in this case, it only says piano (or soft in Italian). But the control system we call musical notation is pretty good, because it conveys a great deal of information in a way we, as humans, are able to make use of quickly and easily. As we shall see when we come to study MIDI, each of these pieces of information is conveyed in a MIDI message too, in the form of a digital code.

Just as written music is suitable for a human machine and a digital code is suitable for a digital synthesiser, it should be no surprise that an analogue synthesiser expects analogue control information. Such information travels in an analogue installation on an interface known as the CV and Gate interface. The control voltage (CV) and gate interface has two fundamental components:

1 A pitch control voltage. A widespread relationship is one volt per octave (or 83.33 mV per semitone).
2 Control pulse to trigger the sound. There exists very little standardisation as to gate-pulse requirements but the 'standard' is a 5 V pulse for the duration of the note. However, some synthesisers expect a short-to-ground action.

Notice the CV and gate interface has a number of very significant drawbacks; it does not transmit dynamic information (how loud or how soft a note should be), it has no standardisation as to control voltage and absolute pitch, and very little standardisation as to gate pulse requirements or connectors. Older CV/gate synthesisers may be interfaced to MIDI systems with the use of commercial interface adapter boxes. Also some 'classic' instruments are now being manufactured once more, complete with MIDI interfaces.

MIDI

For many musicians, developments of the last ten to fifteen years have changed recording forever in that they no longer use a multi-track tape recorder. Instead most orchestral, synthesiser and drum parts can be played on one conventional piano keyboard with the various voice synthesisers linked together using the Musical Instrument Digital Interface or MIDI system as illustrated in Figure 8.2. The data transmitted on the MIDI link specifies not only what note has been played, but also the velocity with which the key has been pressed, the force used to hold the key down and the pedal positions. To recreate the performance, it is only necessary to record the MIDI data rather than the sound signals themselves. This can then be transmitted down the same MIDI link and used to trigger the keyboards and sound modules. An incidental benefit of this approach is that a musician may change voicing after recording and also change the speed of the performance by having the MIDI events transmitted more rapidly or more slowly than they were recorded.

MIDI data containing information on key presses is always preceded by data specifying a 'channel' on which the data is to be received. Most MIDI controlled instruments allow the musician to allocate that particular

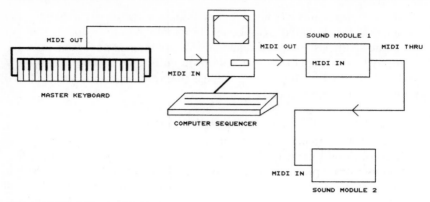

Figure 8.2 *Typical MIDI set-up*

instrument to a particular channel so it is possible to record performances of several musical parts by building up a recording of MIDI event data rather than the musical sounds themselves. This technique has the benefit that it is possible to alter individual parts which, on a conventional multi-track, would be very fiddly. The other boon is that the performance remains distortion, wow, flutter and tape-hiss free no matter how many times it is necessary to record and re-record the individual musical parts. The different products available that allow one to record this MIDI data, and build up the parts one by one, differ greatly from software packages running on IBM PC, Apple or Atari computers to dedicated units. The generic term for any of these devices is sequencer.

The MIDI specification

Like the standards for digital video and digital audio which are explained in Chapter 10, the Musical Instrument Digital Interface (MIDI) standard comes in two parts, a mechanical and electrical specification, and a data-format specification. Part of the success of the MIDI standard certainly issues from its genesis with a manufacturer, in that it was designed for design engineers by design engineers. It is therefore a very practical specification, even to the extent that it suggests particular manufacturers' and part numbers for the interface components! It leaves little to be interpreted – a characteristic sadly lacking from a number of international standards drafted by standards-making bodies.

MIDI is a serial interface protocol which has some similarities to RS232 and other control protocols used in computing. It is an important feature of the specification that, in any MIDI installation, it is only intended that one keyboard or sequencer controller (transmitter) will contribute to the interface although there is no limit to how many receivers (i.e. MIDI controlled sound generators) may 'listen' to the interface.

Mechanically the MIDI standard specifies a 5 pin, 180° DIN socket for the transmission and reception of MIDI data. Cables for the transmission of MIDI data are specified twisted pair with overall screen, terminated at either end with 5 pin, 180° DIN plugs. The maximum physical length of these cables is limited (due to electrical and data coding considerations) to 15 metres.

Electrically, the standard specifies a 'current loop' for the data path. The transmitter has thus a relatively high output impedance and the receiver a relatively low input impedance. The complete MIDI current-loop data path is drawn in Figure 8.3. Logical 0 is equivalent to current ON. This

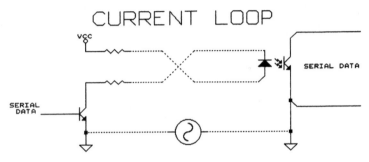

CURRENT LOOP

NOTE IMMUNITY TO EARTH BORNE NOISE

Figure 8.3 *MIDI current-loop data path*

scheme is in many ways excellent. First because it is opto-isolated and therefore provides an interface with no earth path, which might create hum-loop problems in the audio installation, and second because the high output impedance MIDI OUT is inherently protected from short circuits. However, the arrangement creates the minor complication that once one controller (MIDI OUT) has been connected to one receiver (MIDI IN) extra MIDI receivers cannot be driven from the same line. If they are, the drive current divides between the two receivers resulting in unreliable data reception. For this reason the specification suggests the use of a further MIDI port, known as THRU. The MIDI THRU port is effectively an output port which is driven by a replica of the data arriving at the MIDI IN socket as shown in Figure 8.4. Receivers 'down the line' thus receive a buffered version of the data received at the previous instrument or sound generator.

The MIDI interface is designed to use an industry standard universal asynchronous receiver transmitter (UART) integrated circuit. The Harris

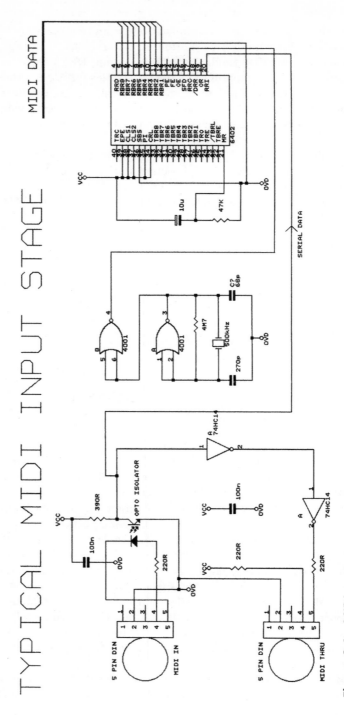

Figure 8.4 *MIDI ports*

semiconductor CDP6402 is commonly used but there are equivalents. The CDP6402 is a CMOS UART designed for interfacing computers and microprocessors to asynchronous serial data channels. It is therefore particularly suited to applications involving the generation of MIDI data from a microcomputer and for reception of that data and its control of sound generators and sound modules. The CDP6402 is designed to provide all the necessary formatting and control for interfacing between serial and parallel channels. The receiver converts serial start, data, parity (if used) and stop bits to parallel data verifying proper code transmission, parity and stop bits. The transmitter converts parallel data (inside the sequencer or controller) into serial form and automatically adds start, parity and stop bits. The data word may be programmed to be 5, 6, 7 or 8 bits in length. Parity may be odd, even or inhibited.

The data format of the MIDI data is specified as 1 start bit, 8 data bits and 1 stop bit with no parity. The baud rate is specified as 31 250 bits/s (±1%). The data format for one byte of MIDI data is illustrated in Figure 8.5.

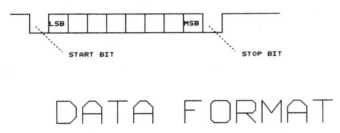

Figure 8.5 *MIDI data format*

Most messages which pass through a MIDI link in the multimedia studio are in the form of 'packages' of two or three bytes. (The two exceptions are system exclusive and real-time messages which are dealt with later.) The format of a multi-byte message is as follows:

Byte 1 A status byte, which indicates to the receiver the type of information that is to be sent in the subsequent two data bytes. The lower nibble of this byte also refers to the channel to which the message refers.

Byte 2 The first data byte.

Byte 3 The second data byte (where used).

A status byte is distinguished from a data byte by the value of its most significant bit (MSB). All status bytes set MSB = 1. All data bytes set MSB

= 0. Messages may be further divided into two groups; MIDI channel messages and MIDI systems messages. Different status bytes reflect these different uses. The format and most common types of MIDI channel message status bytes may be tabulated thus:

MIDI channel message status bytes

MSB	6	5	4	3	2	1	0	Message type
1	0	0	0	One of 16				Note off
1	0	0	1	channel identifier				Note-on
1	0	1	0	nibble				Aftertouch (polyphonic)

The one of 16 channel identifier nibble which forms the lower nibble part of the status byte allows one keyboard or sequencer to control up to 16 sound modules independently on the same link. This multiple channel capability is an essential feature of the MIDI specification and the touchstone of its power over the older synthesiser control interfaces which it has entirely superseded. The multiple channel feature of the MIDI interface allows the possibility of sequences which can be built up channel by channel, yet played back simultaneously, in much the same way as can be done with real instruments using a multi-track tape machine.

Note-on and note-off

Two, 8-bit data bytes follow the note-on/note-off status bytes (their MSBs set to 0). The first data byte refers to the note number and the

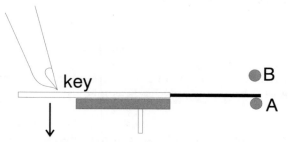

For the note to sound contact A must be broken and contact B made: the *transition time* between these two events is a measure of the *velocity*. A skilled pianist can accomplish about 30 different velocity values; equivalent to a range of 5–35 mS transition time.

Figure 8.5a   *Velocity sensitive keyboard mechanism*

second to the velocity. The note number value relates to the pitch of the note to be played. Note values range from 0 to 127. Each integer relates to a note, one semitone apart. In other words the note values correspond directly to the even-tempered, chromatic scale. Middle C is defined as note 60 (decimal) or 00111100 (binary).[1]

The following, second data byte relates to touch-sensitive keyboards and defines the force (velocity) with which a key is struck. Clearly when playing a real instrument, pressing the key of a piano down gently or pressing it down hard produces not only a different volume of sound, but also a different timbre. Keyboards which feature touch sensitivity encode this tactile information in the form of the 8-bit byte which follows the note-on command. Sound modules may interpret the information in many ways. Some simply reflect the change in the amplitude of the sound output. Other more sophisticated units attempt to emulate the changes in timbre that a player would expect from an acoustic instrument. Low numbers represent low velocity – a lightly played key – and high numbers represent a high velocity – a heavily struck key. Some keyboards go one step further and encode the force with which the key is held down after it has been struck. A pianist, by virtue of the mechanical arrangement within an acoustic piano, has no control over a note once the key has been depressed. Not so a trumpeter or a flautist or a string player each of whom has the ability to change the volume of a note once it has commenced. In order that a piano-style keyboard can be exploited to control sound modules with brass, woodwind and string sound synthesis, better keyboards encode the force with which the key is held down – termed aftertouch – so as to emulate the nuances of real players and send this information to the appropriate sound generator.

Controllers

This concept of the ability to encode the expression or interpretation of a performance over a MIDI link as well as simply the depression and release of the keys themselves is the characteristic which raises MIDI's utility above that of a sort of electronic pianola! However, there exists a bewildering amount of 'expressivity' controllers which may be transmitted via MIDI. These range from the control of the piano damper pedal and sustain pedal to portamento control and breath control. Necessarily, limited as they are to 3 bits in the most significant nibble of the status byte there are simply not enough status bytes to instruct receiving devices which expression parameters are to be changed. So the type of controller is encoded in the 7 bits of the data byte immediately following the status byte. The most common may be tabulated thus:

MIDI channel message status bytes continued

Initial status byte

MSB	6	5	4	3	2	1	0	Message type
1	0	1	1	Channel ident nibble				Control change

Subsequent data byte

0	0	0	0	0	0	0	1	Modulation wheel MSByte
0	0	0	0	0	0	1	0	Breath controller MSByte
0	0	0	0	0	1	0	0	Foot controller MSByte
0	0	0	0	0	1	1	1	Main volume MSByte
0	0	1	0	0	0	0	1	Modulation wheel LSByte
0	0	1	0	0	0	1	0	Breath controller LSByte
0	0	1	0	0	1	0	0	Foot controller LSByte
0	0	1	0	0	1	1	1	Main volume LSByte
0	1	0	0	0	0	0	0	Sustain pedal
0	1	0	0	0	0	1	1	Soft (damper) pedal

In the case of the continuous controllers, the byte following the appropriate identifying data byte is a 7 bit value relating to the most significant (MS) byte or the least significant (LS) byte of the control function. In the case of the soft pedal and sustain pedal the following data byte may only be off or on, values in the range 00000000 to 00111111 are denoted pedal OFF and values on the range 01000000 to 01111111 are denoted pedal ON.

Channel modes

The initial status byte 1011nnnn (where the lower nibble is, as usual to the one of 16 channel identifier) also precedes an important 2 byte sequence which relates to channel modes. Channel mode messages relate to the way an individual sound generator interprets the note-on/note-off commands it receives. It may interpret the channel messages it receives down the MIDI link in one of four ways: it may either act on all note-on/off commands which it receives (irrespective of channel) or it may not. And it may act upon note-on/note-off commands monophonically or polyphonically. In monophonic mode, a sound module will only sound one note at a time (usually the last note received). So, if a keyboard player plays a group of three notes in the form of a chord, a sound module set to sound monophonically will only produce one note of the chord at any one time whereas a polyphonic sound module will produce the sound of the full chord.

There exist three more MIDI channel messages:

Initial status byte

MSB	6	5	4	3	2	1	0	Message type
1	1	0	0	Channel identifier				Program change
1	1	0	1	nibble				Aftertouch (channel)
1	1	1	0					Pitch wheel

The first of these three channel messages, program change, is followed by 1 data byte which informs the sound module which voicing it is to produce. (The term for the name of each voicing produced by a sound generator is patch.) There is an emerging standard for this known as General MIDI, which some newer keyboards and instruments have adopted. (A number of *de facto* standards are in wider circulation so ubiquitous have been the sound modules which have adopted the format.) The standard MIDI voice mapping, into the 127 possibilities offered by the 1 program-change data byte, is given in Table 8.1.

System messages

The second type of MIDI message is the system message. There are three types: common, real-time and exclusive. System messages relate to all devices on the link and are therefore not encoded with channel numbers. System-common commands include Song pointer (status byte 11110010 followed by 2 bytes which indicate a particular position within a song to beat accuracy) and Song select (status byte 11110011 followed by a single byte referencing 1 of 128 possible songs). Both these system-common messages come into play when one MIDI system's information is to be synchronised with another. As do system real-time messages MIDI clock (a byte of unique structure, 11111000 sent 24 times every crotchet) and MIDI timecode.

MIDI timecode

Important for the synchronisation of MIDI information to other sources, MIDI timecode has an important role within a modern studio. Timecode as used in the television and video environment and is a means of encoding a time in hours, minutes, seconds and television frames (of which there are 25 or 30 in a second) into a special code which is recorded along with a video and sound signal onto videotape. This code is applied for identifying particular 'shots' which an editor uses in the process of compiling a television production. As media strands converge, the use of timecode has spread to audio productions so that they may be post-synchronised with video. The appearance of MIDI timecode is an indication of the gathering force of this trend. It takes eight separate

Table 8.1

0 Grand piano; 1 Bright acoustic piano; 2 Electric piano; 4 Rhodes piano; 5 Chorused piano; 6 Harpsichord; 7 Clavichord; 8 Celesta; 9 Glockenspiel; 10 Music box; 11 Vibraphone; 12 Marimba; 13 Xylophone; 14 Tubular bells; 15 Dulcimer; 16 Hammond organ; 17 Percussive organ; 18 Rock organ; 19 Church organ; 20 Reed organ; 21 Accordion; 22 Harmonica; 23 Tango accordion; 24 Acoustic guitar (nylon); 25 Acoustic guitar (steel); 26 Electric guitar (jazz); 27 Electric guitar (clean); 28 Electric guitar (muted); 29 Overdriven guitar; 30 Distortion guitar; 31 Guitar harmonics; 32 Acoustic bass; 33 Electric bass (finger); 34 Electric bass (pick); 35 Fretless bass; 36 Slap bass 1; 37 Slap bass 2; 38 Synth bass 1; 39 Synth bass 2; 40 Violin; 41 Viola; 42 Cello; 43 Contrabass; 44 Tremolo strings; 45 Pizzicato strings; 46 Orchestral harp; 47 Timpani; 48 String ensemble 1; 49 String ensemble 2; 50 Synthstrings 1; 51 Synthstrings 2; 52 Choir aahs; 53 Voice oohs; 54 Synth voice; 55 Orchestral hit; 56 Trumpet; 57 Trombone; 58 Tuba; 59 Muted trumpet; 60 French horn; 61 Brass section; 62 Synth brass 1; 63 Synth brass 2; 64 Soprano sax; 65 Alto sax; 66 Tenor sax; 67 Baritone sax; 68 Oboe; 69 English horn; 70 Bassoon; 71 Clarinet; 72 Piccolo; 73 Flute; 74 Recorder; 75 Pan flute; 76 Bottle blow; 77 Shakuhachi; 78 Whistle; 79 Ocarina; 80 Square lead; 81 Sawtooth lead; 82 Caliope lead; 83 Chiff lead; 84 Charang lead; 85 Voice lead; 86 Fifths lead; 87 Brass and lead; 88 New age pad; 89 Warm pad; 90 Polysynth pad; 91 Choir pad; 92 Bowed pad; 93 Metallic pad; 94 Halo pad; 95 Sweep pad; 96 Rain FX; 97 Soundtrack FX; 98 Crystal FX; 99 Atmosphere FX; 100 Brightness FX; 101 Goblins FX; 102 Echoes FX; 103 Sci-fi FX; 104 Sitar; 105 Banjo; 106 Shamisen; 107 Koto; 108 Kalimba; 109 Bagpipe; 110 Fiddle; 111 Shanai; 112 Tinkle bell; 113 Agogo; 114 Steel drums; 115 Woodblock; 116 Taiko drum; 117 Melodic tom; 118 Synth drum; 119 Reverse cymbal; 120 Guitar fret noise; 121 Breath noise; 122 Sea shore; 123 Bird tweet; 124 Telephone ring; 125 Helicopter; 126 Applause; 127 Gunshot

2-byte, real-time MIDI messages to convey a complete timecode location. Formed by eight occurrences of a unique status byte followed by a data byte where the top (MS) nibble identifies the information occurring in the lower (LS) nibble. In other words, only 4 bits of data are transferred every message:

Status byte	Data Byte MS nibble	LS nibble	
11110001	0000	nnnn	LS frames
	0001	nnnn	MS frames
	0010	nnnn	LS seconds
	0011 etc.		

The complete sequence should be pretty obvious from this excerpt. Like nature itself, timecode only allows for 24 possible hour values. This means that the 8 bits devoted to it are not all necessary, 2 bits being effectively 'spare'. These 2 bits are encoded to denote which type of timecode is in use. The four possible types owe their existence to the different media with which MIDI data may be required to be synchronised. They are: 24 frame (when synchronising to cinema films), 25 frame (when synchronising to PAL television pictures), 30 frame (when synchronising to high definition television pictures and digital audio data) and 30 frame drop-frame (when synchronising to NTSC television pictures). Timecode in music for television applications is covered in Chapter 15.

MIDI system-exclusive messages

The last type of MIDI message is the system exclusive. This message is really a kind of loophole in the standard which allows for almost anything. It was designed so that manufacturers wishing to use the MIDI link could do so for their own devices! Essentially a system exclusive message is, as usual, preceded by a unique status byte. But there the similarity ends, for following this byte there may be as many data bytes as the manufacturer requires for their purpose. The system exclusive message is signalled as terminated by the appearance of another unique status byte:

MSB	6	5	4	3	2	1	0	Message type
1	1	1	1	0	0	0	0	System exclusive start
1	1	1	1	0	1	1	1	System exclusive end

So that one manufacturer's equipment should recognise that it is being sent information during a system exclusive message – and perhaps more importantly that other manufacturers' equipment should realise that it is not being 'spoken to' – the first byte following the unique 'start of system exclusive' status byte 11110000 is a manufacturers' identification byte.[2]

MIDI sequencing

To recap, MIDI (Musical Instrument Digital Interface) allows synthesisers, drum machines and sound generators to be linked together under the control of a master keyboard. The data transmitted on the MIDI link specifies not only what note has been played, but also how the note was played – with what force. To archive and recreate a performance, it is only necessary to record the MIDI data in lieu of the sound signals themselves. This can then be transmitted down the same MIDI link and used to trigger the keyboards and sound modules. Provided everything is configured as it was when the 'recording' was made, the original performance will be exactly reproduced. MIDI offers the musician enormous flexibility via this approach because she may change voicing after recording and also change the speed of the performance (without changing the key) by having the MIDI events transmitted more speedily or more sluggishly than they were recorded. The MIDI sequencer is the given name of the device which records, stores and replays the MIDI data. Originally sequencers were hardware devices but more recently they are nearly always software packages running on desktop personal computers.

Sequencer programs

There are similarities between the MIDI sequencer and a multi-track tape recorder or multi-track hard-disk recorder. First, in the same way a multi-track tape recorder records the signal at its input sockets, so a sequencer program records all the MIDI events which arrive at the computer's MIDI IN socket. Like the tape machine, it stores data detailing each event and when it occurred so that it can reconstruct the exact pattern of events later and, when in replay mode, the program has the computer transmit this data from its MIDI OUT socket. Second, all sequencers allow the possibility of multiple tracking – the process whereby sequentially played input tracks of different instrumental contributions may be played back simultaneously to create a virtual ensemble. A basic MIDI recording set-up might look like Figure 8.2. Note that MIDI by itself doesn't 'sound' at all, so an essential part of the set-up is a sound generator as shown. In record mode, the computer sends a copy of the data it receives to its MIDI output socket. This is termed soft THRU. Without it, the musician would not be able to hear anything played during recording. However, if all a sequencer program could do was to record and replay data in the same form, MIDI sequencing would not have come to dominate multi-track recording as it has done. The power of the sequencer program rests in its ability to alter information once it is recorded.

The first – and most obvious – difference between multi-track tape machines or multi-track disk recorders and MIDI sequencers is the number of tracks available at a given cost. Because of the parsimonious

Table 8.2 *Comparison between the size of various stereo sound files and MIDI files per minute of music. The numbers speak for themselves!*

File type	Sampling frequency	Resolution	Kbytes per minute
WAV	22 kHz	8 bit	2 640
WAV	44.1 kHz	16 bit	10 560
MID	–	–	6

nature of MIDI data, as compared with digital audio data, it is possible to offer very many recording tracks. (Table 8.2 compares the size of various sound files and MIDI files per minute of music.) Numbers range from 32 to 256 depending on the program. But remember, MIDI has only 16 channels available for transmitting data so this vast number of recording tracks is something of an illusion. Still, this abundance is useful in keeping arrangements clear, for instance by splitting a piano keyboard across two tracks – left hand on one track, right hand on the next.

In keeping with the tape recorder analogy, virtually all sequencer programs are controlled by tape machine-style control buttons. These may be clicked, using the mouse, to send the sequencer into play mode, record mode, fast wind and so on – just like a real tape machine. These controls remain a common feature on all program screens. They are clearly visible in Figure 8.6 which illustrates a typical MIDI sequencer program screen presentation.

Also visible in Figure 8.6 is the track/measure window. The left-hand side of this window is given over to track information. Note the reverse video highlight on track 1. Each box in this section of the screen displays information concerning the various parameters which affect the individual track: channel number, voice name, musical transposition, velocity and time offset, MIDI channel allocation, patch name and volume and pan information. The right-hand side of the window (not visible) denotes the track in the form of measures (or bars). Each bar is shown with a 'blob' in it if information exists during that measure. The measure display is useful for viewing entire songs. Cakewalk offers a number of different screens more suitable for judging individual musical parts – these include a piano-roll view, a textual event list and even a representation of the track in conventional musical notation. Several screens are illustrated at once in Figure 8.6. This illustration alone should convince you of the power of the sequencer over a tape machine! The familiar Windows menu bar appears

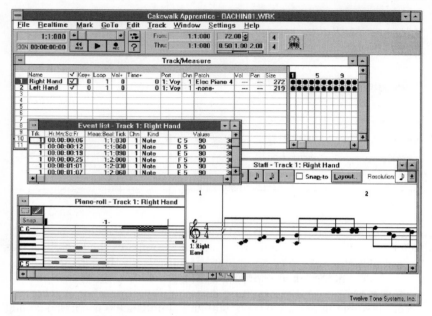

Figure 8.6   *MIDI sequencer screen presentation*

across the top of the screen which allows a host of different functions to be accessed. Unfortunately only a few basic techniques can be covered here.

Sequencer recording

Once the appropriate channel has been selected, by means of clicking on the track number on the extreme left of the track/measure window, recording is initialised by clicking the 'REC' button. All sequencers accept record input on any channel, but soft THRU the information to the channel highlighted. This enables each track to be played without having to change the output channel of the master keyboard each time.

In theory, it is possible to record without regard for the program's tempo, time signature and musical bar structure. However, in order to enjoy the greatest benefit from sequencer recording it is far better to have the program to capture your performance in relation to a predetermined tempo and bar structure. This involves 'telling' it the tempo and time signature of the piece you intend to play and then playing against the program's own generated tempo click. This may sound difficult and unnatural. It may help to think of the sequencer in terms that it lies somewhere between a tape recorder and a musical copyist. A tape machine records real events in real time with no regard for the music

Hymn To The Aten
for soprano, orchestra and pre-recorded tape
By Richard Brice

Figure 8.7 *A full orchestral score created from MIDI data*

being played. A musical copyist on the other hand records symbolic references (crotchets, quavers, minims etc.) in a symbolic time frame of bars, time signatures and bar lines.

Once the recording has been made the advantages of this approach become evident in that we may use the sequencer's intelligent algorithms, and the symbolic nature of MIDI data, to rearrange notes within the time frame. For instance, we can make them more precise with respect to musical tempo and beat structure – a process known as quantising. Or we can adjust the force with which each track, note group, or even individual note, is played by programming velocity offsets. Even the key the music was recorded in can be changed – by adding or subtracting an offset to each note value. Once the desired changes have been made, the sequencer may be spooled back to the start using the fast rewind button and the play button depressed in order to audition the track. The process may continue until the musician is satisfied and with no fear of tape wear. Thereupon the process continues on the next track until the entire multi-track song is built up. Once complete, song files may be stored on the hard drive.

Notes

1 As we saw in Chapter 1, in musical terms there exists a difference between the note D-sharp used in the key of E major and the note E-flat used in the key of B-flat. There exists a similar distinction between the notes of C-sharp and D-flat, of F-sharp and G-flat, of G-sharp and A-flat and of A-sharp and B-flat. Even though these 'non-identical twins' relate to the same mecahnical key on a piano keyboard, violinists and singers may pitch these notes slightly differently. MIDI convention is to ignore these differences as is done in the tuning of an acoustic piano.

2 A manufacturer has to apply to the International MIDI Association for allocation of a unique 'address' value for this byte. Roland's identification byte is, for instance, 01000001; Korg's 01000010; Yamaha's, 01000011; Casio's, 01000100.

Fact Sheet #8: MIDI messages

● Complete list of MIDI messages

Table 8.1   *MIDI 1.0 Specification message summary (Updated 1995 By the MIDI Manufacturers Association)*

Status D7–D0	Data Byte(s) D7–D0	Description
Channel voice messages [nnnn = 0–15 (MIDI Channel Number 1–16)]		
1000nnnn	0kkkkkkk 0vvvvvvv	Note Off event. This message is sent when a note is released (ended). (kkkkkkk) is the key (note) number. (vvvvvvv) is the velocity.
1001nnnn	0kkkkkkk 0vvvvvvv	Note On event. This message is sent when a note is depressed (start). (kkkkkkk) is the key (note) number. (vvvvvvv) is the velocity.
1010nnnn	0kkkkkkk 0vvvvvvv	Polyphonic Key Pressure (Aftertouch). This message is most often sent by pressing down on the key after it 'bottoms out'. (kkkkkkk) is the key (note) number. (vvvvvvv) is the pressure value.
1011nnnn	0ccccccc 0vvvvvvv	Control Change. This message is sent when a controller value changes. Controllers include devices such as pedals and levers. Controller numbers 120–127 are reserved as 'Channel Mode Messages' (below). (ccccccc) is the controller number. (vvvvvvv) is the new value (0–119).
1100nnnn	0ppppppp	Program Change. This message sent when the patch number changes. (ppppppp) is the new program number.
1101nnnn	0vvvvvvv	Channel Pressure (After-touch). This message is most often sent by pressing down on the key after it 'bottoms out'. This message is different from polyphonic after-touch. Use this message to send the single greatest pressure value (of all the current depressed keys). (vvvvvvv) is the pressure value.
1110nnnn	0lllllll 0mmmmmmm	Pitch Wheel Change. This message is sent to indicate a change in the pitch wheel. The pitch wheel is measured by a fourteen bit value. Centre (no pitch change) is 2000 H. Sensitivity is a function of the transmitter. (lllllll) are the least significant 7 bits. (mmmmmmm) are the most significant 7 bits.

Status D7–D0	Data Byte(s) D7–D0	Description

Channel mode messages (see also control change, above)

1011nnnn	0ccccccc 0vvvvvvv	Channel Mode Messages. This is the same code as the Control Change (above), but implements Mode control and special message by using reserved controller numbers 120–127. The commands are: All Sound Off. When All Sound Off is received all oscillators will turn off, and their volume envelopes are set to zero as soon as possible. c = 120, v = 0: All Sound Off Reset All Controllers. When Reset All Controllers is received, all controller values are reset to their default values. (See specific Recommended Practices for defaults). c = 121, v = x: Value must only be zero unless otherwise allowed in a specific Recommended Practice. Local Control. When Local Control is Off, all devices on a given channel will respond only to data received over MIDI. Played data, etc. will be ignored. Local Control On restores the functions of the normal controllers. c = 122, v = 0: Local Control Off c = 122, v = 127: Local Control On All Notes Off. When an All Notes Off is received, all oscillators will turn off. c = 123, v = 0: All Notes Off (See text for description of actual mode commands.) c = 124, v = 0: Omni Mode Off c = 125, v = 0: Omni Mode On c = 126, v = M: Mono Mode On (Poly Off) where M is the number of channels (Omni Off) or 0 (Omni On) c = 127, v = 0: Poly Mode On (Mono Off) (Note: These four messages also cause All Notes Off)

System common messages

11110000	0iiiiiii 0ddddddd 0ddddddd 11110111	System Exclusive. This message makes up for all that MIDI doesn't support. (iiiiiii) is usually a seven-bit Manufacturer's I.D. code. If the synthesizer recognizes the I.D. code as its own, it will listen to the rest of the message (ddddddd). Otherwise, the message will be ignored. System Exclusive is used to send bulk dumps such as patch parameters and other non-spec data.

Status D7–D0	Data Byte(s) D7–D0	Description
		(Note: Real-Time messages ONLY may be interleaved with a System Exclusive.) This message also is used for extensions called Universal Exclusive Messages.
11110001		Undefined. (Reserved)
11110010	0lllllll 0mmmmmmm	Song Position Pointer. This is an internal 14 bit register that holds the number of MIDI beats (1 beat= six MIDI clocks) since the start of the song. l is the LSB, m the MSB.
11110011	0sssssss	Song Select. The Song Select specifies which sequence or song is to be played.
11110100		Undefined. (Reserved)
11110101		Undefined. (Reserved)
11110110		Tune Request. Upon receiving a Tune Request, all analog synthesizers should tune their oscillators.
11110111		End of Exclusive. Used to terminate a System Exclusive dump (see above).

System real-time messages

11111000		Timing Clock. Sent 24 times per quarter note when synchronization is required (see text).
11111001		Undefined. (Reserved)
11111010		Start. Start the current sequence playing. (This message will be followed with Timing Clocks).
11111011		Continue. Continue at the point the sequence was Stopped.
11111100		Stop. Stop the current sequence.
11111101		Undefined. (Reserved)
11111110		Active Sensing. Use of this message is optional. When initially sent, the receiver will expect to receive another Active Sensing message each 300 ms (max), or it will be assumed that the connection has been terminated. At termination, the receiver will turn off all voices and return to normal (non-active sensing) operation.

Status D7–D0	Data Byte(s) D7–D0	Description
11111111		Reset. Reset all receivers in the system to power-up status. This should be used sparingly, preferably under manual control. In particular, it should not be sent on power-up.

Table 8.2　*Expanded status bytes list (Updated 1995 By the MIDI Manufacturers Association)*

	Status byte			Data bytes	
1st Byte Value			Function	2nd Byte	3rd Byte
Binary	Hex	Dec			
10000000 =	80 =	128	Chan 1	Note off	Note	Note
10000001 =	81 =	129	Chan 2	"	Number	Velocity
10000010 =	82 =	130	Chan 3	"	(0–127)	(0–127)
10000011 =	83 =	131	Chan 4	"	see	"
10000100 =	84 =	132	Chan 5	"	Table 4	"
10000101 =	85 =	133	Chan 6	"	"	"
10000110 =	86 =	134	Chan 7	"	"	"
10000111 =	87 =	135	Chan 8	"	"	"
10001000 =	88 =	136	Chan 9	"	"	"
10001001 =	89 =	137	Chan 10	"	"	"
10001010 =	8A =	138	Chan 11	"	"	"
10001011 =	8B =	139	Chan 12	"	"	"
10001100 =	8C =	140	Chan 13	"	"	"
10001101 =	8D =	141	Chan 14	"	"	"
10001110 =	8E =	142	Chan 15	"	"	"
10001111 =	8F =	143	Chan 16	"	"	"
10010000 =	90 =	144	Chan 1	Note on	"	"
10010001 =	91 =	145	Chan 2	"	"	"
10010010 =	92 =	146	Chan 3	"	"	"
10010011 =	93 =	147	Chan 4	"	"	"
10010100 =	94 =	148	Chan 5	"	"	"
10010101 =	95 =	149	Chan 6	"	"	"
10010110 =	96 =	150	Chan 7	"	"	"
10010111 =	97 =	151	Chan 8	"	"	"
10011000 =	98 =	152	Chan 9	"	"	"
10011001 =	99 =	153	Chan 10	"	"	"
10011010 =	9A =	154	Chan 11	"	"	"
10011011 =	9B =	155	Chan 12	"	"	"
10011100 =	9C =	156	Chan 13	"	"	"
10011101 =	9D =	157	Chan 14	"	"	"
10011110 =	9E =	158	Chan 15	"	"	"
10011111 =	9F =	159	Chan 16	"	"	"
10100000 =	A0 =	160	Chan 1	Polyphonic	"	Aftertouch
10100001 =	A1 =	161	Chan 2	Aftertouch	"	amount
10100010 =	A2 =	162	Chan 3	"	"	(0–127)
10100011 =	A3 =	163	Chan 4	"	"	"

Status byte				Data bytes	
1st Byte Value Binary	Hex	Dec	Function	2nd Byte	3rd Byte
10100100 =	A4 =	164	Chan 5	"	"	"
10100101 =	A5 =	165	Chan 6	"	"	"
10100110 =	A6 =	166	Chan 7	"	"	"
10100111 =	A7 =	167	Chan 8	"	"	"
10101000 =	A8 =	168	Chan 9	"	"	"
10101001 =	A9 =	169	Chan 10	"	"	"
10101010 =	AA =	170	Chan 11	"	"	"
10101011 =	AB =	171	Chan 12	"	"	"
10101100 =	AC =	172	Chan 13	"	"	"
10101101 =	AD =	173	Chan 14	"	"	"
10101110 =	AE =	174	Chan 15	"	"	"
10101111 =	AF =	175	Chan 16	"	"	"
10110000 =	B0 =	176	Chan 1	Control/	See	See
10110001 =	B1 =	177	Chan 2	Mode change	Table	Table
10110010 =	B2 =	178	Chan 3	"	3	3
10110011 =	B3 =	179	Chan 4	"	"	"
10110100 =	B4 =	180	Chan 5	"	"	"
10110101 =	B5 =	181	Chan 6	"	"	"
10110110 =	B6 =	182	Chan 7	"	"	"
10110111 =	B7 =	183	Chan 8	"	"	"
10111000 =	B8 =	184	Chan 9	"	"	"
10111001 =	B9 =	185	Chan 10	"	"	"
10111010 =	BA =	186	Chan 11	"	"	"
10111011 =	BB =	187	Chan 12	"	"	"
10111100 =	BC =	188	Chan 13	"	"	"
10111101 =	BD =	189	Chan 14	"	"	"
10111110 =	BE =	190	Chan 15	"	"	"
10111111 =	BF =	191	Chan 16	"	"	"
11000000 =	C0 =	192	Chan 1	Program	Program	NONE
11000001 =	C1 =	193	Chan 2	change	# (0–127)	"
11000010 =	C2 =	194	Chan 3	"	"	"
11000011 =	C3 =	195	Chan 4	"	"	"
11000100 =	C4 =	196	Chan 5	"	"	"
11000101 =	C5 =	197	Chan 6	"	"	"
11000110 =	C6 =	198	Chan 7	"	"	"
11000111 =	C7 =	199	Chan 8	"	"	"
11001000 =	C8 =	200	Chan 9	"	"	"
11001001 =	C9 =	201	Chan 10	"	"	"
11001010 =	CA =	202	Chan 11	"	"	"
11001011 =	CB =	203	Chan 12	"	"	"
11001100 =	CC =	204	Chan 13	"	"	"
11001101 =	CD =	205	Chan 14	"	"	"
11001110 =	CE =	206	Chan 15	"	"	"
11001111 =	CF =	207	Chan 16	"	"	"
11010000 =	D0 =	208	Chan 1	Channel	Aftertouch	"
11010001 =	D1 =	209	Chan 2	Aftertouch	amount	"
11010010 =	D2 =	210	Chan 3	"	(0–127)	"
11010011 =	D3 =	211	Chan 4	"	"	"
11010100 =	D4 =	212	Chan 5	"	"	"
11010101 =	D5 =	213	Chan 6	"	"	"
11010110 =	D6 =	214	Chan 7	"	"	"
11010111 =	D7 =	215	Chan 8	"	"	"

Status byte				Data bytes	
1st Byte Value Binary Hex Dec			*Function*	*2nd Byte*	*3rd Byte*
11011000 = D8 = 216			Chan 9	"	"	"
11011001 = D9 = 217			Chan 10	"	"	"
11011010 = DA = 218			Chan 11	"	"	"
11011011 = DB = 219			Chan 12	"	"	"
11011100 = DC = 220			Chan 13	"	"	"
11011101 = DD = 221			Chan 14	"	"	"
11011110 = DE = 222			Chan 15	"	"	"
11011111 = DF = 223			Chan 16	"	"	"
11100000 = E0 = 224			Chan 1	Pitch	Pitch	Pitch
11100001 = E1 = 225			Chan 2	wheel	wheel	wheel
11100010 = E2 = 226			Chan 3	control	LSB	MSB
11100011 = E3 = 227			Chan 4	"	(0–127)	(0–127)
11100100 = E4 = 228			Chan 5	"	"	"
11100101 = E5 = 229			Chan 6	"	"	"
11100110 = E6 = 230			Chan 7	"	"	"
11100111 = E7 = 231			Chan 8	"	"	"
11101000 = E8 = 232			Chan 9	"	"	"
11101001 = E9 = 233			Chan 10	"	"	"
11101010 = EA = 234			Chan 11	"	"	"
11101011 = EB = 235			Chan 12	"	"	"
11101100 = EC = 236			Chan 13	"	"	"
11101101 = ED = 237			Chan 14	"	"	"
11101110 = EE = 238			Chan 15	"	"	"
11101111 = EF = 239			Chan 16	"	"	"
11110000 = F0 = 240			System Exclusive		**	**
11110001 = F1 = 241			MIDI Time Code Qtr. Frame	-see spec-	-see spec-
11110010 = F2 = 242			Song Position Pointer	LSB	MSB
11110011 = F3 = 243			Song Select (Song #)	(0–127)	NONE
11110100 = F4 = 244			Undefined (Reserved)	?	?
11110101 = F5 = 245			Undefined (Reserved)	?	?
11110110 = F6 = 246			Tune request	NONE	NONE
11110111 = F7 = 247			End of SysEx (EOX)	"	"
11111000 = F8 = 248			Timing clock	"	"
11111001 = F9 = 249			Undefined (Reserved)	"	"
11111010 = FA = 250			Start	"	"
11111011 = FB = 251			Continue	"	"
11111100 = FC = 252			Stop	"	"
11111101 = FD = 253			Undefined (Reserved)	"	"
11111110 = FE = 254			Active Sensing	"	"
11111111 = FF = 255			System Reset	"	"

** Note: System Exclusive (data dump) 2nd byte = Vendor ID (or Universal Exclusive) followed by more data bytes and ending with EOX.

Table 8.3   *Control changes and mode changes (status bytes 176–191) (Updated 1995/1999 By the MIDI Manufacturers Association)*

2nd Byte Value Binary	Hex	Dec	Function	3rd Byte Value	Use
00000000 =	00 =	0	Bank Select	0–127	MSB
00000001 =	01 =	1	Modulation wheel	0–127	MSB
00000010 =	02 =	2	Breath control	0–127	MSB
00000011 =	03 =	3	Undefined	0–127	MSB
00000100 =	04 =	4	Foot controller	0–127	MSB
00000101 =	05 =	5	Portamento time	0–127	MSB
00000110 =	06 =	6	Data Entry	0–127	MSB
00000111 =	07 =	7	Channel Volume (formerly Main Volume)	0–127	MSB
00001000 =	08 =	8	Balance	0–127	MSB
00001001 =	09 =	9	Undefined	0–127	MSB
00001010 =	0A =	10	Pan	0–127	MSB
00001011 =	0B =	11	Expression Controller	0–127	MSB
00001100 =	0C =	12	Effect control 1	0–127	MSB
00001101 =	0D =	13	Effect control 2	0–127	MSB
00001110 =	0E =	14	Undefined	0–127	MSB
00001111 =	0F =	15	Undefined	0–127	MSB
00010000 =	10 =	16	General Purpose Controller #1	0–127	MSB
00010001 =	11 =	17	General Purpose Controller #2	0–127	MSB
00010010 =	12 =	18	General Purpose Controller #3	0–127	MSB
00010011 =	13 =	19	General Purpose Controller #4	0–127	MSB
00010100 =	14 =	20	Undefined	0–127	MSB
00010101 =	15 =	21	Undefined	0–127	MSB
00010110 =	16 =	22	Undefined	0–127	MSB
00010111 =	17 =	23	Undefined	0–127	MSB
00011000 =	18 =	24	Undefined	0–127	MSB
00011001 =	19 =	25	Undefined	0–127	MSB
00011010 =	1A =	26	Undefined	0–127	MSB
00011011 =	1B =	27	Undefined	0–127	MSB
00011100 =	1C =	28	Undefined	0–127	MSB
00011101 =	1D =	29	Undefined	0–127	MSB
00011110 =	1E =	30	Undefined	0–127	MSB
00011111 =	1F =	31	Undefined	0–127	MSB
00100000 =	20 =	32	Bank Select	0–127	LSB
00100001 =	21 =	33	Modulation wheel	0–127	LSB
00100010 =	22 =	34	Breath control	0–127	LSB
00100011 =	23 =	35	Undefined	0–127	LSB
00100100 =	24 =	36	Foot controller	0–127	LSB
00100101 =	25 =	37	Portamento time	0–127	LSB
00100110 =	26 =	38	Data entry	0–127	LSB
00100111 =	27 =	39	Channel Volume (formerly Main Volume)	0–127	LSB
00101000 =	28 =	40	Balance	0–127	LSB
00101001 =	29 =	41	Undefined	0–127	LSB
00101010 =	2A =	42	Pan	0–127	LSB
00101011 =	2B =	43	Expression Controller	0–127	LSB
00101100 =	2C =	44	Effect control 1	0–127	LSB
00101101 =	2D =	45	Effect control 2	0–127	LSB
00101110 =	2E =	46	Undefined	0–127	LSB

2nd Byte Value			Function	3rd Byte	
Binary	Hex	Dec		Value	Use
00101111 = 2F =		47	Undefined	0–127	LSB
00110000 = 30 =		48	General Purpose Controller #1	0–127	LSB
00110001 = 31 =		49	General Purpose Controller #2	0–127	LSB
00110010 = 32 =		50	General Purpose Controller #3	0–127	LSB
00110011 = 33 =		51	General Purpose Controller #4	0–127	LSB
00110100 = 34 =		52	Undefined	0–127	LSB
00110101 = 35 =		53	Undefined	0–127	LSB
00110110 = 36 =		54	Undefined	0–127	LSB
00110111 = 37 =		55	Undefined	0–127	LSB
00111000 = 38 =		56	Undefined	0–127	LSB
00111001 = 39 =		57	Undefined	0–127	LSB
00111010 = 3A =		58	Undefined	0–127	LSB
00111011 = 3B =		59	Undefined	0–127	LSB
00111100 = 3C =		60	Undefined	0–127	LSB
00111101 = 3D =		61	Undefined	0–127	LSB
00111110 = 3E =		62	Undefined	0–127	LSB
00111111 = 3F =		63	Undefined	0–127	LSB
01000000 = 40 =		64	Damper pedal on/off (Sustain)	<63=off	>64=on
01000001 = 41 =		65	Portamento on/off	<63=off	>64=on
01000010 = 42 =		66	Sustenuto on/off	<63=off	>64=on
01000011 = 43 =		67	Soft pedal on/off	<63=off	>64=on
01000100 = 44 =		68	Legato Footswitch	<63=off	>64=on
01000101 = 45 =		69	Hold 2	<63=off	>64=on
01000110 = 46 =		70	Sound Cont. 1 (Sound Variation)	0–127	LSB
01000111 = 47 =		71	Sound Cont. 2 (Timbre/Harmonic Intens.)	0–127	LSB
01001000 = 48 =		72	Sound Cont. 3 (Release Time)	0–127	LSB
01001001 = 49 =		73	Sound Cont. 4 (Attack Time)	0–127	LSB
01001010 = 4A =		74	Sound Cont. 5 (Brightness)	0–127	LSB
01001011 = 4B =		75	Sound Cont. 6 (Decay Time)	0–127	LSB
01001100 = 4C =		76	Sound Cont. 7 (Vibrato Rate)	0–127	LSB
01001101 = 4D =		77	Sound Cont. 8 (Vibrato Depth)	0–127	LSB
01001110 = 4E =		78	Sound Cont. 9 (Vibrato Delay)	0–127	LSB
01001111 = 4F =		79	Sound Cont. 10	0–127	LSB
01010000 = 50 =		80	General Purpose Controller #5	0–127	LSB
01010001 = 51 =		81	General Purpose Controller #6	0–127	LSB
01010010 = 52 =		82	General Purpose Controller #7	0–127	LSB
01010011 = 53 =		83	General Purpose Controller #8	0–127	LSB
01010100 = 54 =		84	Portamento Control	0–127	Source Note
01010101 = 55 =		85	Undefined	0–127	LSB
01010110 = 56 =		86	Undefined	0–127	LSB
01010111 = 57 =		87	Undefined	0–127	LSB
01011000 = 58 =		88	Undefined	0–127	LSB
01011001 = 59 =		89	Undefined	0–127	LSB
01011010 = 5A =		90	Undefined	0–127	LSB
01011011 = 5B =		91	Reverb Send Level	0–127	LSB
01011100 = 5C =		92	Effects 2 Depth (Tremolo Depth)	0–127	LSB
01011101 = 5D =		93	Chorus Send Level	0–127	LSB
01011110 = 5E =		94	Effects 4 Depth (Celeste/Detune Depth)	0–127	LSB
01011111 = 5F =		95	Effects 5 Depth (Phaser Depth)	0–127	LSB

| 2nd Byte Value | | | Function | 3rd Byte | |
Binary	Hex	Dec		Value	Use
01100000 = 60 =		96	Data entry +1		N/A
01100001 = 61 =		97	Data entry −1		N/A
01100010 = 62 =		98	Non-Registered Parameter Number LSB	0–127	LSB
01100011 = 63 =		99	Non-Registered Parameter Number MSB	0–127	MSB
01100100 = 64 =		100	Registered Parameter Number LSB*	0–127	LSB
01100101 = 65 =		101	Registered Parameter Number MSB*	0–127	MSB
01100110 = 66 =		102	Undefined	?	
01100111 = 67 =		103	Undefined	?	
01101000 = 68 =		104	Undefined	?	
01101001 = 69 =		105	Undefined	?	
01101010 = 6A =		106	Undefined	?	
01101011 = 6B =		107	Undefined	?	
01101100 = 6C =		108	Undefined	?	
01101101 = 6D =		109	Undefined	?	
01101110 = 6E =		110	Undefined	?	
01101111 = 6F =		111	Undefined	?	
01110000 = 70 =		112	Undefined	?	
01110001 = 71 =		113	Undefined	?	
01110010 = 72 =		114	Undefined	?	
01110011 = 73 =		115	Undefined	?	
01110100 = 74 =		116	Undefined	?	
01110101 = 75 =		117	Undefined	?	
01110110 = 76 =		118	Undefined	?	
01110111 = 77 =		119	Undefined	?	
01111000 = 78 =		120	All Sound Off	0	
01111001 = 79 =		121	Reset All Controllers	0	
01111010 = 7A =		122	Local control on/off	0=off	127=on
01111011 = 7B =		123	All notes off	0	
01111100 = 7C =		124	Omni mode off (+ all notes off)	0	
01111101 = 7D =		125	Omni mode on (+ all notes off)	0	
01111110 = 7E =		126	Poly mode on/off (+ all notes off)	**	
01111111 = 7F =		127	Poly mode on (incl mono=off +all notes off)	0	

**Note: This equals the number of channels, or zero if the number of channels equals the number of voices in the receiver.

Table F8.3a   *Registered parameter numbers [CC# 65H, 64H]*

CC#65 (MSB) Hex Dec	CC#64 (LSB) Hex Dec	Function
00 = 0	00 = 0	Pitch Bend Sensitivity
00 = 0	01 = 1	Channel Fine Tuning
00 = 0	02 = 2	Channel Coarse Tuning
00 = 0	03 = 3	Tuning Program Change
00 = 0	04 = 4	Tuning Bank Select

9
Got to Get You into My Life – Sound recording

Introduction

Magnetic recording underpins the business of music technology. For all its 'glitz' and glamour, the music business – at its most basic – is concerned with one simple function: the recording of music signals onto tape or on disc for subsequent duplication and sale. Before the widespread advent of computer hardware, this technology was pretty well unique to the music industry. Not that this limitation did anything to thwart its proliferation – the cassette player being the second most commonplace piece of technology after the light bulb! Nowadays, with the massive expansion in data recording products, audio – in the form of digital audio – is just another form of data, to be recorded in formats and distributed via highways originally intended for other media. The long-term advantage for music recording applications is the reduction in price brought about by utilising mass-produced products in high performance applications which previously demanded a precision, bespoke technology.

A sound recording is made onto magnetic tape by drawing the tape past a recording head at a constant speed. The recording head (which is essentially an electromagnet) is energised by the recording amplifier of the tape recorder. The electromagnet, which forms the head itself, has a small gap so the magnetic flux created by the action of the current in the electromagnet's coil is concentrated at this gap. The tape is arranged so that it touches the gap in the record head and effectively 'closes' the magnetic circuit as Figure 9.1 illustrates. Because the tape moves and the energising signal changes with time, a 'record' of the flux at any given time is stored on the tape. Replaying a magnetic tape involves dragging the tape back across a similar (or sometimes identical) electromagnet called the playback head. The changing flux detected at the minute gap in the playback head causes a current to flow in the head's coil. This is applied to an amplifier to recover the information left on the tape.

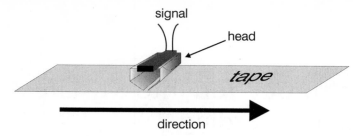

Figure 9.1 *Magnetic tape and the head gap*

In an analogue tape recorder, the pattern stored on the tape is essentially a stored analogue (analogy, see Chapter 10) of the original audio waveform. In a digital recorder the magnetic pattern recorded on the tape is a coded signal which must be decoded by the ensuing operation of the playback electronics. However, at a physical level, analogue and digital recordings using magnetic tape (or discs) are identical.

Magnetic theory

Figure 9.2 illustrates the path of a magnetic tape through the head assembly of a modern analogue tape recorder. The recording tape is fed from the supply reel across an initial erase head by means of the capstan and pinch roller. The purpose of the erase head is to remove any unwanted previous magnetisation on the tape. Next the tape passes the record head, where the audio signal is imprinted upon it, and the playback head, in which the magnetic patterns on the tape are converted

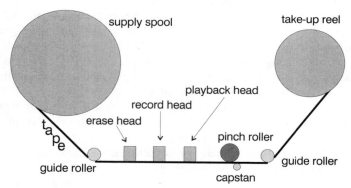

Figure 9.2 *Tape path*

back to an audio signal suitable for subsequent amplification and application to a loudspeaker. Finally, the tape is wound onto the take-up reel. When in playback mode, the erase head and the record head are not energised. Correspondingly, in record mode, the playback head may be used to monitor the signal off-tape to ensure recording levels etc. are correct. Cheaper cassette tape recorders combine the record and playback heads in a composite assembly, in which case off-tape monitoring while recording is not possible.

The physics of magnetic recording

In a tape recording, sound signals are recorded as a magnetic pattern along the length of the tape. The tape itself consists of a polyester-type plastic backing layer, on which is applied a thin coating with magnetic properties. This coating usually contains tiny particles of ferric iron oxide (so called ferric tapes) although more expensive tapes may use chromium dioxide particles or metal alloy particles which have superior magnetic properties (so-called chrome or metal tapes respectively).

The properties of magnetic materials takes place as a result of microscopic magnetic domains – each a tiny bar magnet – within the material. In an unmagnetised state, these domains are effectively randomly aligned so that any overall, macroscopic magnetic external field is cancelled out. Only when the ferrous material is exposed to an external magnetic field do these domains start to align their axis along the axis of the applied field, the fraction of the total number of domains so aligned being dependent on the strength of the externally applied field. Most significantly, after the external field has been removed, the microscopic domains do not altogether return to their pre-ordered state and the bulk material exhibits external magnetic poles.

The relation between the magnetising field (H) and the resultant induction (B) in an iron sample (assumed, initially, to be in a completely demagnetised condition) may be plotted as shown in Figure 9.3. Tracing the path from the origin, note that the first section of the looped curve rises slowly at first (between O and B1), then more rapidly (between B1 and B2), and finally more and more gradually as it approaches a point where there remain only a very few magnetic domains left to be aligned. At this point (B3) the ferrous material is said to be saturated. Significantly, when the magnetising force (H) is reduced, the magnetic induction (B) does not retrace its path along the curve B3–B2–B1–O, instead it falls along a different path, B3–B4, at which point the magnetising force is zero again, but the ferrous material remains magnetised with the residual induction B4. This remaining force is referred to as remnance. For this remnance to be neutralised, an opposite magnetic force must be applied

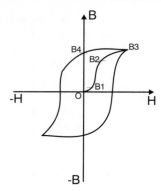

Figure 9.3 *BH curve*

and this accounts for the rest of the looped curve in Figure 9.3. The magnitude of the applied magnetic force required to reduce the remnance to zero is termed coercivity (the ideal magnetic tape exhibiting both high remnance and high coercivity).

Bias

As we saw in earlier chapters, if a sound recording and reproduction system is to perform without adding discernible distortion, a high degree of linearity is required. In the case of tape recording this implies the necessity for a direct relationship between the applied magnetic force and the resultant induction on the tape. Looking again at Figure 9.3, it is apparent that the only linear region over which this relationship holds is between B1 and B2, the relationship being particularly non-linear about the origin. The situation may be compared to a transistor amplifier, which exhibits a high degree of non-linearity in the saturation and cut-off region and a linear portion in between. The essence of good design, in the case of the transistor stage, is appropriately to bias the amplifier in its linear region by means of a steady DC potential. And so it is with magnetic recording. In principle a steady magnetic force may be applied, in conjunction with the varying force dependent on the audio signal, thereby biasing the audio signal portion of the overall magnetic effect into the initial linear region of the *BH* loop. In practice such a scheme has a number of practical disadvantages. Instead a system of ultrasonic AC bias is employed which mixes the audio signal with a high-frequency signal current. This bias signal, as it is known, does not get recorded because the wavelength of the signal is so small that the magnetic domains, resulting from it, neutralise themselves naturally. It acts solely to ensure the audio

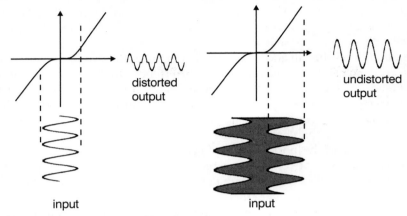

Figure 9.4 *Linearising effect of AC bias*

modulation component of the overall magnetic force influences the tape in its linear region. Figure 9.4 illustrates the mechanism.

It is hardly surprising that the amplitude of the superimposed high-frequency bias signal is important in obtaining the best performance from an analogue tape machine and a given tape. Too high an amplitude and high-frequency response suffers; too low a value and distortion rises dramatically. Different tape formulations differ in their ideal biasing requirements, although international standardisation work (by the International Electrotechnical Commission) has provided recommendations for the formulation of 'standard' tape types.

Equalisation

For a number of reasons, both the signal which is imprinted upon the tape by the action of the record current in the record head and the signal arising as a consequence of the induced current in the playback head are heavily distorted with respect to frequency and must both be equalised. This is an area where standardisation between different manufacturers is particularly important because, without it, tapes recorded on one machine would not be reproducible on another.

In itself, this would not be such a problem were it not for the fact that, due to differences in head geometry and construction, the electrical equalisation differs markedly from manufacturer to manufacturer. The International Electrotechnical Commission (IEC) provided an ingenious solution to widespread standardisation by providing a series of standard pre-recorded tapes on which are recorded frequency sweeps and spot levels. The intention being that these must be reproduced (played back) with a level flat-frequency response characteristic, with the individual manufacturer responsible for choosing the appropriate electrical equalisation to effect this situation. This appears to leave the situation concerning record equalisation undefined but this is not the case because it is intended that the manufacturer chooses record equalisation curves so that tapes recorded on any particular machine must result in a flat-frequency response when replayed using the manufacturer's own IEC standard replay equalisation characteristic.

The issue of the 'portability' should not be overlooked and any serious studio that still relies on analogue recording must ensure its analogue tape equipment is aligned (and continues to remain aligned – usually the duty of the maintenance engineer) to the relevant IEC standards. This, unfortunately, necessarily involves the purchase of the relevant standard alignment tapes.

Tape speed

Clearly another (in fact, the earliest) candidate for standardisation was the choice of linear speed of the tape through the tape path. Without this the signals recorded on one machine replay at a different pitch when replayed on another. While this effect offers important artistic possibilities (see later in this chapter), it is clearly undesirable in most operational circumstances. Table 9.1 lists the standard tape speeds in metric (centimetres per second, cm/s) and imperial measures (inches per second, ips) and their likely applications.

Table 9.1

Tape speed (ips)	cm/sec	Application
30	76	top professional quality
15	38	top professional quality
7.5	19	professional quality (with noise reduction)
3.75	9.5	semi-professional quality (with noise reduction)
1.875	4.75	domestic quality (with noise reduction)

Speed stability

Once standardised, the speed of the tape must remain consistent both over the long and short term. Failure to establish this results in audible effects known, respectively, as wow and flutter. However, these onomatopoeic terms relate to comparatively coarse effects. What is often less appreciated is the action of speed stability upon the purity of audio signals – a fact that is easier to appreciate if speed instability is regarded as a frequency modulation effect. We know (from Chapter 7) that FM modulation results in an infinite set of sidebands around the frequency modulated carrier. The effect of speed instability in an analogue tape recorder may be appreciated in these terms by looking at the output of a pure sine tone recorded and played back analysed on a spectrum analyser, as shown in Figure 9.5. Notice that the tone is surrounded by a 'shoulder' of sidebands around the original tone.

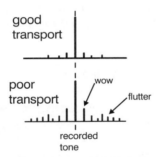

Figure 9.5 *FM sidebands as a result of speed instability*

Happily, the widespread adoption of digital recording has rendered much of the above obsolete, especially in relation to two-track masters. Where analogue tape machines are still ubiquitous (for example, in the case of multi-track recorders) engineering excellence is a necessary by-word, as is the inevitable high cost that this implies. In addition, alignment and calibration to recognised standards must be regularly performed (as well as regular cleaning) in order to ensure multi-track tapes can be recorded and mixed in different studios.

Recording formats – analogue machines

Early tape recorders recorded a single channel of audio across the whole tape width. Pressure to decrease expensive tape usage led to the development of the concept of using 'both sides' of a tape by recording the signal across half the tape width and subsequently flipping over the tape to record the remaining unrecorded half in the opposite direction. The advent of stereo increased the total number of audio tracks to four; two in one direction, two in the other. This format is standard in the familiar analogue cassette. From stereo it is a small conceptual step (albeit a very large practical one) to four, eight, sixteen tracks or more being recorded across the width of a single tape. Such a development demanded various technological innovations, the first was the development of composite multiple head assemblies. Figure 9.6 illustrates the general principle. Given the dimensions, the construction of high quality head assemblies was no mean achievement. The second was the combination of record and replay heads. Without this development, the signal 'coming off' tape would be later than the signal recorded onto the tape, a limitation which would make multi-track recording impossible. In early machines the record head was often made to do temporary duty as playback head during the recording stages of a multi-track session, its less than perfect response characteristic being adequate as a cue track. The optimised playback head was reserved for mixdown only.

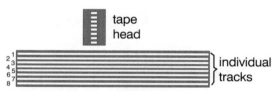

Figure 9.6 *Multiple tape tracks across width of tape*

In spite of this, the number of tracks that it is practical to record across a given width of tape is not governed by head construction limitations only, but by considerations of signal-to-noise ratio. As we saw earlier the signal recorded onto tape is left as a physical arrangement of magnetic domains. Without an audio signal, these domains remain unmagnetised and persist in a random state. These cause noise when the tape is replayed. Similarly at some point, when a strong signal is recorded, all the domains are 'used up' and the tape saturates. A simple rule applies in audio applications: the more domains onto which the signal is imprinted, the better, up to the point just below saturation. This may be achieved in various ways; by running the tape faster and by using a greater tape width for a given number of tracks. Figure 9.7 illustrates this by depicting the saturation levels of a commercial tape at various speeds. This simple principle accounts for the many different tape formats which exist. Each is an attempt to redefine the balance between complexity, sound quality

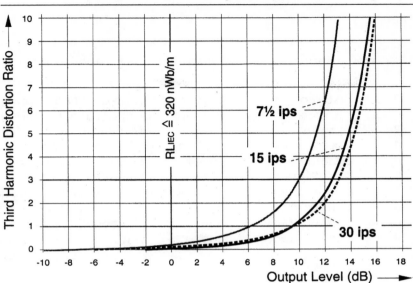

Output Level versus Third Harmonic Distortion Ratio at frequency 1 kHz and tape speeds 30 ips (76.2 cm/s), 15 ips (38.1 cm/s) and 7½ ips (19.05 cm/s). See also Notes, points 2.1 and 2.5.

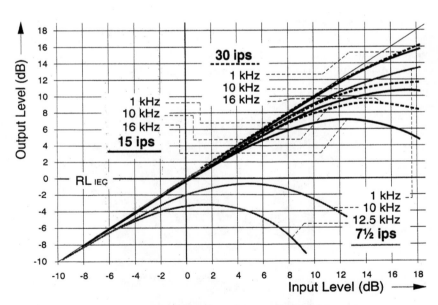

Input Level versus Output Level at frequencies 1 kHz, 10 kHz and 16 kHz (12.5 kHz at 7½ ips) and tape speeds 30 ips (76.2 cm/s), 15 ips (38.1 cm/s) and 7½ ips (19.05 cm/s). See also Notes, point 2.2.

Figure 9.7 *The effects of tape speed on saturation and distortion*

Table 9.2

Tracks	Format	Medium/speed	Application
2	$\frac{1}{2}$" stereo	$\frac{1}{2}$" 7.5–30 ips	High quality mastering
2	$\frac{1}{4}$" stereo	$\frac{1}{4}$" 7.5–30 ips	High quality mastering
2	Cassette	$\frac{1}{8}$" $\frac{15}{8}$" ips	Medium quality replay
4	$\frac{1}{2}$" 4 track	$\frac{1}{2}$" 7.5–30 ips	High quality mastering
4	Cassette	$\frac{1}{8}$" 3.75 ips	Personal multi-track
8	$\frac{1}{4}$" multi-track	$\frac{1}{4}$" 7.5–15 ips	Semi-pro multi-track
16	$\frac{1}{2}$" multi-track	$\frac{1}{2}$" 15–30 ips	Professional multi-track
16	1" multi-track	1" 30 ips	High quality multi-track
16–24	1" multi-track	1" 30 ips	High quality multi-track
24	2" multi-track	2" 30 ips	High quality multi-track

and tape cost appropriate to a certain market sector. Table 9.2 lists some of the major analogue recording formats. Note that the format of a tape relates to its width, specified in inches.

Analogue mastering

Analogue mastering is now very rare, this office having been made the exclusive domain of Digital Audio Tape or DAT. A typical high quality two-track mastering recorder is illustrated in Figure 9.8.

Figure 9.8   *Analogue mastering recorder*

Analogue multi-track tape machines

As mentioned earlier, analogue multi-track machines betray their quality roughly in proportion to the width of the tape utilised for a given number of tracks. A 2-inch tape, 24 track, which utilises a 2-inch width tape drawn across 24 parallel heads, is therefore better than a 1-inch 24 track; but not necessarily better than a $\frac{1}{2}$-inch two track! Not only does a greater head-to-tape contact guarantee higher tape signal-to-noise ratio (i.e. more domains are usefully magnetised) but it also secures less tape dropout. Dropout is an effect where the contact between tape and head is broken microscopically for a small period during which the signal level falls drastically. Sometimes dropout is due to irregularities in the tape, or to the ingress of a tiny particle of dust; whichever, the more tape passing by an individual recording or replay head, the better chance there is of dropouts occurring infrequently. Analogue tape machines are gradually becoming obsolete in multi-track sound recording; however, the huge installed base of these machines means they will be a part of sound recording for many years to come.

Cassette based multi-tracks

Figure 9.9 illustrates a typical analogue cassette-based portable multi-track recorder and mixer combined. This type of low-end 'recording studio in a box' is often termed a Portastudio and these units are widespread as personal recording 'notebooks'. Typically four tracks are available and are recorded across the entire width of the cassette tape

Figure 9.9 *Cassette-based 'notebook' multi-track*

(which is intended to be recorded in one direction only). The cassette tape usually runs at twice normal speed, 3.75 ips. Individual products vary but the mixer of the unit illustrated in Figure 9.9 allows for two (unbalanced) microphone inputs and a further four inputs at line level, of which only two are routed to the tape tracks. Each of the first four inputs may be switched between INPUT, OFF, and TAPE (return). Selecting INPUT will (when the record button is engaged on the tape transport buttons) switch the track to record. The mixer also incorporates two send-return loops and the extra line level inputs mentioned above. In addition an extra monitor mixer is provided, the output of this being selectable via the monitor output. It is thus a tiny split multi-track console (see Chapter 12). Despite the inevitable compromises inherent in such a piece of equipment many portable multi-track units are capable of remarkably high quality and many have been used to record and mix release-quality material. Indeed, so popular has this format proved to be that digital versions have begun to appear, products which offer musical notebook convenience with exemplary sound quality. One such is illustrated in Figure 9.10.

Figure 9.10   *Roland digital multi-track*

Digital tape machines

There exist digital tape formats for both mastering and multi-track. Digital tape recorders may be distinguished by their basic technology pedigree. One type uses stationary heads in the manner of an analogue tape recorder, the so-called Digital Audio Stationary Head (DASH) format. The other uses rotating heads in the manner of videotape machines. The technology is explained in the next chapter. Here consideration is given to the different formats available.

Digital mastering

The cassette-based Digital Audio Tape (DAT) two-track tape format, originally pioneered by Sony but now very widely standardised, is now virtually universal as a mastering format. The DASH method produced mastering formats which permitted the mechanical editing of digital tape, an attribute which was considered important in the early days of digital recording. However, today, with the advent of hard-disk editing, this requirement is no longer required.

DAT technology uses a small videocassette-style tape cassette and a small rotating head assembly. The wrap angle on the tape is, however, very much smaller than that used in most video formats and this has a number of advantages: it reduces tape drag and wear (which makes the format more suitable for portable applications) and it makes the threading process less complicated than that of a videocassette machine. This simplification is possible because, although the bandwidth required for digital audio is, indeed, greater than that required for its analogue counterpart, it is not as high as that required for analogue video. Furthermore because the signal is digital the signal-to-noise requirement is much less stringent too, so Sony took advantage of various simplifications which may be made to the video-style mechanism when designing this bespoke digital audio tape format. DAT achieves a remarkably high data capacity typically consuming 8.15 mm of tape per second, nearly six times slower than the linear tape speed in a standard analogue cassette! Technical details of the DAT format are covered in the next chapter. A DAT cassette is illustrated in Figure 9.11 and a typical semi-professional machine is illustrated in Figure 9.12.

Digital multi-tracks

Digital audio stationary head (DASH) multi-track format

DASH recorders use multi-track head assemblies to write multiple tracks of digital audio onto tape. With the advent of low-cost, rotary-head multi-tracks, DASH is becoming more rare outside of very 'high-end'

Table 9.3 *Digital recording formats*

Tracks	Head	Format	Medium/speed	Application
2	Rotary	1610/1630	60FPS U-matic	Editing/CD mastering
2	Rotary	PCM-F1	NTSC Betamax	Semi-pro recording
2	Rotary	DAT	DAT tape	High quality mastering
2	–	MiniDisc	MiniDisc	High quality replay
2	–	Hard disk	Winchester Disk	Editing/CD mastering
2–4	Stationary	DASH	$\frac{1}{4}$" 7.5ips	High quality mastering
4–16	Stationary	DASH	$\frac{1}{4}$" 30ips	High quality multi-track
8	Rotary	ADAT	S-VHS	High quality multi-track
8	Rotary	DA–88	Hi–8mm videotape	High quality multi-track
24–48	Stationary	DASH	$\frac{1}{2}$" 30ips	High quality multi-track

Figure 9.11   *DAT cassette*

Figure 9.12   *DAT recorder*

Figure 9.13   *DASH machine*

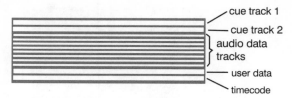

Figure 9.14　*DASH track format*

studios. Figure 9.13 is an illustration of a DASH multi-track machine due to Sony, the PCM–3324S. Figure 9.14 is an illustration of the multi-track nature of the data format on a DASH machine.

Rotary-head digital multi-track formats

The two most common formats for digital multi-track work are the rotary head, videocassette-based Alesis ADAT (Figure 9.15) and the Tascam DA–88. The ADAT format records on readily available S-VHS videocassettes, and up to 16 ADAT recorders can be linked together for 128 tracks with no external synchroniser required and without sacrificing a track to timecode. The new ADAT-XT offers a transport which operates up to four times faster than the original ADAT, with new dynamic braking control for more accurate locates and faster lock times. An onboard digital editor allows musician-engineers to make flawless copy-paste digital edits between tracks on a single unit, or between multiple machines. The XT features Tape Offset, Track Copy, Track Delay, and ten auto-locate buttons for sophisticated flexibility and control over productions. Both +4 dBu balanced and –10 dBV unbalanced connections and the ADAT optical digital interface are provided for flexible hook-up, and a comprehensive vacuum fluorescent display provides all the critical information.

Figure 9.15　*Alesis ADAT*

ADAT-XT specifications

Audio conversion:	Record (A/D): 18-bit linear audio, single converter per channel
Play (D/A):	20-bit linear, 8 times oversampling, single converter per channel
Sample rate:	44.1/48 kHz (selectable); user variable via pitch control
Frequency response:	20 Hz–20 kHz ± 0.5 dB
Dynamic range:	Greater than 92 dB, 20–20 kHz
Distortion:	0.009% THD + noise @ 1 kHz
Crosstalk:	Better than –90 dB @ 1 kHz
Wow and flutter:	Unmeasurable
Input impedance:	10k
Output impedance:	600 Ω
Nominal input levels:	Balanced: +4 dBu unbalanced: –10 dBV
Maximum input level:	Balanced: +19 dBu unbalanced: +5 dBV
Pitch control:	+100 cents/–300 cents (48k), ±200 cents (44.1k)
Digital I/O:	ADAT multi-channel optical digital interface; fibre optic cable
Sync In/Out:	ADAT Synchronisation interface; 9 pin D-subconnector
Foot-switch jacks:	Locate/Play/LRC remote, punch In/Out
Power requirements:	90–250 VAC, 50–60 Hz, 50 W maximum
Dimensions:	(H × W × D): 133.3 mm × 482 mm × 279.5 mm, 3U rack mounting
Weight:	20 lb

DA–88 digital multi-track recorder

The most common alternative to the ADAT is the DA–88, which uses Hi-Band 8 mm compact cassettes (Hi–8 mm) instead of the S-VHS tapes chosen by Alesis. The DA–88's precision 8 mm transport is expressly designed and manufactured by Tascam for digital audio. Tascam chose to create a multi-track transport based on the Hi–8 videocassette format because they believe it has some important advantages.

The DA–88 provides the choice of industry standard 44.1 kHz or 48 kHz sampling rate. This allows transfer in the digital domain from and to samplers, DAT recorders, CD players, digital editing systems and larger format digital multi-track. Since the DA–88 is designed for use in professional recording and production environments, DA–88s can transfer directly between units, also there is an optional AES/EBU digital interface that allows direct connection to other digital audio equipment supporting this standard. Digital audio can be transferred between the

DA–88 and open reel digital recorders, digital mixing consoles, or other digital audio equipment, eliminating the need for D/A and A/D conversion in the transfer process so no loss in sound quality is incurred. SPDIF digital connectors are also provided on the optional IF interfaces so the DA–88 will function within a wide range of systems. In addition to the digital I/O, the DA–88 offers both unbalanced and electronically balanced analogue inputs and outputs for compatibility with the widest possible range of analogue equipment.

The SY–88 optional plug-in board chase-lock synchroniser card provides SMPTE timecode for use as master or slave, plus Video Sync, MMC and Sony 9 pin compatible RS–422 port so you can control the DA–88 directly from a video editor. Because the DA–88 uses the internal system to lock multiple DA–88s together, only one SY–88 chase-lock synchroniser card is necessary in the master DA–88 to synchronise to video.

DA–38 digital multi-track recorder

The lower cost DA–38 uses the same Hi-8 mm format for recording eight tracks of digital audio as the DA–88. Tapes recorded on a DA–38 are 100% compatible with DA–88s and other Hi-8 format machines. The DA–38 handles tape in a same manner and a shuttle-wheel is provided for convenient, accurate locating. The DA–38 also offers a variety of new features, including MIDI, making it ideal for incorporating into a sequencer-based studio. The DA–38 also offers digital track copy capabilities, with an integrated electronic patch bay for super fast assemble editing without having to use multiple tape machines; see Figure 9.16.

Figure 9.16 *Tascam DA–38*

Creative uses of tape

Over and above their original role to capture a live sound as faithfully as possible, tape recorders were soon pressed into service as the tools of sound modification. Especially by the post-war avant-garde composers. To this day, even in a world dominated by electronic music, tape techniques dominate the sound of pop music by way of vocal technique. This section concentrates on some of the tape techniques used by composers, producers and engineers. Note that many of the techniques apply only to analogue tape!

Double tracking

A creative effect originally attributed to Les Paul, double tracking involves two or more vocal (or instrumental) performances by the same musician being combined together in the mix. Confusingly, the term double tracking is sometimes used to refer to two or more takes in which the vocal or instrumental lines are dissimilar (for instance the recording of a harmony track), but in this case the technique is only a special case of a more general multi-track scenario. Double tracking has now come to imply a particular recording technique whereby the original vocal line is doubled at unison on a second track. It is an attractive technique, not so much because the double tracked line inevitably blends so well with the original vocal track, but because (and herein lies the skill which singers must acquire), if the singer doubles the original performance as nearly as possible, the impression is not of two performances but of a single, richer, smoother voice.

ADT (or flanging)

The exacting task of matching a previous performance is surprisingly difficult, especially for an ebullient or nervous performer. Automatic Double Tracking (ADT) was developed by EMI engineer Ken Townsend (Macdonald 1994) and was used originally (and extensively!) by the Beatles. ADT is achieved by taking a vocal signal from the sync head of a multi-track, recording it to another loop of tape which is speed varied with a slow oscillation and recording it back onto the multi-track about a fifth of a second after the original signal. Examples abound on the album *The Beatles* especially (this is the album sometimes referred to as the *White Album*). This technique was originally referred to as flanging (see Chapter 6).

Tape loops

Recordable tape loops originate with the work of Brian Eno and Robert Fripp in the early 1970s. The concept behind tape looping is illustrated in Figure 9.17. Essentially sounds (in the case of Fripp, usually guitar sounds)

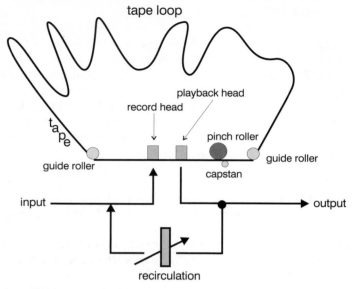

Figure 9.17   *Principle of tape loops*

are recorded over and over onto a loop of magnetic tape on a tape deck which incorporates the crucial modification that the erase head is disabled and an electrical path provided so that sounds may be recirculated in the manner of a tape-echo device (see Figure 6.1). The sounds are therefore recorded 'on top of one another' and one instrument may create vast, dense, musical structures. Importantly, subsequent signals do not simply add because the action of the bias current during recording partially erases the previous sound signal. (This is natural enough, to do otherwise would require an infinite dynamic range!) From an artistic point of view this is extremely valuable because it means, without continually 'fuelling' the process of looping with new signals – or electrically recirculated old signals – the 'sound-scape' gradually dies away. The artist may thereby control the dynamic and tonal 'map' of the piece. Nevertheless the control of this process is not comprehensive and many of the results are partially random. The invention of this process grew out of Eno's long-time interest in self-evolving compositions, a genre which he named 'ambient'. A genre which may be seen as bringing the ideas of John Cage to a pop audience.

Tape 'special effects'

Other tape techniques are employed to produce more spectacular tonal modification. These may be termed tape 'special effects'. Of these temporal distortion is the most common. Temporal distortion covers a wide range of effects from simply implying a speed change, up or down (transposition) to reversing the tape (reversal), thereby producing a sound in which the dynamic envelope is reversed, to the looping of sound snippets. Other effects include the 'chopping up' and reassembly of a sound or the substitution of the transient of one sound with the ongoing sustain of another, for instance a tone generated by joining (splicing) the attack of a French horn onto the sustain and release of a bell, a technique known as 'brassage'. Lastly, by a combination of these means, whole sound collages are possible which derive their constituents from sound 'bites' (perhaps borrowed from radio reportage or from home recording) modified by means detailed above and then combined with other sound (modified and unmodified) to produce a work which is not perhaps instantly recognisable as music but which has, nonetheless, the ability to create a unique and diverting 'sound-scape' with capacity to move and inspire the listener; listen to the Beatles' *Revolution 9* and Luigi Nono's *Non Consumiamo Marx*.

Hard disk recording

The spread of computing technology into all areas of modern life is so obvious as to require no introduction. One consequence of this is the drift towards the recording of sound waveforms (albeit digitally coded) on computer-style hard disks, either within a computer itself, with the machine's operating system dealing with the disk management, or within bespoke recording hardware utilising disk technology. But first a question: why disks and not tape?

Figure 9.18 *Hard-disk drive construction*

The computer industry all but stopped using tape technology many years ago. The reason is simple. While a tape is capable of storing vast quantities of data, it does not provide a very easy mechanism for retrieving that data except in the order that it was recorded. The issue is coined in the computing term access time. To locate a piece of data somewhere on a tape, it may take several minutes even if the tape is wound at high speed from the present position to the desired location. This is really not such an issue for audio entertainment, since music is usually intended to be listened to in the order in which it was recorded. However, it is an issue for recording engineers and musicians because, during a multi-track session, the tape may have to be rewound hundreds, even thousands of times, thereby reducing productivity as well as stifling the creative process. Far better then to enjoy the benefits of computer disks which, because the data is all available at any time spread out as it were 'on a plate' (quite literally – see Figure 9.18), all the recorded signals are available virtually instantaneously.

Disk drive technology

Think of disk drive technology as a mixture of tape and disk technology. In many ways it combines the advantages of both in a reliable, cheap package. In a disk drive, data is written in a series of circular tracks, a bit like a CD or an analogue LP. Not as a wiggly track (as in the case of the LP) or as a series of physical bumps (as in the case of the CD) but as a series of magnetic patterns. As in the case of the CD and the record, this implies that the record and replay head must be on a form of arm which is able to move across the disk's surface. It also implies, and this it has in common with the CD, the presence of a servo-control system to keep the record/replay head assembly accurately tracing the data patterns. As well as a disk operating system to ensure an initial pattern track is written on the disk prior to use (a process known as formatting). Like an LP record, in the magnetic disk the data is written on both sides. In a floppy disk this medium is pliable (floppy in fact!) and feels very much like magnetic tape. The heads, for there are two of them, one below the disk and one above, write to the disk by entering the floppy's body through the two windows which are revealed by drawing back the metal shutter. A process which is undertaken automatically as part of loading a floppy disk. The medium is pliable so that physical contact is ensured between the head and the magnetic medium. The process of formatting a disk involves recording a data pattern onto a new disk so that the heads are able to track this pattern for the purposes of recording new data. The most basic part of this process is breaking the disk into a series of concentric circular tracks. Note that in a disk drive the data is not in the form of a spiral as it is on a CD or an LP record. These concentric circular tracks are known as tracks and are subdivided into sections know as sectors.

In a hard drive, the magnetic medium is not pliable but rigid and is known as a platter. Several platters are stacked together all rotating on a common spindle, along with their associated head assemblies which also move in tandem (see Figure 9.18). Conceptually the process of reading and writing to the disk by means of a movable record/replay head is similar. However, there are a number of important differences. Materially, a hard disk is manufactured to far tighter tolerances than the floppy disk, and rotates some ten times faster. Also the head assembly does not physically touch the disk medium but instead floats on a microscopic cushion of air. If specks of dust or cigarette smoke were allowed to come between the head and the disk, data is lost, the effect being known as a head crash. To prevent this hard drives are manufactured as hermetically sealed units.

Compact disc

During the 1970s, electronics companies were searching for an alternative format to the vinyl record. There was much enthusiasm for a digital format to replace the old analogue system. The winner in the race was the so-called Compact Disc (CD) – a joint development by Philips and Sony. The CD proved to be a great success, to such an extent that the costs of production of CDs dropped quickly; thereby encouraging many other uses for what is, after all, simply a high-capacity data storage medium. From the first specification for Compact Disc in 1982 in the form of a booklet with a red cover (the Red Book standard), these other formats have demanded subsequent standards themselves. These have been published, each with a different colour cover: Yellow Book, Green Book, Orange Book and so on.

Despite their differences, which will be explained below, the physical characteristics and the manner in which all raw data (the 1s and 0s) are recorded is the same for all CDs. All CDs are 1.2 mm thick and have a diameter of 120 mm (or less often 80 mm). Data is stored as a spiral track which begins at the inside edge of the disc (unlike a vinyl record) and extends to the outer edge. Data is represented as a series of microscopic pits in the 'land' of the metallic inner surface of the CD. This inner reflective layer is formed on top of a polycarbonate plastic substrate with good optical properties. The metallic layer is then protected with a protective top layer of plastic. This protective top layer is actually much thinner than the substrate and is the side on which the label is printed. Interestingly, this means that a CD is much more vulnerable to damage on its top label face than it is on the shiny face (Figure 9.19).

The data is read by a laser which scans the surface of the disc under the controlling action of a servo controlled 'sledge' or radial arm. Because the

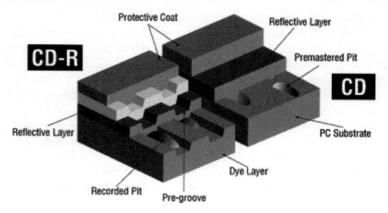

Figure 9.19 *Physical structure of CD and CD-R*

metal surface is protected by the plastic, the fact that the laser has no mechanical interference with the surface and due to the general robust nature of digital storage, CDs are free from the effect of scratches or dust on the surface of the CD.

As the laser scans the spiral track recorded on the CD, light is reflected by the land of the shiny metal surface. But when the laser scans a pit, the light is scattered and the reflection is much less. The land and pits do not represent digital ones and noughts as is sometimes explained. In fact, the change from a land to a pit, or a pit to a land is a logical one: and a continuous pit or land is a logical zero. The received data is compared with a phase-lock controlled, internal 4.3218 MHz clock; this being the average received data-rate.

The CD rainbow

Red Book

Red Book is the standard for Compact Disc-Digital Audio (CD-DA) and was established in 1982. This is still the standard for audio CDs. Audio is recorded as non-compressed 16-bit, PCM audio, sampled at 44.1 kHz; thereby offering a theoretical maximum bandwidth of 22.05 kHz and a maximum theoretical dynamic range of 96 dB. The maximum capacity of a CD is about 750 Mbytes of data, which is equivalent to about 74 minutes of recorded audio in this format. At the time, this was a massive advance over the recording quality available from analogue records. The 16-bit audio bytes are split into 8-bit bytes called 'symbols' in CD parlance. Blocks of 24 symbols are constructed from stereo groups of six, consecutive samples. Two stages of parity symbols are added to this data to enhance further the robustness of the system to data corruption. To this data, a further sub-code byte is added, each bit within this 8-bit byte being referred to as a letter; P, Q, R, S, T, U, V and W. The control bits P and Q are the most important in Red Book and contain the timing information which allows the CD player to cue instantly to the beginning of each selection, display the selection's number and running time, and provide a continuous display of elapsed time. The other bits are used within the other standards and some are still reserved for future uses.

The 8-bit data is now modulated using eight-to-fourteen modulation (EFM). Eight-to-fourteen modulation is, essentially, a look-up table process in which eight-bit bytes are transformed into fourteen-bit bytes; each 14-bit byte especially constructed so that there is never a long group of consecutive 0s or 1s. In other words a code to remove low-frequency components from the recorded data, just as we will see with MADI and with digital video in future chapters. These 14-bit bytes are then serialised and three more bits (merging bits) are added to each group of fourteen bits. These 17-bit words are formed into groups of 33, to make a total of 564-bit strings. These strings are combined with a 24-bit synchronising pattern to create a 588-bit EFM frame.

The information on all compact discs is divided into chunks of uniform size, called sectors, and adjoining sectors are then grouped to form tracks, which are listed in the disc's Table Of Contents (TOC); a special un-numbered track that plays first. Every sector on every type of CD contains 3234 bytes of data, 882 bytes of which are reserved for error detection and correction code and control bytes, leaving 2352 bytes (3234 minus 882) to hold usable data. The difference between CD formats relates to the use of this 2352 bytes of space only. CD-Audio discs (Red-Book) use the space for digitally recorded sound and error-correction data, while other types of CDs contain data of other types. The Red Book standard specifies that a compact disc can have up to 99 tracks of data, with each track containing a single audio selection. Normally, audio discs (Red Book) are 'single session', meaning they have only one lead-in, program area and lead out. In the lead-in area, the Q sub-code contains the Table Of Contents, while in the program area of the disc the P sub-code contains information about where the music starts and ends, and the Q sub-code contains absolute and relative time information. Newer forms of CD can be multi-session. But only multi-session players can read sessions beyond the first.

Yellow Book

Yellow Book is the standard for compact disc-read only memory (CD-ROM) and this extends the Red Book specification by adding an extra layer of error-correction. This is essential because, in computer data applications – even the loss of one bit, is unacceptable. The original CD-

ROM Yellow Book standard (also known as Mode 1) was extended – in 1988 – to include provision for the inclusion of other track types; the resulting disc being known as CD-ROM/XA (for eXtended Architecture), or as Yellow Book, Mode 2. Because CD-ROM is used for computer applications, a file structure had to be defined. This file structure was finalised in the High Sierra Hotel and Casino in Las Vegas and is therefore known informally as the 'High Sierra' file system or – more formally – as ISO9660. This is a universal file system for CD-ROM enabling discs to be read on a number of computer platforms including Apple and PC.

Green Book, White Book and Blue Book

Green Book, White Book and Blue Book are all extensions of Yellow Book. Green Book is the standard for Compact Disc-Interactive (CD-i) and is an extension of Yellow Book to allow discs to contain a mix of video and audio, plus data which the user can control interactively. For this to be possible, Green Book also describes and defines a complete disc operating system based on OS9. The White Book specification was developed to cover the Video CD format, and Blue Book specifies the CD Plus (or Enhanced CD) standard, which was developed as a way to include CD-ROM data on an audio disc. This is achieved by creating two sessions, the first being an audio session, the second a data session. The audio CD player thereby only recognises the audio session but the multi-session CD-ROM drive can read both.

Orange Book

Most importantly for audio recording applications, Orange Book is the standard for recordable compact discs. This specification covers both single (disk-at-once) and incremental multi-session (track-at-once) recording. Multi-session allows you to record a 'session' to part of the disc and then add subsequent sessions at a later date, until the disc is full. Orange Book discs can subsequently be converted to a Red or Yellow Book disc by 'Finalising' the session, to add a final Table Of Contents.

As with a conventional disc, a CD-R has a polycarbonate substrate, a reflective layer and a top protective layer. However, recordable CDs differ from manufactured CDs, in that the data 'pits' are not pressed into the metal but are formed due to the interaction of a writing laser with an organic dye which is sandwiched between the polycarbonate substrate and a gold reflective layer, see Figure 9.19. Unlike a conventional CD, a pre-formed spiral track is used to guide the recoding laser along its coiled trail. (Incidentally, it is the difference in the modulation or 'wobble' of this track which distinguishes CD-Rs for audio-only applications from data CD-Rs.) The recording system may be termed 'heat-mode memory' and therefore differs from practically any other system so far described. Inside the CD recorder the writing laser is focused on the recording dye layer.

When the laser is energised, the minute area heats to over 250 degrees centigrade; enough to decompose the dye and thereby change the reflectivity sufficiently to mimic the effect of the pits in a conventional CD. The result is a CD which is indistinguishable, as far as the player is concerned, from a mass-produced CD.

DVD

Unlike a CD a DVD (digital versatile disk) has the capacity to be double-sided. Two thin (0.6 mm) back-to-back substrates are formed into a single disc that's the same thickness (1.2 mm) as a regular CD but more rigid. Data is represented on a DVD as it is on a CD; by means of physical 'pits' on the disc. But the thinner DVD substrates (and short-wavelength visible light laser) permit the pits to be smaller. In fact, they're roughly half the size, which in turn allows them to be placed closer together. The net effect is that DVDs have the capacity for over four times as many pits per square inch as CDs, totaling some 4.7 GB in a single-sided, single-layer disc.

A DVD's capacity may be further increased by employing more than one physical layer each side of the disc! In this case, the inner layer reflects light from the laser back to a detector through a focusing lens and beam-splitter because the outer layer is only partially reflective. DVD players incorporate novel dual-focus lenses to support two-layer operation, yielding 8.5 GB in a single-sided DVD, or 17 GB in a double-sided disc.

MiniDisc (MD)

MiniDisc is a re-writable, magneto-optical disc which holds 140 MBytes of data. It differs from conventional magneto-optical (MO) drives in that – in the conventional MO writing phase – the magnetic field remains constant while the laser flashes on and off. This requires separate passes for recording and erasure. MiniDisc overcomes this by keeping the laser constant and switching the magnetic field instead, as illustrated in Figure 9.20. With a diameter of 64 mm, MiniDisc can hold only 1/5 of the data of a conventional CD. Therefore, a data compression of 5:1 is employed in order to offer a similar 74 min of playback time. The high-quality audio compression technology used for MiniDisc is called ATRAC.

In the ATRAC encoding process, the signal is divided into three sub-bands by using two stages of digital filters (see Chapter 10 and Fact Sheet 10 for more details on digital filtering). Each band covers 0–5.5 kHz, 5.5–11 kHz, or 11–22 kHz. After that, each of the three sub-bands is

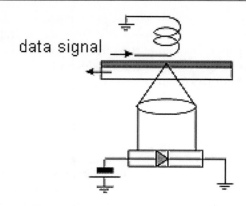

data signal

Figure 9.20   *MiniDisc record system*

transformed into the frequency domain by using the Modified Discrete Cosine Transform (see Chapter 10). The transform block size is adaptively switched between two modes; long mode – 11.6 ms for all frequency bands and short mode and 1.45 ms for the high-frequency band and 2.9 ms for mid- and low-frequency bands. Normally, the long mode is chosen to provide optimum frequency resolution but problems may occur during attack portions of the signal. Specifically, because the quantisation noise is spread over the entire MDCT block time, just before the attack of the sound you hear some 'pre-noise'. Therefore, ATRAC automatically switches to the short mode to avoid this type of noise when it detects a transient.

The transform spectral coefficients are then grouped into 'Blocks' and are quantised using two parameters. One is word length, another is scale factor. The scale factor defines the full-scale range of the quantization and the word length defines the resolution of the scale. The scale factor is chosen from a fixed table and reflects the magnitude of the spectral coefficients and the word length is determined by the bit allocation algorithm. This algorithm divides the available data bits between the various coded blocks. ATRAC does not specify a bit allocation algorithm. The word length of each coded block is stored on the MiniDisc along with the quantised spectra, so that the decoder is independent from the allocation algorithm. This allows for an evolutionary improvement of the encoder without changing the MiniDisc format.

The decoding process is divided into two steps. The decoder first reconstructs the MDCT spectral coefficients from the quantised values, by using the word length and scale factor parameters. The coefficients are transformed back into the time domain by inverse MDCT using either the long mode or the short mode as specified in the parameters. Finally, the three time-domain signals are synthesised into the output signal by QMF synthesis filters. Aside from its consumer acceptance, MiniDisc has found many uses in the hands of the professional; from field recording to radio jingle and TV voice-over playout.

Reference

Macdonald, I. (1994) *Revolution in the Head*, Pimlico.

Fact Sheet #9: Studio data communications

- Data communications in the studio
- RS232
- RS422
- RS485

Data communications in the studio

Since the advent of digital electronics, there has been a need to transfer data from place to place. Sometimes these locations are relatively proximate, other times they are more distant. Usually there is also a requirement for two-way communication. Apart from the MIDI standard (see Chapter 8), the two most widely used standards for communication in the recording studio are RS (for 'Recommended Standard') 232 and 422.

RS232

RS232 has been around since 1962 as a standard for the electrical interface between data terminal equipment (DTE) and data circuit-terminating equipment (DCE). RS232 is what appears on the serial output of your PC! The essential feature of RS-232 is that it is 'single-ended' which is to say the signals are carried as single voltages referred to a common earth. The standard 9-pin D-type pin-out is given in Figure F9.1.
Voltage levels for RS232 are:

Signal > +3 V = 0

Signal < −3 V = 1

The output signal level usually swings between +12 V and −12 V; the 'dead area' between +3 V and −3 V being designed to give some noise immunity.

RS232 pinout

PIN	FUNCTION
1	n/c
2	rx
3	tx
4	n/c
5	0V
6	dtr common
7	rts
8	cts
9	n/c

Figure F9.1 *RS232 pinout*

RS422

RS-232 is simple, universal, well understood and supported everywhere. However, despite RS232's in-built dead-band, noise is always a problem on a RS232 link, except where the length of the cable is less than about 15 metres and the Baud rate is limited to about 56 kbits/s. When communicating at high data rates, or over long distances, single-ended methods are often inadequate. Just as we saw with audio signals, differential data transmission offers superior performance in most applications so, in the RS422 standard, a pair of wires is used to carry each signal. The data is encoded and decoded as a differential voltage between the two lines like this,

$$VA–VB < –0.2\,V = 0$$

$$VA–VB > +0.2\,V = 1$$

Because the signal is differential, the interface is unaffected by differences in ground voltage between sender and receiver. Furthermore, because the balanced lines are close together, they are affected identically by external electromagnetic noise, so these 'common-mode' noise voltages cancel-out. RS422 standard was the inspiration behind the AES/EBU digital audio interface (see Chapter 10). There is no standard pinout for RS422 on a 9-pin D-type connector, however, many pieces of studio equipment adopt the standard illustrated in Figure F9.2 (depending on the role of transmitter or receiver).

RS422 pinouts

PIN	FUNCTION	
	device	controller
1	chassis	chassis
2	rx-	tx-
3	tx+	rx+
4	0V	0V
5	n/c	n/c
6	0V	0V
7	rx+	tx+
8	tx-	rx-
9	chassis	chassis

Figure F9.2 *RS422 pinouts*

In studio situations there is often the necessity to translate between RS232 (usually from a computer serial port) to RS422 (for machine, routing switcher or automation system control). This is usually best accomplished with commercial RS232 to RS422 converters; such as that illustrated in Figure F9.3.

Figure F9.3 *RS232 to RS422 converter*

RS485

Both RS422 and RS485 use twisted-pair cables and they both use the same differential drive with identical voltage swings: 0 to +5 V. RS422 is full-duplex and uses two separate twisted pairs; RS485 is half-duplex. The main difference between RS422 and RS485 is that RS422 has no tri-state capability. Its driver is always enabled, and it is therefore only usable in point-to-point communications. RS485 has tri-state capability and can therefore be used in multidrop systems. It exists in two varieties: 2-wire (which uses a single twisted pair) and 4-wire (which uses two twisted pairs like RS422). RS485 systems are usually 'master/slave'; each slave device has a unique address and it responds only to a correctly addressed 'poll' message from the master. A slave never initiates a dialogue. In a 2-wire system, all devices, including the master, must have tri-state capability. In a 4-wire system, the slaves must have tri-state capability but the master does not need it because it drives a bus on which all other devices merely listen. This is often an advantage because it allows master software and drivers originally developed for RS232 to be used in a multidrop system; provided of course that the master software emits the correct device addresses.

10
Bits 'n' Pieces – Digital Audio

Why digital?

After 332 BC, when the Greeks – under Alexander the Great – took control of Egypt, Greek replaced Ancient Egyptian as the official language of the land of the Nile and gradually the knowledge of how to write and read hieroglyphs was lost. Only in 1799 – after a period of two thousand years – was the key to deciphering this ancient written language found following the discovery of the Rosetta stone. Why was this knowledge lost? It seems pretty inconceivable that a whole culture could just whither away. My guess is, the Egyptian scribes knew they were beaten. Greek writing was based on a written alphabet – a limited number of symbols doing duty for a whole language. Far better then than the seven hundred representational signs of Ancient Egyptian writing. What we are witnessing in today's world is this slice of history repeating itself as analogue systems and signals are being replaced by digital ones. Any analogue system is a representational system – a wavy current represents a wavy sound pressure and so on. Hieroglyphic electronics if you like (Figure 10.1)! The handling and processing of continuous time variable signals (like audio and video waveforms) in digital form has all the advantages of

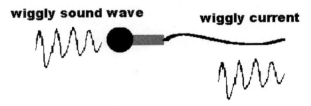

Figure 10.1 *The nature of analogue signals*

243

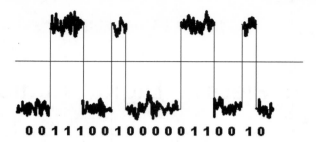

0 0 1 1 1 0 0 1 0 0 0 0 0 1 1 0 0 1 0

Figure 10.2 *Digital signals are relatively immune to noise*

a precise symbolic code (an alphabet) over an older approximate representational code (hieroglyphs).

In practice digital systems do not have to be binary (i.e. use two-levels) but this is preferred, and this choice – in itself – ensures a high resistance to noise and circuit distortion (Figure 10.2). And the advantages don't just stop there. Once represented by a limited number of abstract symbols, a previously undefended signal may be protected by sending special extra codes (parity bits) so that the digital decoder can work out when errors have occurred. For example, if an analogue record is contaminated or damaged, the impulses (in the form of audible 'clicks') will be reproduced by the loudspeakers. This is inevitable because the analogue record-player cannot 'know' what is wanted modulation and what is not. A CD player on the other hand, because the CD data stream contains enough of the right type of extra information, can sort the impulsive interference from wanted signal.

A further advantage which digital systems possess is that the binary symbolic approach facilitates the use of standard electronic memory elements. This allows the storage and manipulation of signals with a facility undreamed of in the days of analogue signal processing. There's a down-side too, of course, and this is the considerably greater capacity, or bandwidth, demanded by digital storage and transmission systems over their analogue counterparts. As we shall see later in the chapter, even this disadvantage is being overcome by gradual advances in data-reduction or compression techniques which make better use of smaller bandwidths.

Sampling theory and conversion

There exist three fundamental differences between a continuous-time, analogue representation of a signal and a digital, pulse code modulation (PCM) description. Firstly, a digital signal is a time-discrete, sampled representation and secondly, it is quantised. Lastly, as we have already noted, it is a symbolic representation of this discontinuous time, quantised signal. Actually it's quite possible to have a sampled analogue signal. (Many exist, for instance film is a temporally sampled system.) And it is obviously quite possible to have a time-continuous, quantised system in which an electrical current or voltage could change state any time it wished but only between certain (allowed) states – the output of a multivibrator is one such circuit. The circuit which performs the function of converting a continuous-time signal with an infinite number of possible states (an analogue signal) into a binary (two state) symbolic, quantised and sampled (PCM) signal is known as an analogue to digital converter (ADC), the reverse process is performed by a digital to analogue converter (DAC).

Theory

The process of analogue to digital conversion and digital to analogue conversion is illustrated in Figure 10.3. As you can see, an early stage of conversion involves sampling. It can be proved mathematically that all the information in a bandwidth-limited analogue signal may be sent in a series of very short, periodic 'snapshots' (samples). The rate these samples need be sent is related to the bandwidth of the analogue signal, the minimum rate required being $1/(2 \times Ft)$, where Ft represents the maximum frequency in the original signal. So, for instance an audio signal (limited – by the filter preceding the sampler – to 15 kHz) will require pulses to be sent every, $1/(2 \times 15000)$ seconds; or 33 microseconds.

The mechanism of sampling

Figure 10.4 illustrates the effect of sampling. Effectively, the analogue signal is multiplied (modulated) by a very short period pulse-train. The spectrum of the pulse-train is (if the pulses were of infinitely short period) infinite and the resulting sampled spectrum (shown too in Figure 10.4) contains the original spectrum as well as images of the spectrum as sidebands around each of the sampling pulse harmonic frequencies. It's very important to realise the reality of the lower diagram in Figure 10.4: The signal carried in a digital system really has this spectrum: discrete-time versions of Fourier analysis prove that all digital signals actually have this form. This, if you are of an intuitive frame of mind, is rather difficult to accept. In fact this effect is termed, even by mathematicians, as the ambiguity of digital signals.

Aliasing

If analogue signals are sampled at an inadequate rate, it results in an effect known as aliasing where the high frequencies get 'folded back' in the frequency domain and come out as low frequencies. Figure 10.5 illustrates the effect which is termed aliasing. Hence the term anti-aliasing

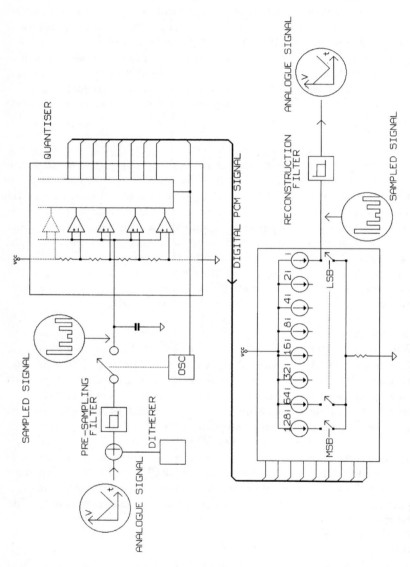

Figure 10.3 *Analogue to digital conversion*

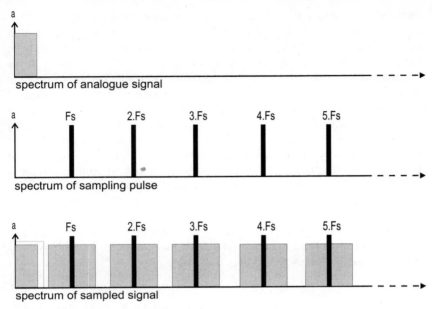

Figure 10.4 *Spectrum of a sampled signal*

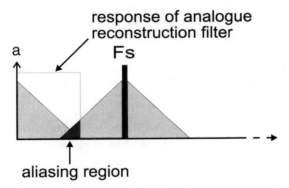

Figure 10.5 *The phenomenon of aliasing*

filter for the first circuit block in Figure 10.3; to remove all frequencies above *Ft*/2.

Quantisation
After sampling, the analogue snapshots pass to the quantiser which performs the function of dividing the input analogue signal range into a number of pre-specified quantisation levels. It's very much as if the circuit

measures the signal with a tape measure with each division of the tape measure being a quantisation level. The important thing to realise is that the result is always an approximation. The finer the metric on the tape measure the better the approximations become. But the process is never completely error free since the smallest increment that can be resolved is limited by the accuracy and fineness of the measure. The errors may be very small indeed for a large signal but for very small signals these errors can become discernible. This quantisation error is inherent in the digital process. Some people incorrectly refer to this quantisation error as quantisation noise. Following the quantiser, the signal is – for the first time – a truly digital signal. However it is often in a far from convenient form. So the last stage in the ADC is the code conversion which formats the data into a binary numerical representation. The choice of the number of quantisation levels determines the dynamic range of a digital PCM system. To a first approximation the dynamic range in dB is the number of digits, in the final binary numerical representation, times six. So, an eight bit signal has $(8 \times 6) = 48\,dB$ dynamic range.

Digital to analogue conversion

The reverse process of digital to analogue conversion (also illustrated in Figure 10.3), involves regenerating the quantised voltage pulses demanded by the digital code which may first have had to pass through a code conversion process. These pulses are then transformed back into continuous analogue signals in the block labelled reconstruction filter. The ideal response of a reconstruction filter is illustrated in Figure 10.6. This has a time-domain performance which is defined by $\{(\sin x)/x\}$. If very short pulses are applied to a filter of this type the analogue signal is 'reconstructed' in the manner illustrated in Figure 10.7.

Jitter

There are a number of things which can adversely affect the action of sampling an analogue signal. One of these is jitter – which is a temporal

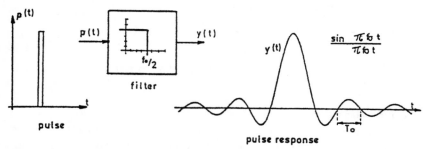

Figure 10.6   *Sin x/x impulse response of a reconstruction filter*

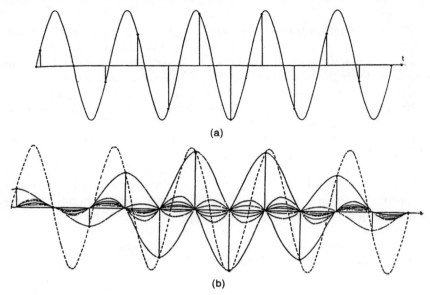

Figure 10.7 *The action of the reconstruction filter*

uncertainty in the exact moment of sampling. On a rapidly changing signal, time uncertainty can result in amplitude quantising errors which in turn lead to noise. (Jitter is discussed more fully in Fact Sheet #10.)

Aperture effect

As we saw, the perfect sampling pulse has a vanishingly short duration. Clearly a practical sampling pulse cannot have an instantaneous effect. The 'moment' of sampling ($t1$) is not truly instantaneous and the converted signal doesn't express the value of the signal at $t1$, but actually expresses an average value between ($t1 - T_o/2$) and ($t1 + T_o/2$) where (T_o) is the duration of the sampling pulse. This distortion is termed aperture effect and it can be shown that the duration of the pulse has an effect on frequency response such that,

$$20 \log \{\text{sinc } (PI/2. \, f/f_n. \, T_s/T_o)\} \text{ dB}$$

where T_s is the duration of the sampling pulse (aperture) and f_n is the Nyquist frequency limit. (Note that sinc is shorthand for sin x/x). Note that the aperture effect is not severe for values of $T_o < 0.2T_s$. Even when $T_o = T_s$, the loss at the band edge (i.e. at the Nyquist frequency) is -3.9 dB. Aperture effect loss is often 'made-up' by arranging the reconstruction filter to have a compensating frequency rise.

Dither

When a quantiser converts a very large signal that crosses many quantisation levels, the resulting errors from literally thousands of very slightly wrong values do indeed create a noise signal which is random in nature. Hence the misnomer quantisation noise. But when a digital system records a very small signal, which only crosses a few quantisation thresholds, the errors cease to be random. Instead the errors become correlated with the signal. And because they are correlated with (or related to) the signal itself, they are far more noticeable than would be an equivalent random source of noise.

In 1984 Vanderkooy and Lipshitz proposed an ingenious and inspired answer to this problem. They demonstrated that it is possible to avoid quantisation errors completely, by adding a very small amount of noise to the original analogue signal prior to the analogue to digital converter integrated circuit. They showed that a small amount of noise can cause low-level information to be coded as a kind of pulse-width modulation of the least significant bit, as illustrated in Figure 10.8. This explains the

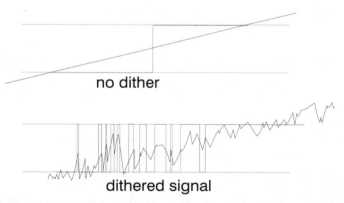

no dither

dithered signal

Figure 10.8   *How 'dither' noise codes low-level information*

block in Figure 10.3 marked dither noise generator which is shown as summing with the input signal prior to the sampler. In the pioneering days of digital audio and video, the design of ADC's and DAC's consumed a vast amount of the available engineering effort. Today's engineer is much luckier. Many 'one-chip' solutions exist which undertake everything but a few ancillary filtering duties. Circuits for a high-quality, commercial audio ADC and DAC are given in Figure 10.9a–c. These utilise one-chip solutions from Crystal Semiconductor. A commercial DAC unit is illustrated in Figure 10.10.

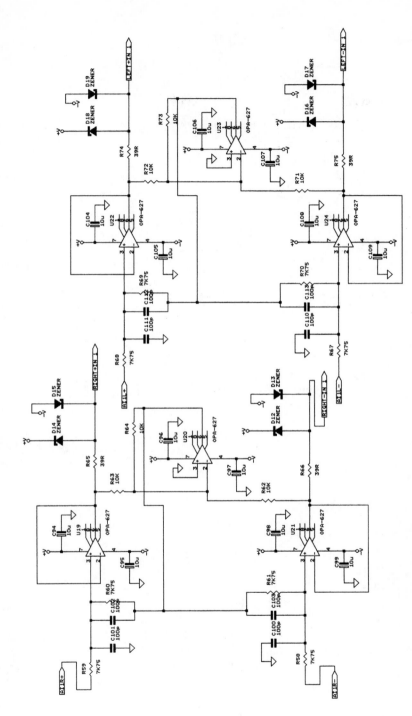

Figure 10.9a *Circuits for ADC and DAC*

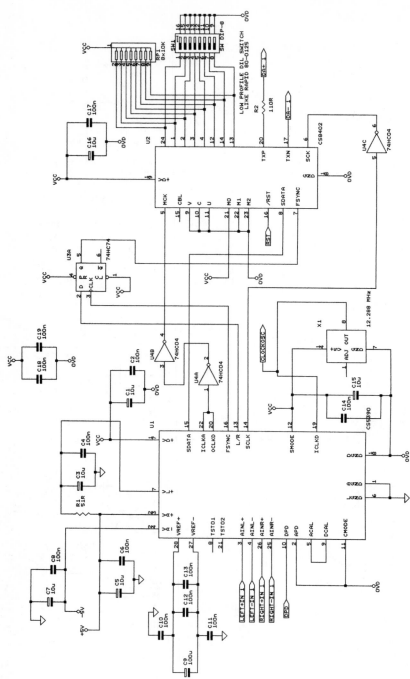

Figure 10.9b

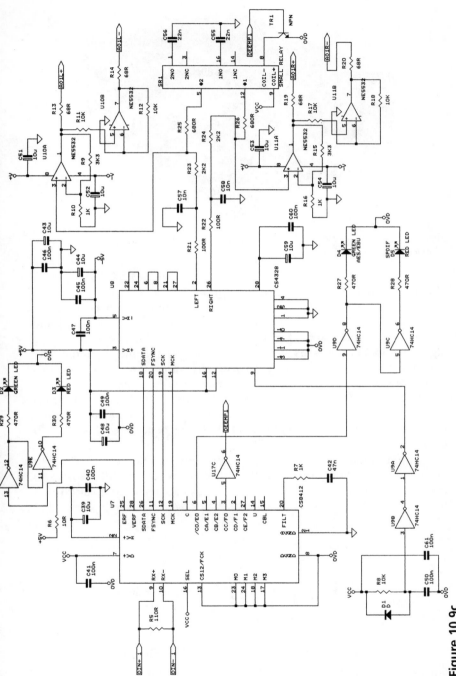

Figure 10.9c

Figure 10.10 *Commercial high-quality digital to analogue converter*

Numerical coding of digital audio signals

Analogue audio signals sit 'around' a zero volt baseline. It is therefore necessary to code this polarity information – inherent in the analogue signal – once the signal has been converted into the digital domain. In the digital audio interfaces used for professional and domestic audio, the digital representation is made in 'two's complement' form. In this numbering system, negative sixteen-bit audio values range from the most negative 1000000000000000 to the least negative, 1111111111111111. And the positive values go from, the least positive, 0000000000000000 to the most positive, 0111111111111111. These codes are illustrated in Figure 10.11.

It's pretty clear that two's complement representation can be thought of as a coding scheme in which the signal is coded from all zeros to all ones, with the first, most-significant bit (MSB) acting as an inverted sign bit; 0 for positive values, 1 for negative values. With this sign inverted, the

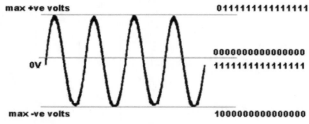

Figure 10.11 *Numerical coding of audio signals*

numerical representation is termed 'signed-integer' format. The addition of an imaginary radix point after the MSB forces the numerical representation into, what is termed 'signed-fractional' format. Signed-integer and signed-fractional formats are the representations most often found inside digital signal processing (DSP) chips; especially signed-fractional. Most fixed-point, DSP operations are optimised for this latter representation.

Digital audio interfaces

Many of the advantages of digital signal processing are lost if signals are repeatedly converted back and forth between the digital and analogue domain. So that the number of conversions could be kept to a minimum, as early as the 1970s, manufacturers started to introduce proprietary digital interface standards enabling various pieces of digital audio hardware to pass digital audio information directly without recourse to standard analogue connections. Unfortunately each manufacturer adopted its own standard and the Sony digital interface (SDIF) and the Mitsubishi interface both bear witness to this early epoch in digital audio technology when compatibility was very poor between different pieces of equipment. It wasn't long before customers were demanding an industry-standard interface so that they could 'mix-and-match' equipment from different manufacturers to suit their own particular requirements. This pressure led to the introduction of widespread, standard interfaces for the connection of both consumer and professional digital audio equipment.

The requirements for standardising a digital interface go beyond those for an analogue interface in that, as well as defining the voltage levels and connector style, it is necessary to define the data format the interface will employ. The two digital audio interface standards described here are:

(1) The two-channel, serial, balanced, professional interface (the so-called AES/EBU or IEC958 type 1 interface) and
(2) The two-channel, serial, unbalanced, consumer interface (the so-called SPDIF or IEC958 type 2 interface).

In fact both these interfaces are very similar, the variation being more due to electrical differences than between differences in data format.

AES/EBU or IEC958 type 1 interface
This electrically balanced version of the standard digital interface, was originally defined in documents produced by the Audio Engineering Society (AES) and the European Broadcasting Union (EBU) and is consequently usually referred to as the AES/EBU standard. This is the

standard adopted mainly by professional and broadcast installations. Mechanically this interface employs the ubiquitous XLR connector and adopts normal convention for female and male versions for inputs and outputs respectively. Electrically, pin 1 is specified as shield and pins 2 and 3 for balanced signal. One of the advantages of the digital audio interface over its analogue predecessor is that polarity is not important, so it is not necessary[1] to specify whether which pin of 2 and 3 is 'hot'. The balanced signal is intended to be carried by balanced, twisted-pair and screen microphone-style cable and voltage levels are allowed to be between 3 V and 8 V pk-pk (EMF, measured differentially). Both inputs and outputs are specified as transformer coupled and earth-free. The output impedance of the interface is defined as 110 ohms and a standard input must always terminate in 110 ohms. A drawing for the electrical standard for this interface is given in Figure 10.12.

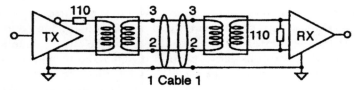

Figure 10.12 *AES/EBU interface*

The SPDIF or IEC985 type 2 interface

This consumer version of the two-channel, serial digital interface is very different electrically from the AES/EBU interface described above. It is a 75 ohm, matched termination interface intended for use with coaxial cable. It therefore has more in common with an analogue video signal interface than with any analogue audio counterpart. Mechanically the connector style recommended for the SPDIF interface is RCA style phono with sockets always being of the isolated type. Voltage levels are defined as 1 V pk-pk when un-terminated. Transformer coupling is by no means always used with this interface but it is recommended at least one end. Figure 10.13 is a drawing of a common implementation of the SPDIF interface.

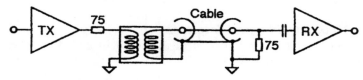

Figure 10.13 *SPDIF interface*

Data

Despite the very considerable electrical differences between the AES/EBU interface and the SPDIF interface, their data formats are very similar. Both interfaces have capacity for the real-time communication of 20 bits of stereo audio information at sampling rates between 32 and 48 kHz as well as provision for extra information which may indicate to the receiving device various important parameters about the data being transferred (such as whether pre-emphasis was used on the original analogue signal prior to digitisation). There is also a small overhead for limited error checking and for synchronisation.

Some of the earlier digital-audio interfaces such as Sony's SDIF and the Mitsubishi interface sent digital audio data and synchronising data clocks on separate wires. Such standards obviously require multi-core cable and multi-way connectors which looked completely different from any analogue interface that had gone before. The intention of the designers of the AES/EBU and SPDIF interfaces was to create standards which created as little 'culture-shock' as possible in both the professional and consumer markets and they therefore chose connector styles that were both readily available and operationally convenient. This obviously ruled out the use of multi-core and multi-way connectors and resulted in the use of a digital coding scheme which buries the digital synchronising signals in with the data signal. Such a code is known as 'serial and self-clocking'. The type of code adopted for AES/EBU and SPDIF is bi-phase mark coding. This scheme is sometimes known as Manchester code and it is the same type of self-clocking, serial code used for SMPTE and EBU timecode. Put at its simplest such a code represents the 'ones and noughts' of a digital signal by two different frequencies where frequency Fn represents a zero and 2Fn represents a one. Such a signal eliminates almost all DC content, enabling it to be transformer coupled and also allows for phase inversion since it is only a frequency (and not its phase) which needs to be detected. The resulting signal has much in common with an analogue FM signal and since the two frequencies are harmonically related (an octave apart) it is a simple matter to extract the bit-clock from the composite incoming data stream.

In data format terms the digital audio signal is divided into frames. Each digital audio frame contains a complete digital audio sample for both left and right channel. If 48 kHz sampling is used, it is obvious the 48 thousand frames pass over the link in every second leading to a final baud rate of 3.072 Mbit/s. If 44.1 kHz sampling is employed, 44 thousand one-hundred frames are transmitted every second, leading to a final baud rate of 2.8224 Mbit/s. The lowest allowable transfer-rate is 2.084 Mbit/s when 32 kHz is used. Just as each complete frame contains a left and right channel sample, so each frame may be further divided into individual audio samples known as sub-frames. A diagram of a complete frame consisting of two sub-frames is given in Figure 10.14.

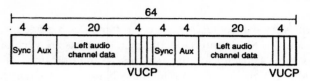

Figure 10.14   *Digital audio data format*

It is manifestly extremely important that any piece of equipment receiving the digital audio signal as shown in Figure 10.14, must know where the boundaries between frames and sub-frames lie. That is the purpose of the 'sync preamble' section of each frame and sub-frame. The sync preamble section of the digital audio signal differs from all the other data sent over the digital interface in that it violates the rules of a bi-phase mark encoded signal. In terms of the FM analogy given above you can think of the sync preamble as containing a third non-harmonically related frequency which, when detected, establishes the start of each sub-frame. There exists a family of three slightly different sync-preambles, one to mark the beginning of a left sample sub-frame and another to mark the start of the right channel sub-frame. The third sync-preamble pattern is used only once every 192 frames (or once every 4 milliseconds in the case of 48 kHz sampling) and is used to establish a 192 bit repeating pattern to the channel-status bit labelled C in Figure 10.14.

The 192 bit repeat pattern of the C bit, builds up into a table of 24 bytes of channel-status information for the transmitted signal. It is in this one bit of data every sub-frame that the difference between the AES/EBU interface data format and the SPDIF data format is at its most significant. The channel status bits in both the AES/EBU format and SPDIF format communicate to the receiving device such important parameters as sample-rate, whether frequency pre-emphasis was used on the recording etc. Channel status data is normally the most troublesome aspect of practical interfacing using the SPDIF and AES/EBU interface – especially where users attempt to mix the two interface standards. This is because the usage of channel status in consumer and professional equipment is almost entirely different. It must be understood that the AES/EBU interface and the SPDIF interface are thus strictly incompatible in data-format terms and the only correct way to transfer data from SPDIF to AES/EBU and AES/EBU to SPDIF is through a properly designed format converter which will decode and re-code the digital audio data to the appropriate standard.

Other features of the data format remain pretty constant across the two interface standards. The Validity bit, labelled V in Figure 10.14 is set to 0 every sub-frame if the signal over the link is suitable for conversion to an analogue signal. The User bit, labelled U, has a multiplicity of uses defined by particular users and manufacturers. It is most often used over the domestic SPDIF interface. And the Parity bit, labelled P, which is set such that the number of ones in a subframe is always even. It may be used to detect individual bit errors but not conceal them.

It's important to point out, both the AES/EBU interface and its SPDIF brother are designed to be used in an error free environment. Errors are not expected over digital links and there is no way of correcting for them.

Practical digital audio interface

There are many ways of constructing a digital audio interface and variations abound from different manufacturers. Probably the simplest consists of an HC-family inverter IC, biased at its mid-point with a feedback resistor and protected with diodes across the input to prevent damage from static or over-voltage conditions. (About the only real merit of this circuit is simplicity!) Transformer coupling is infinitely preferred. Happily, whilst analogue audio transformers are complex and expensive items, digital audio – containing as it does no DC component and very little low-frequency component – can be coupled via transformers which are tiny and cheap! So, it represents a false economy indeed to omit them in the design of digital interfaces. Data-bus isolators manufactured by Newport are very suitable. Two or four transformers are contained within one IC-style package. Each transformer costs about 2 dollars – a long way from the 20 or so dollars required for analogue transformers. Remember too, that 'in digits' only one transformer is required to couple both channels of the stereo signal. You'll notice, looking at the circuit diagrams (Figure 10.15), RS422 (RS485) receiver-chips buffer and re-slice the digital audio data. The SN75173J, is a quad receiver in a single 16 pin package costing a few dollars. The part has the added advantage that, to adapt the interface between SPDIF and AES, all that is required is to change the value of the terminating resistor on the secondary side of the input transformer. SPDIF digital output can be derived by inverters driving in tandem. If AES/EBU output is required it is best performed by an RS422 driver IC.

TOSlink optical interface

In many ways an optical link seems to be the ideal solution for joining two pieces of digital audio equipment together. Obviously a link that has no electrical contact cannot introduce ground-loop, hum problems. Also, because the bandwidth of an optical link is so high, it would appear from a superficial inspection that an optical link would provide the very fastest (and therefore 'cleanest') signal path possible. However the optical TOSLink is widely regarded as sounding a little less crisp than its coaxial, electrical counterpart. There are a number of possible reasons for this: In

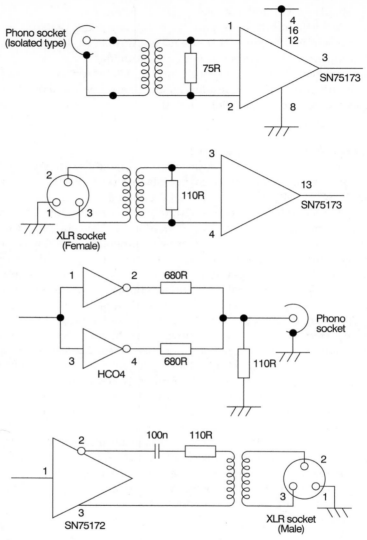

Figure 10.15 *Practical digital audio interfaces*

the first place, the speed of the link is compromised by the relatively slow light emitting diode transmitter and photo-transistor receiver housed within the connector shells. Secondly, cheap optical fibres, which allow the optical signal more than one direct path between transmitter and receiver (the technical term is multimodes), causes a temporal smearing of the audio pulses, resulting in an effect known as modal dispersion. This

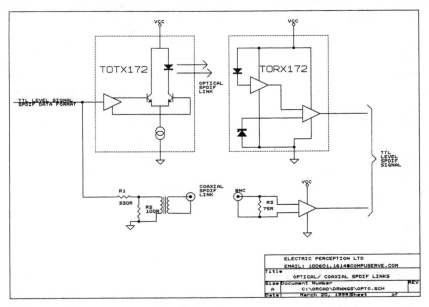

Figure 10.16   *Optical digital audio interface and adaption to coaxial SPDIF*

can cause a degree of timing instability in digital audio circuits (jitter) and this can affect sound quality. The only advantage the optical link confers therefore, is its inherent freedom from ground-path induced interference signals such as hum and RF noise. Yet at digital audio frequencies, ground isolation – if it is required – is much better obtained by means of a transformer. If you want to modify a piece of equipment with an optical interface to include SPDIF coaxial output, a modification is shown in Figure 10.16.

Transmission of AES3 formatted data by unbalanced coaxial cable

In October 1995, the AES produced an information document (AES-3id-1995) relating to the transmission of digital audio information (utilising the professional data format) over an electrical interface which has much in common with the interconnection standards employed in analogue video. Limitations of AES data travelling on twisted pairs and terminated in XLRs include poor RF radiation performance and a limitation of maximum transmission distance to 100 metres. The proposed unbalanced interface is suitable for transmission distances of up to 1000 metres. Furthermore by a prudent choice of impedance and voltage operating

Table 10.1

General

Transmission data format	Electrically equivalent to AES
Impedance	75 ohms
Mechanical	BNC connector
Signal Characteristics	
Output voltage	1 V, measured when terminated in 75 ohms
DC Offset	<50 mV
Rise/Fall time	30 to 44 nS
Bit width (at 48 kHz)	162.8 nS

---

level coupled with a sensible specification of minimum rise-time, the signal is suitable for routing through existing analogue video cables, switchers and distribution amplifiers.

The salient parts of the signal and interface specification are given in Table 10.1.

MADI (AES10–1991) serial multi-channel audio digital interface

The MADI standard is a serial transmission format for multi-channel linearly represented PCM audio data. The specification covers transmission of 56, mono, 24-bit resolution channels of audio data with a common sampling frequency in the range of 32 kHz to 48 kHz. Perhaps, this is more easily conceived of in terms of 28 stereo 'AES' audio channels (i.e. of AES3–1985 data) travelling on a common bearer; as illustrated in Figure 10.17. The MADI standard is not a 'networking' standard; in other words, it only supports point-to-point interconnections.

Data format

The MADI serial data stream is organised into frames which consist of 56 channels (numbered 0–55). These channels are consecutive within the frame and the audio data remains, just as it is in the original digital audio interface, in linearly-coded, 2's-complement form, although, this is scrambled as described below. The frame format is illustrated in Figure 10.17. Each channel 'packet' consists of 32 bits (as shown in Figure 10.17

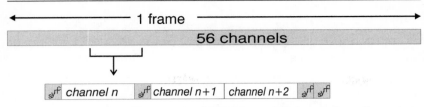

Figure 10.17   *Data structure of MADI, multi-channel audio interface*

too), in which 24 are allocated to audio data (or possibly non-audio data if the non-valid flag is invoked) and four bits for the validity (V), user (U), channel-status (C) and parity (P) bits as they are used in the AES3–1985 standard audio interface. In this manner the structure and data within contributing dual-channel AES bitstreams can be preserved intact when travelling in the MADI multi-channel bitstream. The remaining 4 bits per channel (called, confusingly mode-bits) are used for frame synchronisation on the MADI interface and for preserving information concerning A/B pre-ambles and start of channel-status block within each of the contributing audio channels.

Scrambling and synchronisation

Serial data is transmitted over the MADI link in polarity-insensitive (NRZI) form. However, before the data is sent it is subjected to a 4-bit to 5-bit encoding, as defined in Table 10.2. MADI has a rather unusual synchronisation scheme in order to keep transmitter and receiver in step. The standard specifies that the transmitter inserts a special synchronising sequence (1100010001) at least once per frame. Note that this sequence cannot be derived from data, as specified in Table 10.2. Unusually, this sync signal need not appear between every frame as Figure 10.17 illustrates. This sync signal is simply repeated wherever required in order to regulate the final data rate of 100 megabits/second specified in the standard.

Electrical format

MADI travels on a coaxial cable interface with a characteristic impedance of 75 ohms. Video-style BNC connectors are specified. Because the signal output is practically DC free, it may be AC coupled and must sit around 0 V ± 100 mV. This signal is specified to have a peak-to-peak amplitude of 300–600 mV when terminated – this choice of amplitude being determined by the practical consideration that the signal could be directly derived from the output of an ECL gate (see Chapter 15).

Table 10.2

Input data sequence	Output data sequence
0000	11110
0001	01001
0010	10100
0011	10101
0100	01010
0101	01011
0110	01110
0111	01111
1000	10010
1001	10011
1010	10110
1011	10111
1100	11010
1101	11011
1110	11100
1111	11101

Fibre-optic format

Oddly, the MADI standard did not define a fibre implementation, despite the fact that the copper implementation was based on a widely used fibre interface known as FDDI (ISO 9314). It is this standard, which pre-dates the MADI, which specified the 4-bit to 5-bit mapping defined in Table 10.2! This lack of standardisation has resulted in a rather disorganised situation regarding MADI over fibre. The AES's own admission is simply that 'any fibre-system could be used for MADI as long as the basic bandwidth and data-rate can be supported . . . However, adoption of a common implementation would be advantageous'.

Digital tape recorders

The bandwidth of a digital audio PCM signal is much higher than its analogue counterpart; several Megahertz in fact. In many ways, the digital audio PCM signal has more in common with an analogue television signal than it has with an analogue audio signal. It is therefore not surprising that several of the first attempts to record a PCM digital audio signal appropriated video technology in order to capture the PCM signal. Video tape recorders had always been faced with the need to record a wide bandwidth signal, because a television signal extends from DC to perhaps four or five Megahertz[2]. The DC component in a television signal exists to represent overall scene brightness whereas in a PCM digital audio signal, near zero-frequency must be preserved in order to capture long strings of similar symbols (0s or 1s).

There exist two fundamental limitations to the reproducible bandwidth from an analogue tape recorder of the type considered so far. The first is due to method of induction of an output signal; which is – in turn – due to the rate of change of flux in the tape head. Clearly a zero frequency signal can never be recorded and reproduced because, by definition there would exist no change in flux and therefore no output signal. In fact, the frequency response of an un-equalised tape recorder varies linearly with frequency; the higher the frequency, the faster the rate of change of flux and the higher the induced electrical output. This effect is illustrated in Figure 10.18. In audio tape recorders the intrinsic limitation of an inability

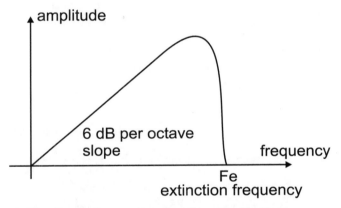

Figure 10.18 *Frequency response of magnetic tape*

to record zero frequency is not important because, usually, 20 Hz is regarded as the lowest frequency required to be reproduced in an audio system. Similarly, the changing frequency response is 'engineered around' by the application of complementary equalisation. But in video tape recorders, where the DC component must be preserved, this is achieved by the use of frequency modulation; a modulation scheme in which a continuous modulating frequency is present at the tape heads even if there is little or no signal information, or where the signal information changes very slowly.

The second limitation is a function of recorded wavelength and head gap. Essentially, once the recorded wavelength approaches the dimension of the head gap, the response of the record–replay system falls sharply as is illustrated in Figure 10.18. (The response reaches total extinction at the frequency at which the recorded wavelength is equal to that of the head gap.) Of course, recorded wavelength is itself a function of linear tape speed, the faster the tape travels, the longer the recorded wavelength, so theoretically the bandwidth of a tape recorder can be extended indefinitely by increasing the tape speed. It's pretty clear however that there exist some overwhelming practical, and commercial, obstacles to such an approach.

The alternative approach developed first by Ampex in the VR-1000 video tape recorder was to spin a number of heads in a transverse fashion across the width of the tape; thereby increasing the head to tape writing speed without increasing the linear tape speed. This video technology was named by Ampex, Quadruplex, after the four heads which rotated on a drum across a 2″ wide tape. Each video field was written in one rotation of the drum, so that each video field was split into four sections. This led to one of the problems which best 'Quad', as this tape format is often called, where the picture appears to be broken into four discrete bands due to a differing response from each of the heads. During the 1950s many companies worked on variations of the Ampex scheme which utilised the now virtually universal helical recording format; a scheme – illustrated in Figure 10.19 – in which the tape is wrapped around a

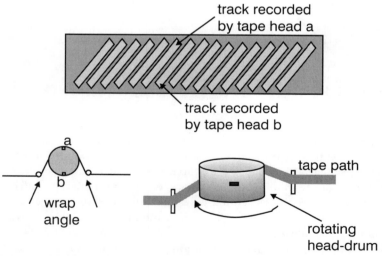

Figure 10.19   *Helical recording format*

rotating drum which contains just two-heads. One head writes (or reads) a complete field of video in a slanting path across the width of the head. By this means head switching can be made to happen invisibly, just before vertical-blanking interval (see Chapter 15). Virtually all video tape formats and contemporary digital audio recording formats employ a variation of this technique.

Spinning-head technology is eminently suitable for digital audio applications; not the least because of the cost-effective adoption of mature video technology. So suitable in fact that early PCM audio recorders were nothing more than slightly modified video cassette recorders and an associated video encoder–decoder (see Figure 10.20).

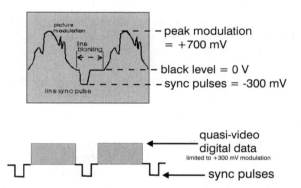

Figure 10.20 *Coding of digital audio as an analogue video signal*

Two famous examples are the SONY 1610/1630 and the PCM-F1. The first was (and still is!) the standard mastering format for CDs and is essentially a CODEC for assembling the PCM audio signal into a quasi-video format suitable for recording on a slightly modified semi-professional U-Matic tape machine. The second, the PCM-F1, was a 'consumer' unit but, like so much other SONY equipment failed in its original marketplace only to become much loved by professionals, which was a CODEC suitable for marrying with a consumer Betamax tape recorder.

Despite the ubiquity of video equipment, its adoption as the basis for audio recording brings with it a number of significant complications. Firstly, the tape is not mechanically editable as is standard with analogue tape formats (video tapes are never spliced together). Instead editing is achieved by writing the new signal (which must be provided by a second video recorder) onto the tape at precisely the right point, this process being governed with reference to extra, linear control-tracks which are recorded along the length of the tape in the region where the transverse heads do not sweep the tape – for fear of snagging. These tracks are used to control the moment of switching and in order to keep the VCRs running at precisely the same speed (in synchronism). Any digital audio system which utilises video technology must employ a similarly complicated procedure for editing. Indeed SONY 1610/1630 editing involved nothing more than modified video editors. Secondly the rotating head arrangement in most video tape recorders is rather complicated and heavy and suffers from various gyroscopic effects if the tape recorder is to be used in a portable role. Nevertheless modified video technology remains the technological basis which underpins digital audio.

Digital audio stationary head (DASH) format

An alternative approach (and an approach adopted in the development of an early BBC prototype PCM audio recorder) utilised an analogue audio multi-track recorder in which, each of the individual digital words was written across the width of the tape; one bit of the complete word to each head. This approach dispenses with the need for complex coding and decoding circuits (although error protection is still necessary) and even dispenses with the requirement for ultra-sonic bias because it is only necessary to record the change of state of each bit's signal as Figure 10.21 illustrates. With careful design, error-correction circuitry can be constructed to account for the damage caused by even the most careful of razor blades thereby permitting the mechanical editing of digital tape. Regarded as very important in the early days of digital recording, nowadays – and especially with the advent of hard-disk editing – this requirement is no longer paramount and this type of digital tape recorder is less and less common.

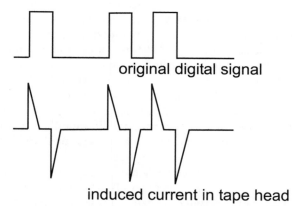

original digital signal

induced current in tape head

Figure 10.21 *Digital signals recorded on magnetic tape*

Data compression techniques

Data reduction or compression techniques are important because universal laws put a premium on information. You couldn't read all the books in the world, neither could you store them. You might make a start on reading every book by making it a team effort. In other words, you might tackle the problem with more than one brain and one pair of eyes. In communication theory terms this approach is known as increasing the channel capacity by broadening the bandwidth. But you wouldn't have an infinite number of people at your disposal unless you had an infinite amount of money to pay them! Likewise no one has an infinite channel capacity or an infinite bandwidth at their disposal. The similar argument applies to storage. Stated axiomatically: Information, in all its forms, is using up valuable resources, so the more efficiently we can send it and store it the better. That's where compression comes in.

If I say to you, 'Wow, I had a bad night, the baby cried from three 'til six!' You understand perfectly what I mean because you know what a baby crying sounds like. I might alternatively have said, 'Wow, I had a bad night, the baby did this; wah, bwah, bwah, wah . . .' and continue doing it for three hours. Try it. You'll find you lose a lot of friends because nobody needs to have it demonstrated. Most of the three hour impersonation is superfluous. The second message is said to have a high level of redundancy in the terms of communication theory. The trick performed by any compression system is sorting out the necessary information content – sometimes called the entropy – from the redundancy. (If, like me you find it difficult to comprehend the use of entropy in this context consider this: Entropy refers here to a lack of pattern; to disorder. Everything in a signal which has a pattern is, by definition, predictable and therefore redundant. Only those parts of the signal which possess no pattern are unpredictable and therefore represent necessary information.)

All compression techniques may be divided between lossless systems and lossy systems. Lossless compression makes use of efficiency gains in the manner in which the data is coded. All that is required to recover the original data exactly is a decoder which implements the reverse process performed by the coder. Such a system does not confuse entropy for redundancy and hence dispense with important information. However, neither does the lossless coder divide perfectly entropy from redundancy. A good deal of redundancy remains and a lossless system is therefore only capable of relatively small compression gains. Lossy compression techniques attempt a more complete distinction between entropy and redundancy by relying on a knowledge of the predictive powers of the human perceptual systems. This explains why these systems are referred to as implementing perceptual coding techniques. Unfortunately, not only are these systems inherently more complicated, they are also more likely to get things wrong and produce artefacts.

Lossless compression

Consider the following contiguous stream of luminance bytes taken from a bitmap graphic:

00101011
00101011
00101011
00101011
00101011
00101011
00101100
00101100
00101100
00101100
00101100

There must be a more efficient way of coding this! 'Six lots of 00101011 followed by five lots of 00101100' springs to mind. Like this:

00000110
00101011
00000101
00101100

This is the essence of a compression technique known as run-length encoding (RLE). RLE works really well but it has a problem. If a data file is comprised of data which is predominantly non-repetitive data, RLE actually makes the file bigger! So RLE must be made adaptive so that it is only applied to strings of similar data (where redundancy is high) and, when the coder detects continuously changing data (where entropy is high), it simply reverts back to sending the bytes in an uncompressed form. Evidently it also has to insert a small information overhead to instruct the decoder when it is (and isn't) applying the compression algorithm.

Another lossless compression technique is known as Huffman coding and is suitable for use with signals in which sample values appear with a known statistical frequency. The analogy with Morse code is frequently drawn, in which letters that appear frequently are allocated simple patterns and letters that appear rarely are allocated more complex patterns. A similar technique, known by the splendid name of the Lempel-Ziv-Welch (LZW) algorithm is based on the coding of repeated data chains or patterns. A bit like Huffman's coding, LZW sets up a table of common patterns and codes specific instances of patterns in terms of 'pointers' which refer to much longer sequences in the table. The algorithm doesn't use a pre-defined set of patterns but instead builds up a table of patterns which it 'sees' from the incoming data. LZW is a very effective technique – even better than RLE. But for the really high compression ratios, made necessary by the transmission and storage of high-quality audio down low bandwidth links, different approaches are required, based on an understanding of human perception processes.

In principle the engineering problem presented by low-data rates, and therefore in reduced digital resolution, is no different to the age-old analogue problems of reduced dynamic range. In analogue systems, noise reduction systems (either complementary – i.e. involving encoding and complementary decoding like Dolby B and dbx1, or single ended such as that described in detail in Chapter 6) have been used for many years to enhance the dynamic range of inherently noisy transmission systems like analogue tape. All of these analogue systems rely on a method called 'compansion' a word derived from the contraction of compression and expansion. The dynamic range being deliberately reduced (compressed) in the recording stage processing and recovered (expanded) in the playback electronics. In some systems this compansion acts over the whole frequency range (dbx is one such type). Others work over a selected frequency range (Dolby A, B, C and SR). We shall see that the principle of compansion applies in just the same way to digital systems of data reduction. Furthermore the distinction made between systems which act across the whole audio frequency spectrum and those which act selectively on ranges of frequencies (sub-bands) is true too of digital implementations. However, digital systems have carried the principle of sub-band working to a sophistication undreamed of in analogue implementations.

Intermediate compression systems

Consider the 8 bit digital values: 00001101, 00011010, 00110100, 0110100 and 11010000. (Eight bit examples are used because the process is easier to follow but the principles below apply in just the same way to digital audio samples of 16 bits or, indeed, any word length.) We might just as correctly write these values thus:

```
00001101  =  1101 * 1
00011010  =  1101 * 10
00110100  =  1101 * 100
01101000  =  1101 * 1000
11010000  =  1101 * 10000
```

If you think of the multipliers 1, 10, 100 and so on as powers of two then it's pretty easy to appreciate that the representation above is a logarithmic description (to the log of base two) with a four-bit mantissa and a three-bit exponent. So already we've saved one bit in eight (a 20% data reduction). We've paid a price of course because we've sacrificed accuracy in the larger values by truncating the mantissas to four bits. However this is possible in any case with audio because of the principle of masking which underlies the operation of all noise reduction systems (see Chapter 2). Put at its simplest, masking is the reason we strain to listen to a conversation on a busy street and why we cannot hear the clock ticking when the television set is on: Loud sounds mask quiet ones. So the logarithmic representation makes sense because resolution is maintained at low levels but sacrificed at high levels where the programme signal will mask the resulting, relatively small, quantisation errors.

NICAM

Further reductions may be made because real audio signals do not change instantaneously from very large to very small values so the exponent value may be sent less often than the mantissas. This is the principle behind the stereo television technique of NICAM which stands for near instantaneous companded audio multiplex. In NICAM 782, 14 bit samples are converted to 10 bit mantissas in blocks of 32 samples with a common three bit exponent. This is an excellent and straightforward technique but it is only possible to secure relatively small reductions in data throughput of around 30 per cent.

Psychoacoustic masking systems

Wideband compansion systems view the phenomenon of masking very simply; and rely simply on the fact that programme material will mask system noise. But actually masking is a more complex phenomena. Essentially it operates in frequency bands and is related to the way in which the human ear performs a mechanical Fourier analysis of the incoming acoustic signal. It turns out (see Chapter 2), a loud sound only masks a quieter one when the louder sound is lower in frequency than the quieter, and only then, when both signals are relatively close in frequency. It is due to this effect that all wideband compansion systems can only achieve relatively small gains. The more data we want to discard the more subtle must our data reduction algorithm be in its appreciation of the human masking phenomena. These compression systems are termed psychoacoustic systems and, as you will see, some systems are very subtle indeed.

MPEG layer 1 compression (PASC)

It's not stretching the truth too much to say that the failed Philips' Digital Compact Cassette (DCC) system was the first non-professional digital audio tape format. As we have seen, other digital audio developments had ridden on the back of video technology. The CD rose from the ashes of Philips Laserdisc and DAT machines use the spinning-head tape recording technique originally developed for B and C-Format one-inch video machines, later exploited in U-Matic and domestic videotape recorders. To their credit then that, in developing the Digital Compact Cassette, Philips chose not to follow so many other manufacturers down the route of modified video technology. Inside a DCC machine, there's no head-wrap, no spinning head and few moving precision parts. Until DCC, it had taken a medium suitable for recording the complex signal of a colour television picture to store the sheer amount of information needed for a high quality digital audio signal. Philips' remarkable technological breakthrough in squeezing two high quality, stereo digital audio channels into a final data rate of 384 kBaud was accomplished by, quite simply, dispensing with the majority (75%) of the digital audio data! Philips named their technique of bit-rate reduction or data-rate compression, Precision Adaptive Sub-band Coding (PASC). PASC was adopted as the original audio compression scheme for MPEG video/audio coding (layer 1).

In MPEG layer 1 or PASC audio coding, the whole audio band is divided up into 32 frequency sub-bands by means of a digital wave filter. At first sight, it might seem that this process will increase the amount of data to be handled tremendously – or by 32 times anyway! This, in fact, is not the case because the output of the filter bank, for any one frequency band, is at 1/32nd of the original sampling rate. If this sounds counter intuitive, take another look at the Fourier transform in Chapter 2, where the details of the discrete Fourier transform are given, and note that a very similar process is being performed here. Observe that when a periodic waveform is sampled n times and transformed the result is composed of n frequency components. Imagine computing the transform over a 32 sample period: Thirty-two separate calculations will yield 32 values. In other words the data rate is the same in the frequency domain as it is in the time domain. Actually, considering both describe exactly the same thing with exactly the same degree of accuracy, this shouldn't be surprising. Once split into sub-bands, sample values are expressed in terms of a mantissa and exponent exactly as explained above. Audio is then grouped into discrete time periods and the maximum magnitude in each block is used to establish the masking 'profile' at any one moment and thus predict the mantissa accuracy to which the samples in that sub-band can be reduced, without their quantisation errors becoming perceivable (see Figure 10.22).

Despite the commercial failure of DCC, the techniques employed in PASC are indicative of techniques now widely used in the digital audio industry. All bit-rate reduction coders have the same basic architecture, pioneered in PASC: however details differ. All systems accept PCM dual-

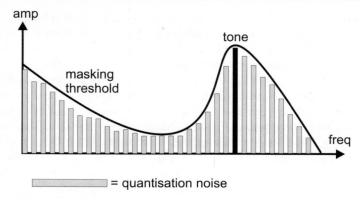

Figure 10.22　*Sub-band quantisation and how it relates to masking profile*

channel, digital audio (in the form of one or more AES pairs) is windowed over small time periods and transformed into the frequency domain by means of sub-band filters or via a transform filter bank. Masking effects are then computed based on a psychoacoustic model of the ear. Note that blocks of sample values are used in the calculation of masking. Because of the temporal – as well as frequency dependent – effects of masking, it's not necessary to compute masking on a sample by sample basis. However, the time-period over which the transform is performed and the masking effects computed are often made variable so that quasi-steady-state signals are treated rather differently to transients. If coders do not include this modification, masking can be incorrectly predicted resulting in a rush of quantisation noise just prior to a transient sound. Subjectively this sounds like a type of pre-echo. Once the effects of masking are known, the bit allocation routine apportions the available bit-rate so that quantisation noise is acceptably low in each frequency region. Finally, ancillary data is sometimes added and the bitstream is formatted and encoded.

Intensity stereo coding

Because of the ear's insensitivity to phase response above about 2 kHz (see Chapter 11), further coding gains can be achieved by sending by coding the derived signals (L + R) and (L – R) rather than the original left and right channel signals. Once these signals have been transformed into the frequency domain, only spectral amplitude data is coded in the HF region, the phase component is simply ignored.

The discrete cosine transform

The encoded data's similarity to a Fourier transform representation has already been noted. Indeed, in a process developed for a very similar

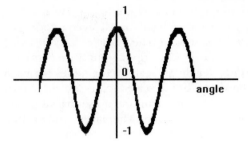

Figure 10.23 *Cosine function*

application, Sony's compression scheme for MiniDisc, actually uses a frequency domain representation utilising a variation of the DFT method known as the Discrete Cosine Transform. The DCT takes advantage of a distinguishing feature of the cosine function which is illustrated in Figure 10.23; that the cosine curve is symmetrical about the time origin. In fact, it's true to say that any waveform which is symmetrical about an arbitrary 'origin', is made up of solely cosine functions. Difficult to believe, but consider adding other cosine functions to the curve illustrated in Figure 10.23. It doesn't matter what size or what period waves you add, the curve will always be symmetrical about the origin. Now, it would obviously be a great help, when we come to perform a Fourier transform, if we knew the function to be transformed was only made up of cosines because that would cut down the maths by half (see Chapter 2). This is exactly what is done in the DCT. A sequence of samples from the incoming waveform are stored and reflected about an origin. Then one half of the Fourier transform performed. When the waveform is inverse transformed, the front half of the waveform is simple ignored, revealing the original structure.

MPEG layer 2 audio coding (MUSICAM)

The MPEG layer 2 algorithm is the preferred algorithm for European DTV and includes a number of simple enhancements of layer 1 (or PASC). Layer 2 was originally adopted as the transmission coding standard for the European digital radio project (Digital Audio Broadcasting or DAB) where it was termed MUSICAM. The full range of bit rates for each layer is supported, as are all three sampling frequencies, 32, 44.1 and 48 kHz. Note that MPEG decoders are always backward compatible; i.e. a layer 2 decoder can decode layer 1 or layer 2 bitstreams, however a layer 2 decoder cannot decode a layer 3 encoded stream.

MPEG layer 2 coding improves compression performance by coding data in larger groups. The layer 2 encoder forms frames of 3 by 12 by 32

= 1152 samples per audio channel. Whereas layer 1 codes data in single groups of 12 samples for each sub-band, layer 2 codes data in 3 groups of 12 samples for each sub-band. The encoder encodes with a unique scale factor for each group of 12 samples only if necessary to avoid audible distortion. The encoder shares scale factor values between two or all three groups when the values of the scale factors are sufficiently close or when the encoder anticipates that temporal noise masking (see Chapter 2) will hide the consequent distortion. The layer 2 algorithm also improves performance over layer 1 by representing the bit allocation, the scale factor values, and the quantised samples with a more efficient code. Layer 2 coding also added 5.1 multi-channel capability. This was done in a scaleable way, so as to be compatible with layer 1 audio.

MPEG layers 1 and 2 contain a number of engineering compromises. The most severe concerns the 32 constant-width sub-bands which do not reflect accurately the equivalent filters in the human hearing system (the critical bands). Specifically, the bandwidth is too wide for the lower frequencies so the number of quantiser bits cannot be specifically tuned for the noise sensitivity within each critical band. Furthermore, the filters have insufficient Q so that signal at a single frequency can affect two adjacent filter bank outputs. Another limitation concerns the time-frequency-time domain transformations achieved with the wave filter. These are not transparent so, even without quantisation, the inverse transformation would not perfectly recover the original input signal.

MPEG layer 3

The layer 3 algorithm is a much more refined approach. Layer 3 is finding its application on the Internet where the ability to compress audio files by a large factor is important in download times. In layer 3, time to frequency mapping is performed by a hybrid filter bank composed of the 512-tap polyphase quadrature mirror filter (used in layers 1 and 2) followed by an 18-point modified cosine transform filter. This produces a signal in 576 bands (or 192 bands during a transient). Masking is computed using a 1024-point FFT: once again more refined than the 512 point FFT used in layers 1 and 2. This extra complexity accounts for the increased coding gains achieved with layer 3, but increases the time delay of the coding process considerably. Of course, this is of no account at all when the result is an encoded .mp3 file.

Dolby AC-3

The analogy between data compression systems and noise reduction has already been drawn. It should therefore come as no surprise that one of the leading players in audio data compression should be Dolby, with that company's unrivalled track-record in noise reduction systems for analogue magnetic tape. Dolby AC-3 is the adopted coding standard for terrestrial

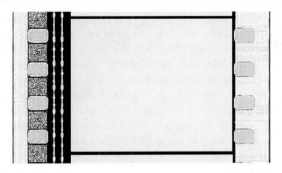

Figure 10.24 *Dolby Digital as originally coded on film stock*

digital television in the US, however, it was actually implemented for the cinema first, where it was called Dolby Digital. It was developed to provide multi-channel digital sound with 35 mm prints. In order to retain an analogue track so that these prints could play in any cinema, it was decided to place the new digital optical track between the sprocket holes, as illustrated in Figure 10.24. The physical space limitation (rather than crude bandwidth) was thereby a key factor in defining its maximum practical bit-rate. Dolby Labs did a great deal of work to find a channel format that would best satisfy the requirements of theatrical film presentation. They discovered that 5 discrete channels; left (L), right (R), centre (C), left surround (LS), right surround (RS) set the right balance between realism and profligacy! To this they added a limited (1/10th) bandwidth sub-woofer channel; the resulting system being termed 5.1 channels. Dolby Digital provided Dolby Labs a unique springboard for consumer formats for the new DTV (ATSC) systems.

Like MPEG, AC-3 divides the audio spectrum of each channel into narrow frequency bands of different sizes optimised with respect to the frequency selectivity of human hearing. This makes it possible to sharply filter coding noise so that it is forced to stay very close in frequency to the frequency components of the audio signal being coded. By reducing or eliminating coding noise wherever there are no audio signals to mask it, the sound quality of the original signal can be subjectively preserved. In this key respect, a perceptual coding system like AC-3 is essentially a form of very selective and powerful Dolby-A type noise reduction! Typical final data-rate applications include 384 kb/s for 5.1-channel Dolby Surround Digital consumer formats, and 192 kb/s for two-channel audio distribution.

Dolby E

Dolby E is a digital audio compression technology designed for use by TV broadcast and production professionals which allows an AES/EBU audio pair to carry up to eight channels of digital audio. Because the coded audio frame is arranged to be synchronous with the video frame, encoded audio can be edited without mutes or clicks. Dolby E can be recorded on a studio-level digital VTR and switched or edited just like any other AES digital audio signal, as long as some basic precautions are observed. The data must not be altered by any part of the system it passes through. That's to say, the gain must not be changed, the data must not be truncated or dithered, neither must the sample rate be converted. Dolby E technology is designed to work with most popular international video standards. In its first implementation Dolby E supported 29.97 fps, 20-bit word size, and 48 kHz audio. Newer versions will support 25 fps, 24 fps, and 16-bit and 24-bit audio.

DTS

DTS, the rival to Dolby Digital in the cinema, uses an entirely different approach to AC-3 coded data on the film-stock. In DTS, the digital sound (up to 10 channels and a sub-woofer channel) are recorded on CDs, which are synchronised to the film by means of a timecode. Because of the higher data rate available (CD as against optical film-track), DTS uses a relatively low 4:1 compression ratio.

MPEG AAC

MPEG-2 advanced audio coding (AAC) was finalised as a standard in 1997 (ISO/IEC 13818–7). AAC constitutes the coding algorithms of the new MPEG-4 standard.

MPEG-4

MPEG-4 will define a method of describing objects (both visual and audible) and how they are 'composited' and interact together to form 'scenes'. The scene description part of the MPEG-4 standard describes a format for transmitting the spatio-temporal positioning information that describes how individual audio-visual objects are composed within a scene. A 'real world' audio object is defined as an audible semantic entity recorded with one microphone – in case of a mono recording – or with more microphones, at different positions, in case of a multi-channel recording. Audio objects can be grouped or mixed together, but objects can not easily be split into sub-objects. Applications for MPEG-4 audio might include 'mix minus 1' applications in which an orchestra is recorded minus the concerto instrument; allowing a musician to play along with her instrument at home. Or where all effects and music tracks in a feature film are 'mix minus the dialogue' allowing very flexible multilingual applications because each language is a separate audio object and can be selected as required in the decoder.

In principle, none of these applications is anything but straightforward; they could be handled by existing digital (or analogue) systems. The problem, once again, is bandwidth. MPEG-4 is designed for very low bitrates and this should suggest that MPEG have designed (or integrated) a number of very powerful audio tools to reduce necessary data throughput. These tools include the MPEG-4 Structured Audio format which uses low bitrate algorithmic sound models to code sounds. Furthermore MPEG-4 includes the functionality to use and control post production panning and reverberation effects at the decoder as well as the use of a SAOL signal-processing language enabling music synthesis and sound-effects to be generated, once again, at the terminal; rather than prior to transmission.

Structured audio

We have already seen how MPEG (and Dolby) coding aims to remove perceptual redundancy from an audio signal; as well as removing other simpler representational redundancy by means of efficient bit-coding schemes. Structured Audio (SA) compression schemes compress sound by, first, exploiting another type of redundancy in signals – structural redundancy.

Structural redundancy is a natural result of the way sound is created in human situations. The same sounds, or sounds which are very similar, occur over and over again. For example, a performance of a work for solo piano consists of many piano notes. Each time the performer strikes the 'middle C' key on the piano, very similar sound is created by the piano's mechanism. To a first approximation, we could view the sound as exactly the same upon each strike; to a closer one, we could view it as the same except for the velocity with which the key is struck and so on. In a PCM representation of the piano performance, each note is treated as a completely independent entity; each time the 'middle C' is struck, the sound of that note is independently represented in the data sequence. This is even true in a perceptual coding of the sound. The representation has been compressed, but the structural redundancy present in re-representing the same note as different events has not been removed.

In structured coding, we assume that each occurrence of a particular note is the same, except for a difference which is described by an algorithm with a few parameters. In the model-transmission stage we transmit the basic sound (either a sound sample or another algorithm) and the algorithm which describes the differences. Then, for sound transmission, we need only code the note desired, the time of occurrence, and the parameters controlling the differentiating algorithm.

SAOL

SAOL (pronounced 'sail') stands for 'Structured Audio Orchestra Language' and it falls into the music-synthesis category of 'Music V' languages. Its fundamental processing model is based on the interaction of oscillators running at various rates. Note that this approach is different from the idea (used in the multimedia world) of using MIDI information to drive synthesis chips on soundcards. This latter approach has the disadvantage that, depending on IC technology, music will sound different depending on which soundcard it is realised. Using SAOL (a much 'lower-level' language than MIDI) realisations will always sound the same.

At the beginning of an MPEG-4 session involving Structured Audio (SA), the server transmits to the client a stream information header, which contains a number of data elements. The most important of these is the orchestra chunk, which contains a tokenised representation of a program written in Structured Audio Orchestra Language. The orchestra chunk consists of the description of a number of instruments. Each instrument is a single parametric signal-processing element that maps a set of parametric controls to a sound. For example, a SAOL instrument might describe a physical model of a plucked string. The model is transmitted through code which implements it, using the repertoire of delay lines, digital filters, fractional-delay interpolators, and so forth that are the basic building blocks of SAOL.

The bitstream data itself, which follows the header, is made up mainly of time-stamped parametric events. Each event refers to an instrument described in the orchestra chunk in the header and provides the parameters required for that instrument. Other sorts of data may also be conveyed in the bitstream; tempo and pitch changes for example.

Unfortunately, as at the time of writing (and probably for some time beyond!) the techniques required for automatically producing a Structured Audio bitstream from an arbitrary, pre-recorded sound are beyond today's state of the art, although they are an active research topic. These techniques are often called 'automatic source separation' or 'automatic transcription'. In the mean time, composers and sound designers will use special content creation tools to directly create Structured Audio bitstreams. This is not considered to be a fundamental obstacle to the use of MPEG-4 Structured Audio, because these tools are very similar to the ones that contemporary composers and editors use already; all that is required is to make their tools capable of producing MPEG-4 output bitstreams. There is an interesting corollary here with MPEG-4 for video. For, whilst we are not yet capable of integrating and coding real-world images and sounds, there are immediate applications for directly synthesised programmes. MPEG-4 audio also foresees the use of text-to-speech (TTS) conversion systems; these are covered in greater detail in Brice (1999).

Audio scenes

Just as video scenes are made from visual objects, audio scenes may be usefully described as the spatio-temporal combination of audio objects. An 'audio object' is a single audio stream coded using one of the MPEG-4 coding tools, like Structured Audio. Audio objects are related to each other by mixing, effects processing, switching, and delaying them, and may be panned to a particular 3-D location. The effects processing is described abstractly in terms of a signal-processing language – the same language used for Structured Audio.

Digital audio production

We've already looked at the technical advantages of digital signal processing and recording over its older analogue counterpart. We now come to consider the operational impact of this technology, where it has brought with it a raft of new tools and some new problems.

Digital audio workstations (DAWs)

When applied to a digital audio application, a computer hardware platform is termed a Digital Audio Workstation or DAW. The two ubiquitous standards in the audio arena are the Apple Macintosh computer family (or 'Macs') which use Motorola processors and the IBM PC and compatibles (PCs), which use Intel-based processors. The Macintosh was always 'audio ready' because it was designed with an audio capacity beyond the PC's dumb 'beep'. Other small personal computers (especially Atari ST) became very popular in music applications. The Atari ST computer (Figure 10.25) was another 68000 based

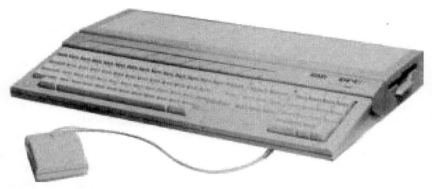

Figure 10.25 *Atari ST computer*

computer (like the Apple Mac). Including a powerful ROM based operating system and a desktop metaphor very like that now common-place in Windows; the Atari was pretty much ahead of its time in the early 1980s. However, the Atari ST owes its tremendous success, and continuing long-overdue existence in many recording studios to the decision of the designers to include MIDI IN and OUT sockets on the side of the machine. This made the Atari the only ready-to-go, 'plug and play' MIDI sequencer platform. A fact reflected in the number of software products designed for it.

PowerPCs and PowerMacs are machines built around a reduced instruction set computing (RISC) processor developed jointly by IBM, Apple and Motorola. They are designed to run operating system software which supports both PC and Mac applications and are designed to be especially good at handling large data files typical of media components. RISC technology is especially noted in workstation computers. Three computers designed and manufactured by the American high-end computer company Silicon Graphics Inc. (SGI) make extensive use of RISC technology. SGI's subsidiary, MIPS Technologies, Inc., design the RISC processor technology inside SGI machines. MIPS' new R5000 MIPS RISC processor delivers a peak of 480 million floating point operations per second (MFLOPS) – up to twice as fast as Intel's 200 MHz Pentium Pro and over seven times as fast as a 133 MHz Pentium! Workstations from Silicon Graphics Inc. are also finding their way into high-end audio production.

SGI is the leading manufacturer of high-performance visual comput-ing systems. The company delivers interactive three-dimensional graph-ics, digital media and multiprocessing super-computing technologies to technical, scientific and creative professionals. Silicon Graphics Inc. manufacture some of the best tools for multimedia creation as well as white-heat video, audio and graphics standalone packages. They also provide specialist tools for HTML, hypermedia page creation and serving for the creation of multimedia creations on the Internet/World Wide Web (WWW). Silicon Graphics has its headquarters in Mountain View, California. SGI's products include the Indy, which is a 'value' RISC workstation utilising 64 bit system architecture and MIPS pro-cessors. On the audio side it has digital audio I/O as well as analogue ports. The Indigo 2, which is aimed as a cost-effective desktop alternative to older style dedicated video production hardware. The Onyx which is a super-computer with a graphics bias! SGI also manufacture the CHALLENGE Media Server for the broadcast television environment. Table 10.3 is the Audio Subsystem specification for Onyx and CHALLENGE and it represents the typical digital audio performance from desktop audio. The option also provides for microphone input and headphone output but these figures are not quoted here.

Table 10.3

Number of channels	4 analogue (16 bit), 2 digital (24 bit)
Input analogue route	
Input Z	20 k ohms
Input amplitude	from 0.63 V pp to 8.4 V pp for full scale modulation (this level is adjustable under software control)
Frequency response	±0.81 dB 20 Hz to 20 kHz
THD + Noise	less than 0.007% 20 Hz to 20 kHz
Residual noise	−86 dB unweighted
Crosstalk	−82 dB at 1 kHz, −67 dB at 20 kHz
ADC type	16 bit Delta-Sigma
Output analogue route	
Output Z	600 ohms
Output level	4.7 V pp (4.4 dBV) for 0 dBFS
Sampling rates	32 kHz, 44.1 kHz, 48 kHz or divisors selectable
Frequency response	±1.2 dB 20 Hz to 20 kHz
THD + Noise	less than 0.02% 20 Hz to 20 kHz
Residual noise	−81 dB unweighted
Crosstalk	−80 dB at 1 kHz, −71 dB at 20 kHz
Digital Serial I/O	
Type	coaxial only
Input Z	75 ohms, transformer coupled
Input level	0.5 V pp into 75 ohms
Sample rates	30 kHz to 50 kHz
Output Z	75 ohms, transformer coupled
Output level	0.5 V when terminated in 75 ohms
Coding	AES-3, IEC 958.

Note the input ports for audio (both analogue and digital) conform to consumer levels and practice even though SGI themselves refer to the digital inputs as AES/EBU.

Hard-disk editing

Not long ago, most recordings were mastered on two-track analogue tape. Whether the performance was by a soloist, small classical or rock ensemble or full orchestra, the good 'takes' of a recording session were separated from the bad and joined together using razor-blade editing. Using this technique the tape was physically cut with a razor-blade and joined using a special sticky tape. With the high tape speeds employed in professional recording, accurate editing was possible using this technique and many fine recordings were assembled this way. Any engineer who has been involved with razor-blade editing will know that it is a satisfying skill to acquire but it is tricky and it is always fraught. The reason being that a mis-timed or misjudged edit is difficult to put right once the first 'incision' has been made! So much so that a dub or copy of the original master tapes was sometimes made for editing lest the original master should be irreparably damaged by a poor edit. Unfortunately because analogue recording is never perfect this meant that editing inevitably meant one tape-generation of quality loss before the master tape had left the studio for production. The advent of digital audio has brought about a new vista of possibility in sound editing. Apart from the obvious advantages of digital storage, that once the audio signal is in robust digital form it can be copied an infinite number of times thus providing an identical tape 'clone' for editing purposes, the arrival of the ability to process digital audio on desktop PCs has revolutionised the way we think about audio editing providing a flexibility undreamed of in the days of analogue mastering machines.

Editing digital audio on a desktop microcomputer has two major advantages:

(1) an edit may be made with sample accuracy, by which is meant, a cut may be made with a precision of 1/40 000th of a second!

(2) An edit may be made non-destructively, meaning that when the computer is instructed to join two separate takes together, it doesn't create a new file with a join at the specified point, but instead records two pointers which instruct on subsequent playback to vector or jump to another data location and play from the new file at that point.

In other words, it 'lists' the edits in a new file of stored vector instructions. Indeed this file is known as an edit decision list. (Remember that the hard-disk doesn't have to jump instantaneously to another location because the computer holds a few seconds of audio data in a RAM cache memory.) This opens the possibility of almost limitless editing in order to assemble a 'perfect' performance. Edits may be rehearsed and auditioned many times without ever 'molesting' the original sound files.

Low-cost audio editing

Most audio capture cards like the Creative Labs Soundblaster come bundled with a primitive sound-file editing software. Usually this permits a 'butt-join' edit (the exact analogy of a razor blade splice) between different sound files or between data loaded onto the clipboard in rather the same way as a word processor works on text files. In fact Creative Labs' Wave Studio utility is quite powerful and affords some manipulations (such as the ability to operate on left and right channels of a stereo signal separately) which exceed the capabilities of some low-end professional editing software. However the big disadvantage with Wave Studio is that it does not allow for non-destructive editing. An inexpensive and truly excellent package is authored by Minnetonka Software Inc. and is known as FastEddie. This 'value' package permits non-destructive editing to sample accuracy, pre-audition of edit points, the ability to mix files, time 'stretch' or compress the WAV file and normalise gain – so that a WAV file may be increased in gain just enough so that the largest excursion on the file is brought almost to within clipping level – thus maximising dynamic range. The utility can also be used to generate audio effects such as echo and reverb and reversal of the sound file, so that it plays backwards. In order to facilitate the judgement of edit points, most editing software provides an on-screen waveform display. This may be zoomed at will; IN to examine individual samples, OUT to reveal musical sections or the whole file.

An example is shown in Figure 10.26, which is an off-screen capture of a FastEddie window. The highlighted sections in the lower, read-only part of

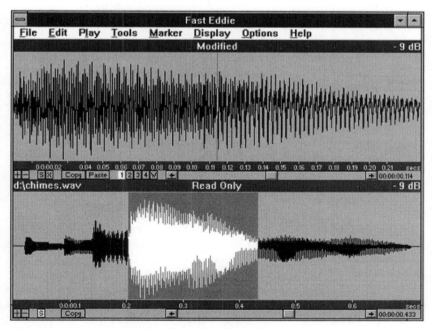

Figure 10.26 *FastEddie digital audio editor*

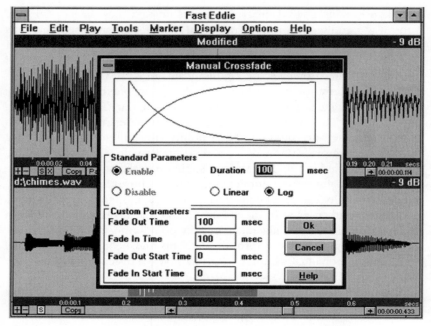

Figure 10.27   *Audio editing with cross-fades for edit points*

the window, are ready for editing in the top editing part of the window. Given its price, FastEddie has a number of very professional features, amongst these is the ability to cross-fade at edit points rather than to produce a butt-join (Figure 10.27). The advantage of this feature is due to the complex nature of musical sounds. Even if an edit is made at a precise point in a musical score a discontinuity is produced in the audio waveform. This discontinuity usually manifests itself as an audible click or 'bump'. The use of cross-fading ameliorates these effects. FastEddie also provides facilities for producing fade-outs and fade-ins on complete sound-file data.

Professional audio editing

For professional music editing, there exist still most sophisticated editing systems, ranging from the reasonable to the very expensive. At the higher end, the platforms are predominantly Mac based. The crème-de-la-crème in this field being the Sonic Solutions – the choice of music editing system for most classical music producers. The high end systems mostly show up with their own proprietary hardware and it is here, as much as in software, that these systems score over desktop PC systems using a 16 bit sound card. The professional units typically offer better quality A to D and D to A conversion and more transparent signal processing.

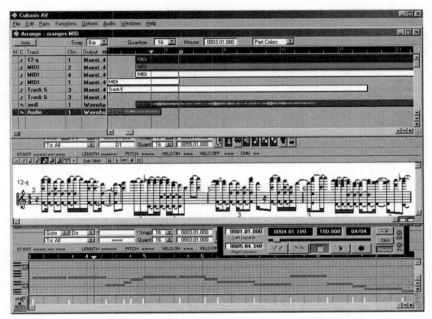

Figure 10.28 *Multi-track audio recording combined with MIDI sequencing*

Muti-track hard-disk recording

Hard-disk capacity and computer speeds are not so high that multi-track recording on a DAW is commonplace. A typical home-studio application (Cubase AV) is illustrated in Figure 10.28, illustrating the combination of multi-track audio and MIDI data, all within one screen environment.

Plug-ins

As gradually more and more audio production is taking place on computer platforms; or on the desktop, hardware manufacturers are now producing software programs which imitate their outboard hardware in a DSP program. This software interfaces with mixing and editing programs (via mutually agreed data-exchange standards) so as to provide the metaphor of 'outboard' facilities on the desktop. These programs are known as plug-ins.

Audio data files

Digital representations of sound, when stored on computer, are stored just like any other kind of data; as files. There exist a number of different file formats in common usage. Most sound files begin with a header consisting of information describing the format of that file. Characteristics such as word length, number of channels, and sampling frequency are specified so that audio applications can properly read the file. One very common type of file format is the WAV (or Wave) format. This is a good example because it demonstrates all the typical features of a typical audio file.

WAV files

WAV files are a version of the generic RIFF file format. This was co-developed by Microsoft and IBM. RIFF represents information in pre-defined blocks, preceded by a header that identifies exactly what the data is. This format is very similar to the AIFF format developed by Apple (see below) in that it supports monaural and multi-channel samples and a variety of sample rates. Like AIFF, WAVE files are big and require approximately 10 Mbytes per minute of 16-bit stereo samples with a sampling rate of 44.1 kHz. Here is a hexadecimal representation of the first 128 bytes of a WAV file.

```
26B7:0100  52 49 46 46 28 3E 00 00-57 41 56 45 66 6D 74 20  RIFF(>..WAVEfmt
26B7:0110  10 00 00 00 01 00 01 00-22 56 00 00 22 56 00 00  ....."V.."V..
26B7:0120  01 00 08 00 64 61 74 61-04 3E 00 00 80 80 80 80  ...data.>.........
26B7:0130  80 80 80 80 80 80 80 80-80 80 80 80 80 80 80 80  ............
26B7:0140  80 80 80 80 80 80 80 80-80 80 80 80 80 80 80 80  ............
26B7:0150  80 80 80 80 80 80 80 80-80 80 80 80 80 80 80 80  ............
26B7:0160  80 80 80 80 80 80 80 80-80 80 80 80 80 80 80 80  ............
26B7:0170  80 80 80 80 80 80 80 80-80 80 80 80 80 80 80 80  ............
```

The header provides Windows with all the information it needs. First off, it defines the type of RIFF file, in this case, WAVEfmt. Notice the bytes which are shown underlined. The first two; 22 and 56 relate to the audio sampling frequency. Their order needs reversing to read; 5622 hexadecimal, which is equivalent to 22 050 in decimal – in other words, 22 kHz sampling. The next two inform the Media Player the sound file is 1 byte per sample (mono) and 8 bits per sample.

AU files

AU (or μ-law – pronounced mu-law), files utilise an international standard for compressing audio data. It has a compression ratio of 2:1. The compression technique is optimised for speech (in the United States it is a standard compression technique for telephone systems; in Europe, a-law is used). This file format is most frequently found on the Internet where it is used for '.au' file formats, alternately know as 'Sun audio' or 'NeXT' format. Even though it's not the highest quality audio file format available, it's non-linear, logarithmic coding scheme results in a relatively small file size; ideal for applications where download time is a problem.

AIFF & AIFC

The Audio Interchange File Format (AIFF) allows for the storage of monaural and multi-channel sample sounds at a variety of sample rates. AIFF format is frequently found in high-end audio recording applications. Originally developed by Apple, this format is used predominantly by Silicon Graphics and Macintosh applications. Like WAV, AIFF files can be quite large; one minute of 16-bit stereo audio sampled at 44.1 kHz usually takes up about 10 megabytes. To allow for compressed audio data, Apple introduced the new AIFF-C, or AIFC, format, which allows for the storage of compressed and uncompressed audio data. AIFC supports compression ratios as high as 6:1. Most of the applications that support AIFF playback also support AIFC.

MPEG

The International Standard Organisation's Moving Picture Expert Group is responsible for one of the most popular compression standards in use on the Internet today. Designed for both audio and video file compression, MPEG-I audio compression specifies three layers, and each layer specifies its own format. The more complex layers take longer to encode but produce higher compression ratios while keeping much of an audio file's original fidelity. Layer 1 takes the least amount of time to compress, but layer 3 yields higher compression ratios for comparable quality files.

VOC

Creative Voice (.voc) is the proprietary sound file format that is recorded with Creative Lab's Sound Blaster and Sound Blaster Pro audio cards. This format supports only 8-bit mono audio files up to sampling rates of 44.1 kHz, and stereo files up to 22 kHz.

Raw PCM data

Raw Pulse Code Modulated data is sometimes identified with the .pcm, but it sometimes has no extension at all. Since no header information is provided in the file, you must specify the waveform's sample rate, resolution and number of channels to the application to which it is loaded.

Sound cards

Available in a bewildering array of different guises, for serious audio work, 16 bit cards only are suitable; and even then beware of very poor noise

Figure 10.29 *Creative Labs' Creative Mixer utility*

levels. Computers differ widely in their suitability as devices for high-quality audio. The Creative Technology Ltd. Sound Blaster card family are some of the most widespread sound cards used in the PC world. Supplied standard with a four operator FM sound generator chip for sound synthesis, Creative Labs offer a wavetable based upgrade. Sound Blaster ships with sound file editing software. The card comes with a utility program for controlling the analogue mixer on the card where the various sound sources are combined and routed. This utility is called Creative Mixer and it's illustrated in Figure 10.29. Notice that fader style controls are implemented in software so as to provide a familiar user interface. Control over CD, MIDI synthesiser, WAV file replay as well as line and microphone inputs are provided. Global (all sources) equalisation is also provided.

PCI bus versus ISA bus

Most PCs, until the arrival of the Pentium, were provided with a PC/AT bus (or ISA bus) for connecting peripherals (such as sound cards, frame grabbers and so on). The ISA bus operates with 16 bit data bus and a 16 bit address bus and operates with a divided clock. The ISA bus limits real-world transfer rates to around 1–2 Mbytes/s which is just enough for high-quality, dual-channel audio. The Peripheral Component Interconnect (PCI) bus is incorporated in newer Pentium-based IBM PCs. PCI is a local bus, so named because it is a bus which is much 'closer' to the CPU. Local buses run at a much higher rate and PCI offers considerable performance advantages over the traditional ISA bus allowing data to be transferred at between 5 and 70 Mbytes/s; allowing the possibility of real-time, multi-track audio applications. PCI bus is a processor-independent bus specification which allows peripheral boards to access system memory directly (under the aegis of a local bus controller) without directly using the CPU, employing a 32 bit data bus and a 64 bit address bus at full clock speed. Installation and configuration of PCI bus plug-in cards is much simpler than the equivalent installation on the ISA Bus. Commonly referred to as the 'plug-and-play' feature of the PCI Bus, this user-transparency is achieved by having the PCs BIOS configure the plug-in card's base address and interrupt level at power-up. Because all cards are automatically configured, conflicts between them are eliminated. A process which can only be done manually with cards on the ISA Bus. The PCI Bus is not limited to PCs; it is the primary peripheral bus in the PowerPC and PowerMacs from Apple. Incorporation of the PCI Bus is planned for other RISC-based processor platforms.

Disks and other peripheral hardware

Read/write compact disk (CD-R) drives are now available at a price within the reach of the small recording studio, and the recordable media less than $2. CD-R drives are usually SCSI based so PCs usually have to have an extra expansion card fitted to provide this interface (see below). Recordable CDs rely on a laser-based magneto-optical system to 'burn' data into the recorded medium. Once written the data cannot be erased. Software exists (and usually comes bundled with the drive) which enables the drive to be used as a data medium or an audio carrier (or sometimes as both). There exist a number of different variations of the standard ISO-9600 CD-ROM. The two most important are the (HFS/ISO) Hybrid disk which provides support for CD-ROM on Mac and PC using separate partitions and the Mixed mode disk which allows one track of either HFS (Mac) or ISO-9600 information and subsequent tracks of audio.

A number of alternative removable media are available and suitable for audio use; some based on magnetic storage (like a floppy disk or a Winchester hard-drive) and some on magneto-optical techniques – nearer to CD technology: Bernoulli cartridges are based on floppy disk, magnetic storage technology. Disks up to 150 MByte are available. Access times are fast; around 20 milli seconds. SyQuest are similar. Modern SyQuest cartridges and drives are now available in up to 1.3 GByte capacity and 11 millisecond access times, making SyQuest the nearest thing to a portable hard-drive. Magneto-optical drives use similar technology to CD, they are written and read using a laser (Sony is a major manufacturer of optical drives). Sizes up to 1.3 GBytes are available with access times between 20 and 30 milliseconds.

Hard drive interface standards

There are several interface standards for passing data between a hard disk and a computer. The most common are: the SCSI or Small Computer System Interface, the standard interface for Apple Macs; the IDE or Integrated Drive Interface, which is not as fast as SCSI; and the Enhanced IDE interface which is a new version of the IDE interface that supports data transfer rates comparable to SCSI.

IDE drives

The Integrated Drive Electronics interface was designed for mass storage devices, in which the controller is integrated into the disk or CD-ROM drive. It is thereby a lower cost alternative to SCSI interfaces in which the interface handling is separate from the drive electronics. The original IDE interface supports data transfer rates of about 3.3 Mbytes per second and has a limit of 538 Mbytes per device. However, a recent version of IDE, called enhanced IDE (EIDE) or Fast IDE, supports data transfer rates of about 12 Mbytes per second and storage devices of up to 8.4 Gbytes. These numbers are comparable to what SCSI offers. But, because the interface handling is handled by the disk-drive, IDE is a very simple interface and does not exist as an inter-equipment standard; i.e. you cannot connect an external drive using IDE. Due to demands for easily up-gradable storage capacity, and for connection with external devices such as recordable CD players, SCSI has become the preferred bus standard in audio applications.

SCSI

An abbreviation of Small Computer System Interface and pronounced 'scuzzy', SCSI is a parallel interface standard used by Apple Macintosh computers (and some PCs) for attaching peripheral devices to computers. All Apple Macintosh computers starting with the Macintosh Plus come with a SCSI port for attaching devices such as disk drives and printers. SCSI interfaces provide for fast data transmission rates; up to 40 Mbytes per second. In addition SCSI is a multi-drop interface which means you can attach many devices to a single SCSI port.

Although SCSI is an ANSI standard, unfortunately, due to ever higher demands on throughput, SCSI comes in a variety of 'flavours'! Each is used in various studio and mastering applications and, as a musician engineer, you will need to be aware of the differences. The following varieties of SCSI are currently implemented:

SCSI-1: Uses an 8-bit bus, and supports data rates of 4 Mbytes/s.

SCSI-2: Same as SCSI-1, but uses a 50-pin connector instead of a 25-pin connector. This is what most people mean when they refer to plain SCSI.

Fast SCSI: Uses an 8-bit bus, and supports data rates of 10 Mbytes/s.

Ultra SCSI: Uses an 8-bit bus, and supports data rates of 20 Mbytes/s.

Fast Wide SCSI: Uses a 16-bit bus and supports data rates of 20 Mbytes/s.

Ultra Wide SCSI: Uses a 16-bit bus and supports data rates of 40 Mbytes/s, this is also called SCSI-3.

Fibre channel

Fibre channel is a data transfer architecture developed by a consortium of computer and mass storage device manufacturers. The most prominent Fibre Channel standard is Fibre Channel Arbitrated Loop (FC-AL) which was designed for new mass storage devices and other peripheral devices that require very high bandwidth. Using an optical fibre to connect devices, FC-AL supports full-duplex data transfer rates of 100 MBit/s. This is far too high a transfer rate to be relevant as an audio-only standard. However in multi-channel applications and in multi-media applications (with video for example) Fibre Channel may well find its way into the modern studio. So much so that FC-AL is expected to eventually replace, SCSI for high-performance storage systems.

Firewire (IEEE 1394) interface

The 'Firewire' (IEEE 1394 interface) is an international standard, low-cost digital interface that is intended to integrate entertainment, communication, and computing electronics into consumer multimedia. Originated by Apple Computer as a desktop LAN, Firewire has been developed by the IEEE 1394 working group. Firewire supports 63 devices on a single bus (SCSI supports 7, SCSI Wide supports 15) and allows buses to be bridged (joined together) to give a theoretical maximum of thousands of devices. It uses a thin, easy to handle cable that can stretch further between devices than SCSI which only supports a maximum 'chain' length of 7 metres (20 feet). Firewire supports 64 bit addressing with automatic address selection and has been designed from the ground up as a 'plug and play' interface. Firewire can handle 10 Mbytes per second of continuous data with improvements in the design promising continuous throughput of 20–40 Mbytes per second in the very near future and a long term potential of over 100 Mbytes/s. Much like LANs and WANs, IEEE 1394 is defined by the high level application interfaces that use it, not a single physical implementation. Therefore as new silicon technologies allow high higher speeds, longer distances, IEEE 1394 will scale to enable new applications.

Digital noise generation – chain-code generators

The binary sequence generated by a chain code generator appears to have
no logical pattern; it is, to all intents and purposes, a random sequence of
binary numbers. The code is generated by a shift register which is clocked
at a predetermined frequency and whose input is derived from a network
which develops a function of the outputs from the register.

A basic form of chain code generator is illustrated in Figure 10.30,
which consists of a four-bit shift register whose input is derived from the
output of an Exclusive-OR gate, itself fed from the penultimate and last
output of the shift register. The output from the chain code generator may
be taken in serial form (from any of the data-latch outputs) or in parallel
form (from all the data latch outputs). In operation, imagine that the
output of stage B starts with a 1 as power is applied, but that all the other
outputs start with a 0. Note that the output of the XOR gate will only
equal 1 when its inputs are of a different state (i.e. non-identical). We can
now predict the ensuing pattern which results as the shift register is
clocked:

State	Output (A,B,C,D)
0 (initial)	0,1,0,0
1	0,0,1,0
2	1,0,0,1
3	1,1,0,0
4	0,1,1,0
5	1,0,1,1
6	0,10,1
7	1,01,0
8	1,1,0,1
9	1,1,1,0
10	1,1,1,1
11	0,1,1,1
12	0,0,1,1
13	0,0,0,1
14	1,0,0,0
15	0,1,0,0
16	0,0,1,0
17	1,0,0,1
18	1,1,0,0
etc.	

Note that, at state 15, the pattern starts to repeat. This sequence is
known as the maximum-length sequence. The fact that the outputs states
are predictable illustrates that the output of the code generator is not

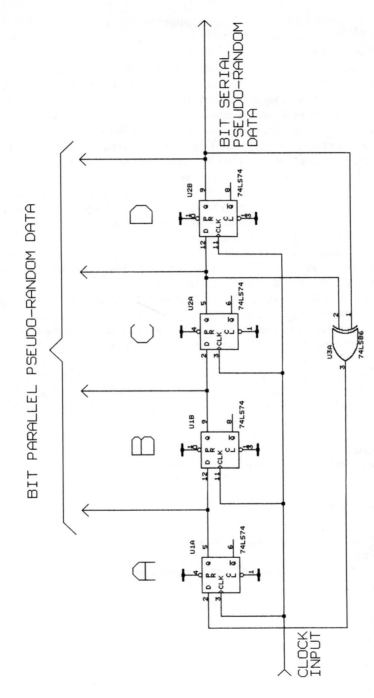

Figure 10.30 *Chain-code generator*

really random at all but is a pseudo-random binary sequence (PRBS). The sequence does, however, have some very 'random' qualities – like a very nearly equal number of 1s and 0s (think of it a coin-tossing machine)! Practically, this lack of true randomness, does not matter provided the sequence is long enough to appear random in any particular application. In every case of an n-stage, chain-code generator, the longest (maximal-length) sequence of 1s and 0s repeats after (2e n – 1) states. Note that, as illustrated in Figure 10.30, a pathological condition can occur if all outputs power up in an identical 0 state – in which case 0s will propagate indefinitely around the chain-code generator, resulting in no output. Practical circuits have to include provision to prevent this situation from ever occurring. Indeed it is precisely because of the necessity to avoid this 'forbidden' all zeros state that the output of the chain code generator illustrated in Figure 10.30 consists of a cycle of fifteen (rather than the more intuitively expected 16) states.

It can be shown mathematically, that the output binary sequence from the chain-code generator has a frequency spectrum extending from the repeat frequency of the entire sequence up to the clock frequency and beyond. The noise is effectively flat (within 0.1 dB) to about 0.12 of the clock frequency (F_c). The noise source is –3 dB at $0.44 F_c$ and falls off rapidly after that. For most applications (audio included), simple low-pass filtering of the digital maximal-length sequence results in white Gaussian noise – provided the breakpoint of the low-pass filter is well below the clock frequency of the register (say $0.05 F_c$ to $0.1 F_c$) that is, in the region where the spectrum of the digital noise is constant. The analogue filter may be a simple 6 dB/octave RC circuit, but usually a sharper active-filter is desirable.

References
Brice, R. (1999) *Newnes Guide to Digital Television*. Newnes.

Notes

1 Nevertheless the AES recommends pin 3 is 'phase' or 'hot'; and pin 2 is 'return' or 'cold'.
2 Note that a description of video signals and concepts is given in Chapter 15.

Fact Sheet #10: A digital-audio jitter-filter

- Jitter
- Further sampling theory
- Phase-locked loop receivers
- Asynchronous sample-rate conversion
- Practical circuit

Jitter

Jitter on a digital audio signal is known to cause appreciable signal degradation. All the more irksome then, that its elimination is extremely difficult by means of classic PLL style digital audio interface receivers; especially when the modulation is at a relatively low frequency, such as that caused by power-supply induced coupling. This fact sheet describes a practical circuit for a digital interface unit which may be used to remove LF jitter from a digital audio signal. Its use between a CD output and an external converter is described. The unit has a number of useful ancillary provisions which allow it to be modified to transcode between the SPDIF consumer interface and the various AES/EBU interfaces and also to strip copy-code allowing direct digital copies to be made: this makes the design useful in a number of studio interfacing roles.

Further sampling theory

As we saw in the last chapter, the quality of digital audio is mathematically definable in terms of the sampling frequency employed, the bit 'depth', the sampling-pulse aperture and time uncertainty. Expressions for the first two are well known. The effect of the latter two parameters is less well appreciated. Sampling pulse width (as a proportion of sampling period) simply has an effect on frequency response as defined in the expression,

$$20 \log \{ \mathrm{sinc}(PI/2 \ . \ f/f_n \ . \ T_s/T_o) \} \ \mathrm{dB} \tag{1}$$

where, T_s is the duration of the sampling pulse (aperture) and f_n is the Nyquist frequency limit. (Note that sinc is shorthand for sin x/x.) This is termed aperture effect and is actually relatively benign.

As Table F10.1 indicates, even when the sampling pulse width is equal to the sampling period, the loss, at the band

Table F10.1   *Aperture effect*

T_s/T_o	Attenuation at pass-band edge
1	3.9 dB
0.5	0.9 dB
0.25	0.2 dB
0.2	0.1 dB
0.1	0.04 dB

edge, is only −3.9 dB. Provided $T_s < 0.2T_o$, the effect is pretty negligible. In any case, frequency response 'droop' can always be made up in the design of the reconstruction filter following the DAC (where it is often referred to as sin *x*/*x* correction).

The effect of sampling-pulse time uncertainty or 'jitter' is much more destructive. Because all signals change their amplitude with respect to time, the effect of a slightly misplaced sampling point has the effect of superimposing a distortion on the original waveform; effectively reducing available dynamic range. Equation (2) is an expression which defines the limit of sampling uncertainty (dT) for a digital system of n bits,

$$dT/T_o = 1 / (PI . 2n-1) \qquad (2)$$

Working through an example – a sixteen-bit audio system, with 48 kbit/s sampling must have a jitter performance of less than 200 ps in order to preserve the theoretical dynamic range available from the 16-bit system. In other words the jitter must be just 0.001% of the sampling period!

Even if this requirement has been met in the recording stage, for absolute fidelity to be preserved, this value must be 'recreated' in any subsequent conversion to analogue for playback.

Phase-locked loop receivers

Most digital audio converters rely on a phase-locked loop front-end to extract clock from the self-clocking AES/EBU or SPDIF digital audio interface and to use this in the reconstruction of the analogue signal. Several very good chips exist for this purpose, one of the most famous being the CS8412 from

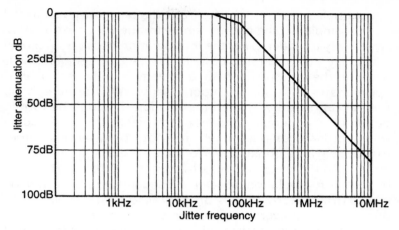

Figure F10.1 *PLL digital audio receiver response – note that only HF jitter is attenuated*

Crystal Semiconductor. Should there be any high-frequency jitter on the interface, the PLL type receiver, does a very good job in rejecting it. But, at low frequencies, it has no effect whatsoever, as Figure F10.1 shows.

This is unfortunate for the audiophile because jitter very often exists at much lower frequencies, usually due to the interaction of other analogue or digital signals or to power-supply induced effects. Experiments have shown that the effect of substantially monotonic jitter, even on modern over-sampling ADCs and DACs, indicate that the limits defined in equation (2) still apply.

Asynchronous sample-rate conversion

The construction of high-frequency phase-locked loops with low-frequency rejection is no mean task. Effectively the circuit must behave like a resonant circuit with a Q of thousands; a design constraint which usually compromises lock-time and centre frequency variability without recourse to complicated multi-stage designs. Fortunately there exists an alternative, in the form of a family of chips from Analog Devices based upon asynchronous sample-rate conversion (ASRC) technology. There are more ways than one to describe the nature of asynchronous sample rate conversion. The easiest to understand is the interpolation-decimation model in which the input signal is over-sampled to a much higher rate, digitally low-pass filtered and re-sampled at the output sample frequency. Unfortunately, whilst easy to understand, the interpolation-decimation model is actually not a suitable basis for a practical system. This is because the output of such a system is only the nearest appropriate value in a temporal sense. Whilst there is no theoretical reason why the interpolation shouldn't be carried out at a fast enough rate to make this viable, there exist some very good practical reasons why it is not. For instance, in order to achieve a reasonable performance (and this means, to achieve 16-bit levels of THD+N across the 0 to 20 kHz audio band) the interpolation up-sample frequency would need to be over 3 GHz! Clearly, this is an impractical rate at which to operate a low-power IC, so the Analog Devices' chips use a less commonly known method of sample rate conversion called polyphase filtering.

In the polyphase filter ASRC, the digital audio sample sequence is over-sampled (but at a manageable rate of MHz) and applied to a digital FIR low pass filter in which the required impulse response (20 kHz cut-off) is itself highly over-sampled. The filter is 'over-sampled' in the sense that it comprises many times the required number of coefficient sample taps to satisfy the Nyquist criterion. This means that, at any one moment, only a sparsely sampled subset of coefficients of this filter need be chosen to process the input samples. These subsets of coefficients, create a kind of 'sub-filter', each possessing an identical 0 to 20 kHz magnitude response but with a fractionally different group delay (hence the term 'polyphase'). It is as if the input signal was being applied to a very great number (thousands) of digital delay-lines; each with a slightly differing delay as shown (greatly simplified) in Figure F10.2. The sample-rate conversion process works like this; if a request for an output sample occurs immediately after an input sample has arrived, a polyphase filter is chosen which imposes a short group delay. If a request for an output sample occurs late in the input sample period, a polyphase filter is chosen which imposes a long group delay. In this fashion, the amplitude of the output sample is precisely computed at the desired output sample frequency.

Looking at the output commutator in Figure F10.2, it's possible to imagine that, provided the relationship between the input and output frequencies is not completely random, there will be a pattern to the switch selection when looked at over a certain period of time. Indeed, provided the input and

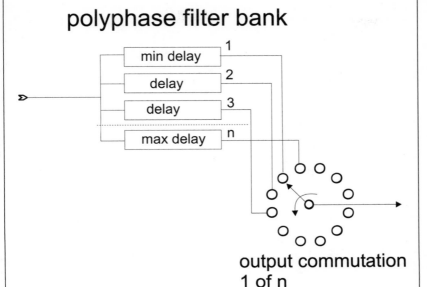

Figure F10.2   *Architecture of a polyphase filter*

output frequency are relatively stable, you can imagine the commutator revolving at the computed difference frequency between the input and output sample frequency. This process is controlled, within the Analog Devices' parts, by an on-chip, digital servo control system which bases its commutation decisions, not on an instantaneous measurement, but rather a digitally filtered ratio. It is the effect of this powerful, low-pass filtering mechanism that greatly reduces any jitter which may be present on the sample clocks; even when the jitter frequency is just a few tens of Hz.

Practical circuit

Figure F10.3 is a practical implementation of the AD1892 used as a jitter rejection device to be utilised between the output of a CD player and the input of an outboard DAC. The AD1892 is not just an ASRC, it is also an AES/SPDIF interface receiver too, so the circuit implementation is very simple.

The AD1892 has some limitations, the most severe of which is that it only retains its performance over a limited range of upward sample-rate conversion and a very limited range of

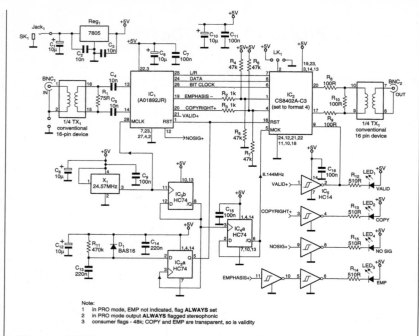

Figure F10.3   *Sample-rate converter schematic*

downward rate conversion. For this design, I decided to use an up-conversion from 44.1 kHz to 48 kHz. The part works well at these two rates and the master oscillator (which must be 512 times output sample rate; 24.576 MHz) is relatively easy to source at this frequency.

The SPDIF signal arrives at TX1 – one part of a 16 pin, four transformer data-bus isolator and is terminated, on the far side of the transformer, by R1. The signal is applied directly to the AD1892 part via coupling caps, C4 and C5. The master output clock is derived from a small 24 MHz crystal oscillator. The composite AES/SPDIF signal, having been broken down into separate clocks and data by the Analog Devices part, is put back together again by the Crystal Semiconductor CS8402 transmitter chip. This too requires a master clock, but at one quarter of the frequency of the AD1892, hence the inclusion of the divide-by-two bistables IC3 and IC4. SPDIF output is via transformer TX1, which is another part of the same data-bus isolator used for the input. Note resistors R8, 9, 10, these produce an output impedance of 75 ohms at a level of about 2 V EMF. This is above that specified for SPDIF (1 V EMF) and is therefore a bit 'non-standard'. I made the choice for two reasons: Firstly, I have found that outboard DACs like to have a bit more level; and secondly, the circuit may be used (by changing the position of LK1) to encode a digital signal to the unbalanced form of the professional AES/EBU digital interface which requires the higher output level. This provision makes the circuit useful if you need to interface a non-professional CD player in a digital studio. The output is also quite suitable for driving symmetrical a 110 ohm, AES-style interface, mutatis mutandis.

The circuit includes several user LEDs to indicate; validity, copy-code, pre-emphasis and signal loss which are derived and decoded by the AD1892, these are driven by an HC14 and are primarily there for amusement since no user intervention is required. Emphasis state and Copyright prohibit are decoded and re-coded by the CS8402. Pull-up, pull-down resistor positions are provided here to allow for various options; the most useful being the removal of R3 and R6 which strips copy-code and allows direct digital copies to be made.

Figure F10.4 illustrates a view of the completed SRC-1 PCB and a finished production unit is shown in Figure F10.5. Note that the signal inputs and outputs are on BNC as I prefer this connector enormously to the RCA phono alternative. The PSU input is 'squeezed' between the input and output and the whole circuit is enclosed in a little anodised, aluminium extrusion box, no bigger than a household box of matches, ideally suited for sitting on top of a CD player or DAC.

Figure F10.4 *PCB of sample-rate converter*

Figure F10.5   *Production sample of SRC-1*

Although it's unwise to be adamant in this area, everyone who has listened to the circuit has been amazed by the improvement in quality that it yields; especially in definition at the bass-end of the spectrum.

11
Space Odyssey – Stereo and spatial sound

Stereo

When listening to music on a two-channel stereo audio system, a sound 'image' is spread out in the space between the two loudspeakers. The reproduced image thus has some characteristics in common with the way the same music is heard in real life – that is, with individual instruments or voices each occupying, to a greater or lesser extent, a particular and distinct position in space. Insofar as this process is concerned with creating and re-creating a 'sound event', it is woefully inadequate. First, the image is flat and occupies only the space bounded by the loudspeakers. Second, even this limited image is distorted with respect to frequency. (There exists an analogy with chromatic aberration in optics.) Happily there exist relatively simple techniques for both the improvement of existing stereophonic images and for the creation of synthetic sound fields in a 360° circle around the listening position. The basic techniques of stereophony and these newer techniques, are covered later on in this chapter, but first there is a brief review of spatial hearing – our capacity to localise (determine the position of) a sound in our environment.

Spatial hearing

Sound localisation in humans is remarkably acute. As well as being able to judge the direction of sounds within a few degrees, experimental subjects are sometimes able to estimate the distance of a sound source as well. Consider the situation shown in Figure 11.1, where an experimental subject is presented with a source of steady sound located at some distance from the side of the head. The two most important cues the brain uses to determine the direction of a sound are due to the physical nature

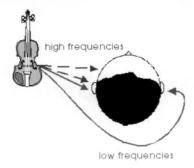

high frequencies

low frequencies

Figure 11.1 *Head and sound source*

of sound and its propagation through the atmosphere and around solid objects. We can make two reliable observations:

1 at high frequencies, the relative loudness of a sound at the two ears is different since the nearer ear receives a louder signal compared with the remote ear; and
2 at all frequencies, there is a delay between the sound reaching the near ear and the further ear.

It can be demonstrated that both effects aid the nervous system in its judgement as to the location of a sound source: at high frequencies, the head casts an effective acoustic 'shadow' which acts like a low-pass filter and attenuates high frequencies arriving at the far ear, thus enabling the nervous system to make use of interaural intensity differences to determine direction. At low frequencies, sound diffracts and bends around the head to reach the far ear virtually unimpeded. So, in the absence of intensity-type directional cues, the nervous system compares the relative delay of the signals at each ear. This effect is termed interaural delay difference. In the case of steady-state sounds or pure tones, the low-frequency delay manifests itself as a phase difference between the signals arriving at either ear. The idea that sound localisation is based upon interaural time differences at low frequencies and interaural intensity differences at high frequencies has been called Duplex theory and it originates with Lord Rayleigh at the turn of the century.[1]

Binaural techniques

In 1881, Monsieur Clement Ader placed two microphones about eight inches apart (the average distance between the ears known technically as the interaural spacing) on stage at the Paris Opera where a concert was being performed and relayed these signals over telephone lines to two telephone earpieces at the Paris Exhibition of Electricity. The amazed listeners were able to hear, by holding one earpiece to each ear, a remarkably lifelike impression that they too were sitting in the Paris Opera audience. This was the first public demonstration of binaural stereophony, the word binaural being derived from the Latin for two ears.

The techniques of binaural stereophony, little different from this original, have been exploited many times in the century since the first demonstration. However, psychophysicists and audiologists have gradually realised that considerable improvements can be made to the simple spaced microphone system by encapsulating the two microphones in a synthetic head and torso. The illusion is strengthened still more if the dummy head is provided with artificial auricles (external ears or pinnae – see Chapter 2). The binaural stereophonic illusion is improved by the addition of an artificial head and torso and external ears because it is now known that sound interacts with these structures before entering the ear canal. If, in a recording, microphones can be arranged to interact with similar features, the illusion is greatly improved in terms of realism and accuracy when the signals are relayed over headphones. This is because headphones sit right over the ears and thus do not interact with the listener's anatomy on the final playback.

The most significant effect brought about by the addition of artificial auricles is the tendency for the spectrally modified sound events to localise outside of the listener's head as they do in real life. Without these additions the sound image tends to lateralise – or remain artificially inside the head. And they play another role: we saw in Figure 11.1 that the main cues for spatial hearing lay in the different amplitudes and phases of the signals at either ear. Such a system clearly breaks down in its ability to distinguish from a sound directly in front from one directly behind (because, in both cases, there are neither amplitude nor phase differences in the sounds at the ears – see Figure 11.2). A little thought reveals that the

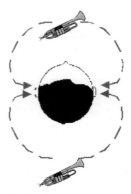

Figure 11.2 *Front-back ambiguity*

same thing goes for a sound from any direction – it is always possible to confuse a sound in the forward 180° arc from its mirror image in the rear 180° arc. A common failing of binaural stereophonic recordings (made without false pinnae) is the erroneous impression, gained on playback, that the recorded sound took place behind the head, rather than in front. Manifestly, the pinnae appear to be involved in reducing this false impression and there exist a number of excellent experimental papers describing this role, known as investigations into the front–back impression (Blauert 1983).

Two-loudspeaker stereophony

If the signals from a dummy head recording are replayed over two loudspeakers placed in the conventional stereophonic listening arrangement (as shown in Figure 11.3), the results are very disappointing. The

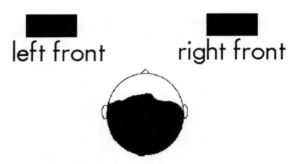

Figure 11.3 *'Classic' stereophonic arrangement*

reason for this is the two unwanted crosstalk signals: The signal emanating from the right loudspeaker which reaches the left ear; and the signal emanating from the left loudspeaker which reaches the right ear (shown in Figure 11.4). These signals result in a failure to reproduce the correct interaural time delay cues at low frequencies. Furthermore, the filtering effects of the pinnae (so vital in obtaining out-of-the-head localisation when listening over headphones) are not only superfluous, but impart to the sound an unpleasant and unnatural tonal balance. Because of these limitations binaural recordings are made only in special circumstances. A different technique is used in the production of most stereo records and CDs. A system was invented in 1928 by Alan Blumlein – an unsung British genius. Unfortunately Blumlein's solution is so complete – and so elegant – that it is still widely misunderstood or regarded as simplistic.

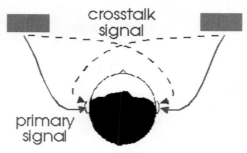

Figure 11.4 *Binaural crosstalk signals*

Summing localisation

Consider the arrangement shown in Figure 11.3 again. A moment's thought will probably lead you to some fairly obvious conclusions: if all the sound comes out of the left loudspeaker, the listener will clearly experience the sound 'from the left'. Similarly with the right. If both loudspeakers reproduce identical sounds at identical intensity, it is reasonable to assume that the listener's brain will conclude the existence of a 'phantom' sound, coming from directly in front, because in nature that situation will result in the sound at both ears being identical. And indeed it does, as experiments have confirmed. Furthermore proportionally varied signal intensities result in a continuum of perceived 'phantom' image positions between the loudspeakers. But how does a system which works on intensity alone fool the brain into thinking there is a phantom sound source other than in the three rather special positions noted above? While it is fairly obvious that interchannel intensity differences will reliably result in the appropriate interaural intensity differences at high frequencies, what about at low frequencies where the sound can diffract around the head and reach the other ear? Where does the necessary low-frequency time-delay component come from?

When two spaced loudspeakers produce identically phased low-frequency sounds at different intensities, the soundwaves from both loudspeakers travel the different distances to both ears and arrive at either ear at different times. Figure 11.5 illustrates the principle involved: the louder signal travels the shorter distance to the right ear and the longer distance to the left ear. But the quieter signal travels the shorter distance to the left ear and the longer distance to the right ear. The result is that the sounds add vectorially to the same intensity but different phase at each ear. The brain interprets this phase information in terms of interaural delay. Remember that stereophonic reproduction from loudspeakers requires only that stereo information be carried by interchannel intensity difference. Despite a huge body of confused literature to the contrary,

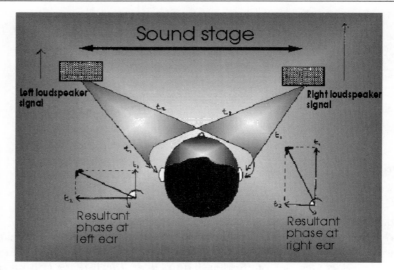

Figure 11.5 *Vectorial addition results in phase differences at the ears*

there is no requirement to encode interchannel delay difference. If this were not the case, the pan control, which the sound engineer uses to 'steer' instruments into position in the stereo 'sound stage', would not be the simple potentiometer control described in the next chapter.

FRANCINSTIEN stereophonic image enhancement technique

So far we have only considered individual sounds; however, stereo works by eliciting, within the mind of the listener, a continuum of simultaneous and coexistent stereo images for each voice and/or instrument. As we have seen, at high frequencies, the stereo image is largely intensity derived and, at low frequency the image is largely delay derived. Unfortunately, conventional intensity-derived two loudspeaker stereo cannot create a perfect illusion, the problem being that the simultaneous stereo images (one set predominantly high frequency and the other predominantly low frequency) are not in exact perceptual spatial register. In this section, circuitry is described which electrically manipulates the signals of an existing stereophonic recording, to bring about an improvement in the realism of two-speaker stereophonic reproduction. A review of the literature reveals that the technique is not new. Nevertheless the implementation described is both radical and novel and side-steps many of the problems and awkwardness of the previous attempts to solve this frequency dependent 'smearing' of the stereo image described here.

The important qualitative fact to appreciate is that, for a given interchannel intensity difference, the direction of the perceived auditory event is further from a central point between the loudspeakers when a high-frequency signal is reproduced than when a low frequency is reproduced. Since music is itself a wideband signal, when two loud-speakers reproduce a stereo image from an interchannel intensity derived stereo music signal, the high-frequency components of each instrument or voice will subtend a greater angle than will the low-frequency components. This problem was appreciated even in the very early days of research on interchannel intensity related stereophony and, through the years, a number of different solutions have been proposed.

The Shuffler

Blumlein mentioned in his 1931 patent application that it was possible to control the width of a stereo image by matrixing the left and right signal channels into a sum and difference signal pair and controlling the gain of the difference channel prior to rematrixing back to the normal left and right signals. He further suggested that, should it be necessary to alter the stereo image width in a frequency dependent fashion, all that was needed was a filter with the appropriate characteristics to be inserted in this difference channel. After his untimely death, the post-war team working at EMI on a practical stereo system and attempting to cure this frequency dependent 'smearing' of the stereo picture implemented just such an arrangement and introduced a filter of the form:

$$H(s) = A[(1 + s/a^2) / (1 + s/a^1)]$$

into the difference channel. Figure 11.6 is an illustration of their practical Shuffler circuit (as they termed it) and its implementation in the difference channel. Unfortunately this circuit was found to introduce distortion and tonal colouring and was eventually abandoned. (It has been demonstrated that the time constants and attenuation ratios used in the original Shuffler circuit were not, in any case, well chosen.) Nevertheless it is well worth investigating the Shuffler technique because it demonstrates particularly clearly the requirement for any mechanism designed to tackle the image blurring problem.

Manifestly, there is only one particular filter characteristic which will equalise the interchannel intensity difference signal in the appropriate fashion. If we rewrite the above equation, the characteristics for such a filter become easier to conceptualise:

$$H(s) = A\{a^1/a^2 \ [(a^2 + s) / (a^1 + s)]\}$$

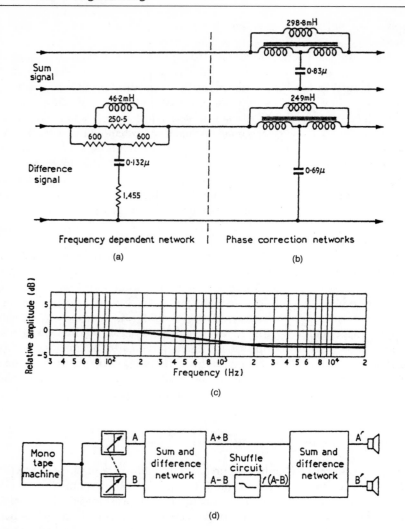

Figure 11.6   *Practical Shuffler circuit*

This is because the variable A represents overall difference-channel gain, and demonstrates how overall image width may be manipulated, and further, since

$[(a^2 + s) / (a^1 + s)]$ tends to 1 as s tends to infinity,

the term a^1/a^2 defines the image narrowing at high frequency and demonstrates the filtering necessary to match the interaural intensity derived high-frequency image with the interaural delay derived image at low frequency. One might therefore call a^1 and a^2, psychoacoustic constants. It was precisely these constants which were at fault in the original Shuffler circuit.

Other derivatives of the Shuffler, using operational amplifier techniques, have appeared, but the act of matrixing, filtering and rematrixing is fraught with problems since it is necessary to introduce compensating delays in the sum channel which very exactly match the frequency dependent delay caused by the filters in the difference channel if comb filter coloration effects are to be avoided. Furthermore the very precise choice of the correct constants is crucial. After all, existing two-loudspeaker stereo is generally regarded as being a tolerably good system, the required signal manipulation is slight and failure to use the correct constants, far from improving stereo image sharpness, can actually make the frequency dependent blurring worse! Others have taken a more imaginative and unusual approach.

Edeko

It is a fundamental characteristic of the blurring problem that the brain perceives the high-frequency intensity derived image as generally wider than the low-frequency, delay derived image. With this in mind Dr Edeko conceived of a way of solving the problem acoustically (and therefore of side-stepping the problems which beset electronic solutions).

Edeko (1988) suggested a specially designed loudspeaker arrangement as shown in Figure 11.7 where the angle between the high frequency loudspeaker drive units subtended a smaller angle at the listening position than the mid-range drive units and these, in turn, subtended a smaller angle than the low-frequency units. This device, coupled with precise designs of electrical crossover network enabled the image width to be manipulated with respect to frequency.

Improving image sharpness by means of interchannel crosstalk

There is a much simpler technique which may be used to narrow a stereo image at high frequencies and that is by the application of periodic interchannel crosstalk (Brice 1997). Interestingly investigations reveal that distortion mechanisms in reproduction from vinyl and other analogue media may indeed be just those required to bring about an improvement in the realism of the reproduced stereo image.[2] This suggests there may be something in the hi-fi cognoscenti's preference for vinyl over CD and for many recording musicians' preference for analogue recording over the, apparently better, digital alternative – though not, as they invariably suppose, due to digital mysteriously taking something away but due to the analogue equipment adding beneficial distortion. This crosstalk technique is exploited in the FRANCINSTIEN range of stereophonic image enhance-

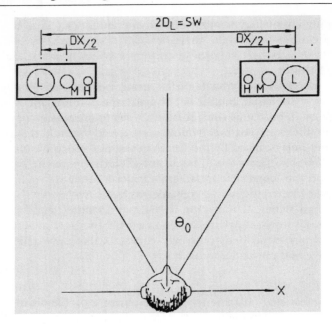

Figure 11.7   *Edeko's loudspeaker arrangement*

Figure 11.8   *Domestic FRANCINSTIEN unit*

ment systems developed by Perfect Pitch Music Ltd. Commercial units for use in hi-fi systems and recording studios are illustrated in Figures 11.8 and 11.9. The simplest possible implementation of the FRANCINSTIEN effect is achieved very easily. The schematic is given in Figure 11.10. This circuit may be included in the record or replay chain but not both.

Figure 11.9 *Studio FRANCINSTIEN unit*

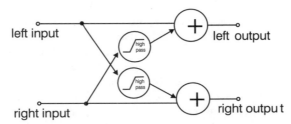

Figure 11.10 *FRANCINSTIEN system*

3D sound fields

Perhaps the crudest method for generating audible spatial effects is to provide more channels of audio and more loudspeakers!

Dolby Surround

Walt Disney Studio's *Fantasia* was the first film ever to be shown with a stereo soundtrack. That was in 1941. Stereo in the home has been a reality since the 1950s. Half a century on, it is reasonable that people might be looking for 'something more'. With the advent of videocassette players, watching film at home has become a way of life. Dolby Surround was originally developed as a method of bringing part of the cinema experience to the home where a similar system named Dolby Stereo has been in use since 1976. Like Dolby Stereo, Dolby Surround is essentially a four-channel audio system encoded or matrixed into the standard two-stereo channels. Because these four discrete channels are encoded within the stereo channels extra hardware is required both at the production house and in the home. Decoders are now very widespread because of the take-up of home cinema systems. The extra hardware required, in addition to normal stereo, is a number of extra loudspeakers (ideally three), a decoder and an extra stereo power amplifier. Some manufacturers supply decoder and four power amplifiers in one AV amplifier unit. In addition a sub-woofer channel may be added. (A sub-woofer is a loudspeaker unit devoted to handling nothing but the very lowest audio frequencies – say below 100 Hz.) Frequencies in this range do add a disproportionate level of realism to reproduced sound. In view of the very small amount of information (bandwidth) this is surprising. However, it is likely humans infer the scale of an acoustic environment from these subsonic cues. (Think about the low, thunderous sound of the interior of a cathedral.)

In order for an audio, multimedia or VR studio to produce Dolby Surround material, and claim that it is such, permission has to be granted by Dolby who provide the encoding unit and a consultant on a daily basis. (You have, of course, also to set up your studio so that it has the necessary decoding facilities by buying extra loudspeakers, decoder and amplifiers.) This is fine if you're a Hollywood film company but less good if you're an independent producer of video, audio or multimedia material. Fortunately, while you will not be able to mark your studio productions as Dolby Surround encoded there is no reason why you may not encode information so that you can exploit the effects of Dolby Surround decoding. In fact encoding is much simpler than decoding and this is discussed in this section. But first a description of the Dolby Surround process.

A typical surround listening set-up is illustrated in Figure 11.11. Note the extra two channels, centre and surround, and the terminology for the final matrixed two channels signals Lt and Rt; standing for left-total and right-total respectively. The simplest form of decoder (which most certainly does not conform to Dolby's criteria but is nevertheless reasonably effective) is to feed the centre channel power amplifier with a sum signal (Lt + Rt) and the surround channel amplifier with a difference signal (Lt – Rt). This bare-bones decoder works because it complements (to a first approximation) the way a Dolby Surround encoder matrixes the four channels onto the left and right channel: centre channel split

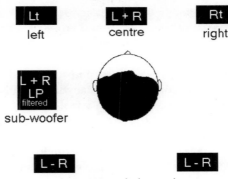

Figure 11.11   *Typical surround-sound listening arrangement*

between left and right, surround channel split between left and right with one channel phase reversed. If we label the original left/right signals L and R we can state the fundamental process formally:

Input channels
Left (sometimes called left music channel):   L
Right (sometimes called right music channel):   R
Centre channel (sometimes called dialogue channel):   C
Surround channel (for carrying atmosphere sound
effects etc.).   S

Output channels (encoding process)

$$Lt = i(L + jC + kS)$$
$$Rt = i(R + jC - kS)$$

where i, j and k are simply constants. And the decoding process yields:

Left (L′) $= e(Lt)$
Right (R′) $= f(Rt)$
Centre (C′) $= u(Lt + Rt) = u[i(L + jC + kS + R + jC - kS)]$
          $= u[i(L + R + 2jC)]$
Surround (S′) $= v(Lt - Rt) = v[i(L + jC + kS - R - jC + kS)]$
          $= v[i(L - R + 2kS)]$

where e and f and u and v are constants.

Which demonstrates this is far from a perfect encoding and decoding process. However a number of important requirements are fulfilled even by this most simple of matrixing systems and to some extent the failure mechanisms are masked by operational standards of film production.

Dolby have cleverly modified this basic system to ameliorate the perceptible disturbance of these unwanted crosstalk signals. Looking at the system as a whole – as an encode and decode process, first, and most important, note that no original centre channel (C) appears in the decoded rear, surround signal (S'). Also note that no original surround signal (S) appears in the decoded centre channel (C'). This requirement is important because of the way these channels are used in movie production. The centre channel (C) is always reserved for mono dialogue. This may strike you as unusual but it is absolutely standard in cinema audio production. Left (L) and Right (R) channels usually carry music score. Surround (S) carries sound effects and ambience. Therefore, considering the crosstalk artefacts, at least no dialogue will appear in the rear channel – an effect which would be most odd! Similarly, although centre channel information (C) crosstalks into left and right speaker channels (L' and R'), this only serves to reinforce the centre dialogue channel. The most troublesome crosstalk artefact is the $v(iL - iR)$ term in the S' signal which is the part of the left/right music mix which feeds into the decoded surround channel – especially if the mix contains widely panned material (with a high interchannel intensity ratio). Something really has to be done about this artefact for the system to work adequately and this is the most important modification to the simple matrix process stated above which is implemented inside all Dolby Surround decoders. All decoders delay the S' signal by around 20 ms which, due to an effect known as the law of the first wavefront or the Hass effect, ensures that the ear and brain tend to ignore the directional information contained within signals which correlate strongly with signals received from another direction but at an earlier time. This is certainly an evolutionary adaptation to avoid directional confusion in reverberant conditions and biases the listener, in these circumstances, to ignore unwanted crosstalk artefacts. This advantage is further enhanced by band limiting the surround channel to around 7 kHz and using a small degree of high-frequency expansion (as explained in Chapter 6). Dolby Pro Logic enhances the system still more by controlling the constants written e, f, u and v above dynamically, based on programme information. This technique is known as adaptive matrixing.

One very important point to remember regarding Dolby Surround is that it does not succeed in presenting images at positions around the listening position. Instead the surround channel is devoted to providing a diffuse sound atmosphere or ambience. While this is effective, the system is not concerned with the creation of realistic (virtual) sound fields. Nevertheless, Dolby Surround systems with ProLogic are becoming a widespread feature in domestic listening environments and the wise musician-engineer could do worse than to exploit the emotive power this technology most certainly possesses. But how?

DIY surround mixing

Our understanding of Dolby replay has made it clear that encoding (for subsequent decoding by Dolby Surround decoders) can be undertaken quite simply. The golden rule, of course, is to set up a good surround monitoring system in the studio in the first place. Systems are available quite cheaply. With the monitoring system in place and adjusted to sound good with known material mixes can be undertaken quite simply: pan music mixes as normal – but avoid extreme pan positions. Ensure all narration (if appropriate) is panned absolutely dead centre. Introduce surround effects as mono signals fed to two channels panned hard left and right. Invert the phase of one of these channels. (Sometimes this is as simple as switching the channel phase invert button – if your mixer has one of these.) If it hasn't you will have to devise a simple inverting operational amplifier stage with a gain of one. Equalise the 'rear' channels to roll off around 7 kHz, but you can add a touch of boost around 5 kHz to keep the sound crisp despite the action of the high-frequency expansion undertaken by the decoder.

Ambisonics

When mono recording was the norm, the recording engineer's ideal was expressed in terms of the recording chain providing an acoustic 'window' at the position of the reproducing loudspeaker, through which the listener could hear the original acoustic event – a 'hole in the concert hall wall' if you like (Malham 1995). It is still quite common to see explanations of stereo which regard it as an extension of this earlier theory, in that it provides two holes in the concert wall! In fact such a formalisation (known as wavefront-reconstruction theory) is quite inappropriate unless a very great number of separate channels are employed. As a result, two-channel loudspeaker stereophony based on this technique – two wide-spaced microphones feeding two equally spaced loudspeakers – produces a very inaccurate stereo image. As we have seen, Blumlein took the view that what was really required was the 'capturing' of all the sound information at a single point and the recreation of this local sound field at the final destination – the point where the listener is sitting. He demonstrated that for this to happen, it required that the signals collected by the microphones and emitted by the loudspeakers would be of a different form to those we might expect at the listener's ears, because we have to allow for the effects of crosstalk. Blumlein considered the recreation of height information (periphony) but he did not consider the recreation of phantom sound sources over a full 360° azimuth (pantophony). The recording techniques of commercial quadraphonic[3] systems (which blossomed in the 1970s) were largely based on a groundless extension of the already flawed wavefront-reconstruction stereo techniques and hence derived left-front, left-back signals and so on.

Not so Ambisonics – brainchild of Michael Gerzon which although too a child of the 1970s and, in principle, a four-channel system, builds upon Blumlein's work to create a complete system for the acquisition, synthesis and reproduction of enveloping sound fields from a limited number of loudspeakers.

Consider a sound field disturbed by a single sound source. The sound is propagated as a longitudinal wave which gives rise to a particle motion along a particular axis drawn about a pressure microphone placed in that sound field. Such a microphone will respond by generating an output voltage which is proportional to the intensity of the sound, irrespective of the direction of the sound source. Such a microphone is called omnidirectional because it cannot distinguish the direction of a sound source. If the pressure microphone is replaced with a velocity microphone, which responds to the particle motion and is therefore capable of being directional, the output is proportional to the intensity of the sound multiplied by cos I, where I is the angle between the angle of the incident sound and the major axis of the microphone response.

But there is an ambiguity as to the sound source's direction if the intensity of the signal emerging from a velocity microphone is considered alone. As I varies from 0° to 359°, the same pattern of sensitivity is repeated twice over. On a practical level this means the microphone is equally responsive in two symmetrical lobes, known as the figure-of-eight response that we saw in Chapter 3. Mathematically put, this is because the magnitude of the each half-cycle of the cosine function is identical. But not in sign; the cosine function is negative in the second and third quadrant. So, this extra directional information is not lost, but is encoded differently, in phase information rather than intensity information. What is needed to resolve this ambiguity is a measure of reference phase to which the output of the velocity microphone can be compared. Just such a reference would be provided by a pressure microphone occupying a position, ideally coincident but practically very close to, the velocity type. (This explanation is a rigorous version of the qualitative illustration, in Figure 3.1, of the combination of a figure-of-eight microphone and an omnidirectional microphone 'adding together' to make a unidirectional, cardioid directional response.)

More intuitively stated: a velocity microphone sited in a sound field will resolve sound waves along (for instance) a left–right axis but not be able to resolve a sound from the left differently from a sound from the right. The addition of a pressure microphone would enable the latter distinction to be made by either subtracting or adding the signals from one another. Now consider rotating the velocity microphone so that it faced front–back. This time it would resolve particle motion along this new axis but would be equally responsive regardless of whether the sound came from in front or behind. The same pressure microphone would resolve the situation. Now contemplate rotating the microphone again, this time so it faced up-down, the same ambiguity would arise and would once more be resolved by the suitable addition or subtraction of the pressure microphone signal.

Consider placing three velocity microphones each at right angles to one another (orthogonally) and, as nearly as possible in the same position in space. The combination of signals from these three microphones, coupled with the response of a single omnidirectional, phase-reference microphone, would permit the resolution of a sound from any direction. Which is the same thing as saying, that a sound in a unique position in space will translate to a unique combination of outputs from the four microphones. These four signals (from the three orthogonal velocity microphones and the single pressure microphone) are the four signals which travel in the four primary channels of an Ambisonics recording. In practical implementations, the up-down component is often ignored, reducing the system to three primary channels. These three signals may be combined in the form of a two-channel, compatible stereo signal in a process called UHJ coding, although this process is lossy. Ambisonics provides for several different loudspeaker layouts for reproduction.

If we consider the *de facto* four-speaker arrangement and limit the consideration of Ambisonics recording to horizontal plane only (three-channel Ambisonics), it is possible to consider rotating the velocity microphones so that, instead of facing front–back, left–right as described above, they face as shown in Figure 11.12. An arrangement which is identical to a 'Blumlein' crossed pair.

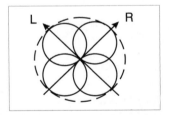

Figure 11.12   *Ambisonic microphone arrangement*

If we label one microphone L, the other R, and the third pressure microphone P as shown, by a process of simple addition and subtraction, four signals are obtainable from the combination of these three microphones:

(L + P), (L – P), (R + P) and (R – P)

each one equivalent to a cardioid microphone facing diagonally left-front, diagonally left-back, right front and right rear, in other words into the four cardinal positions occupied by the loudspeakers on replay. This approach is equivalent to the BBC's policy on quadraphonic recording (Nisbett 1979). Ambisonics theory has enabled the construction of special Ambisonics panpots which enable sounds to be artificially positioned anywhere in a 360° sound stage based on simple channel intensity ratios.

Roland RSS system and Thorn EMI's Sensaura

Anyone who has listened to a good binaural recording on headphones will know what an amazing experience it is, far more realistic than anything experienced using conventional stereo on loudspeakers. As noted earlier binaural recordings 'fail' on loudspeakers due to the crosstalk signals illustrated in Figure 11.4. A modern solution to this difficulty, known as crosstalk cancellation, was originally proposed by Schroeder and Atal (Begault 1994). The technique involves the addition, to the right-hand loudspeaker signal, of an out-of-phase version of the left channel signal anticipated to reach the right ear via crosstalk, and the addition, to the left-hand loudspeaker signal, of an out-of-phase version of the right-hand channel signal expected to reach the left ear via crosstalk. The idea is that these extra out-of-phase signals cancel the unwanted crosstalk signals, resulting in the equivalent of the original binaural signals only reaching their appropriate ears. Unfortunately, without fixing the head in a very

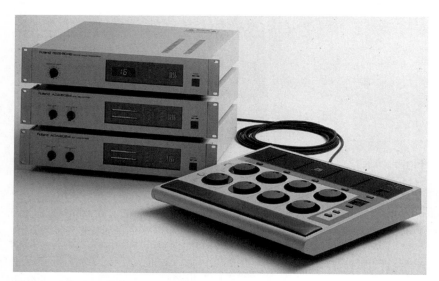

Figure 11.13   *Roland's RSS 3D sound system*

precise location relative to the loudspeakers, this technique is very difficult to put into practice (Blauert 1985; Begault 1994) although several commercial examples exist, including the Roland RSS System and Thorn EMI's Sensaura.

The Roland Corporation's RSS system (Roland Sound Space) was introduced around 1990 and was primarily targeted at the commercial music industry. The original system consisted of a desktop control unit and a rack mount processor. The RSS system, illustrated in Figure 11.13, allowed up to four mono signals to be panned in 360° azimuth and elevation using the eight controls on the control unit. In essence the unit implemented a two-stage process. The first stage consisted of the generation of synthetic binaural signals followed by a second stage of binaural crosstalk cancellation. The unit also provided MIDI input so that the sound sources could be controlled remotely. Roland utilised a derivative of this technology in various reverb units (see Figure 11.14).

Figure 11.14 *A later development of the unit shown in Figure 11.13*

Thorn EMI's Sensaura system also implemented a binaural crosstalk cancellation system, this time on signals recorded from a dummy head. There exists a paucity of technical information concerning this implementation and how it differs from that originally proposed by Schroeder and Atal (Begault 1994).

OM 3D sound processor

The RSS System and the Sensaura are intriguing systems because they offer the possibility of sound from all directions (pantophony) without the use of extra loudspeakers. The OM 3D sound system, developed by Perfect Pitch Music Ltd in Farnham, England, was developed with a similar goal in mind. The approach to the generation of two loudspeaker 'surround sound' taken in the OM system was an extension of summing stereophony. The OM system provided four mono inputs, two of which could be panned in a 180° arc which extended beyond the conventional loudspeaker boundary. Two further channels were provided which permitted sound sources to be panned 300°; from loudspeaker to loudspeaker in an arc to the rear of the stereo listening position. Figure 11.15 illustrates the two panning regimes. The unit was ideally integrated in a mixer system with four auxiliary sends. The resulting 3D-panned signals were summed on a common bus and output as a single stereo pair; the idea being that the output signal could be routed back to the mixer as a single stereo return. No provision was provided for adjusting the input sensitivity of any of the four input channels but VU meters allowed these

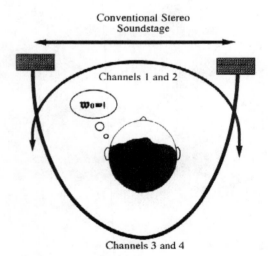

Figure 11.15 *Perfect Pitch Music OM 3D sound panner; pan effects*

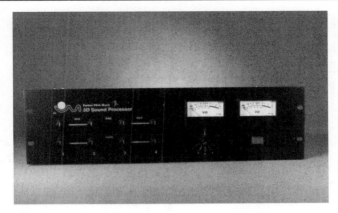

Figure 11.16   *OM hardware*

levels to be monitored so that they could be set in the mixer send circuitry. Figure 11.16 is an illustration of the OM hardware and Figure 11.17 is a system diagram. Figure 11.18 illustrates the system configuration for a 3D mix. Looking first at channels 1 and 2; how was the 180° frontal pan achieved? OM was developed using a spatial hearing theory based upon Duplex theory. This approach suggests that any extension of the generation of spatial sound is best broken down in terms of high and low frequencies. Let's take the low frequencies first.

Figure 11.19 illustrates the maximum interaural time difference which may be generated by a hard-right panned signal in a conventional stereophonic arrangement. The signal is shown at the right ear delayed by t_1 and at the left ear delayed by t_2. The problem is trying to make a sound appear as if it is at position P. If it was, the signals would be as shown, with the left ear experiencing a signal delayed by t_3. Figure 11.10 illustrates how this may be achieved. With the addition of an out-of-phase version of the signal at the left-hand loudspeaker, the interaural phase difference may be increased, thus generating the required perception. Unfortunately, if high frequencies were allowed to be emitted by the left-hand loudspeaker these would tend to recentralise the image because the ear is insensitive to phase for signals above about 700 Hz. So another approach has to be made. This may be illustrated by comparison with the monaural transfer functions of a sound approaching from 30° from the front (the true loudspeaker position) and the desired position (at 90° to the front). The monaural transfer functions for these two directions in the horizontal plane are given in Figure 11.21.

The required spectral modification (which must be made to the high-frequency signals emitted from the right loudspeaker alone – the other channel remaining mute) is derived by subtracting one response from the

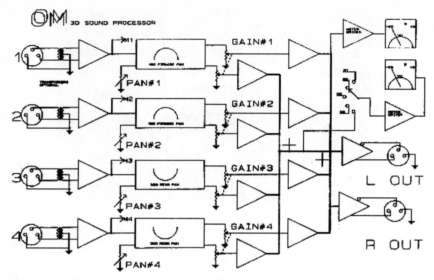

Figure 11.17 *OM system diagram*

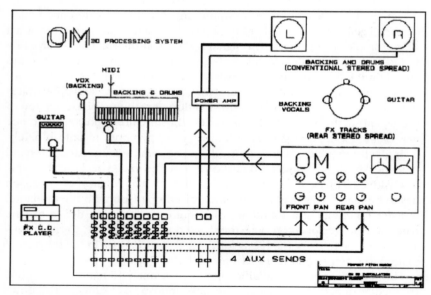

Figure 11.18 *Typical configuration for 3D mix*

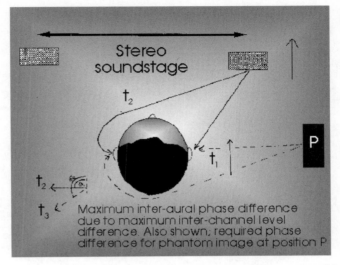

Figure 11.19 *The maximum interaural time difference which may be generated by a hard-right panned signal in a conventional stereophonic arrangement*

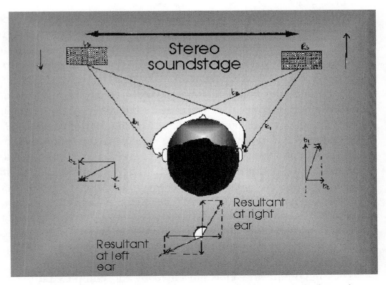

Figure 11.20 *The addition of an out-of-phase signal at the 'mute' loudspeaker can be used to increase the apparent interaural phase difference*

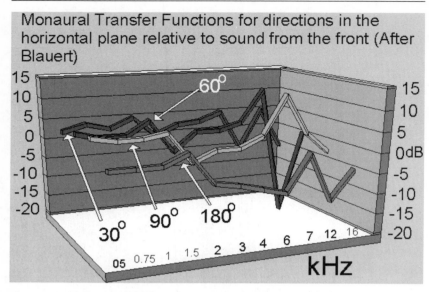

Figure 11.21   *Monaural transfer functions (after Blauert)*

other. The clearest component is around 7 kHz which shows a pronounced dip at 30° and a marked peak at 90°. Note that at 60°, an intermediate level peak at the same frequency exists – suggesting that linear interpolation is justifiable. Furthermore there is a gradual rise in HF at the 90 degree position and another high-Q effect at about 10 kHz. Obviously conditions in the example given are generalisable to extreme left positions *mutatis mutandis*.

To implement such a technique electronically we require a panning circuit which operates over much of its range like an existing pan control and then, at a certain point, splits the signal into high- and low-frequency components and feeds out-of-phase signals to the loudspeaker opposite the full pan and spectrally modifies the signals emitted from the speaker nearest the desired phantom position. This is relatively easy to implement digitally but an analogue approach was taken in the development of OM so that the unit could be easily installed in existing analogue mixing consoles. The circuit for the OM panner is given in Figure 11.22. The circuit may be broken down into four elements: a variable equaliser stage; a gyrator stage; a conventional pan control (see next chapter) and a low-frequency differential mode amplifier.

The signal entering A1 is amplified by the complex ratio $1 + (R_x/Z_1)$. Clearly this amplification is related to the setting of VR1a, which presents its maximum resistance in the centre position and its minimum at either end of its travel. Due to the impedance presented at the slider of VR1$_A$, the

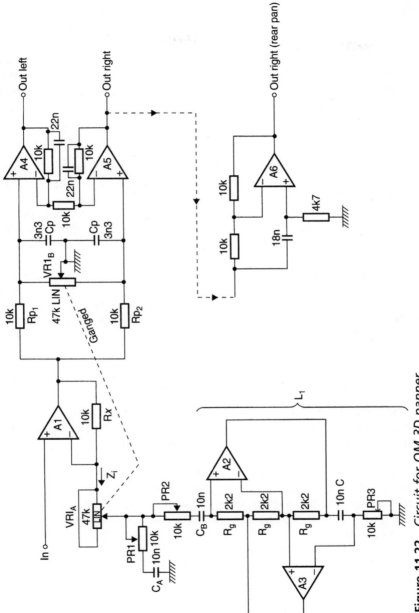

Figure 11.22 *Circuit for OM 3D panner*

frequency response of this circuit changes according to the control position. The function of C_A is to provide a general high-frequency rise. Were it not for the imposition of the preset resistance PR1, the circuit would become unstable when $VR1_A$ was at either end of its travel. PR1 allows the degree of boost to be set as $VR1_A$ is moved from its centre position. C_B, A2 and A3 and supporting components form a resonant acceptor circuit which resonates at the frequency the simulated inductance gyrator (L1) has the same reactance magnitude as C_B. And where L1 is given by the expression:

$$C \times R_g \times PR3$$

PR2 sets, in the same way as PR1 did above, the degree of lift at the resonant frequency of C_B and L1; this set to 7 kHz during test – as explained above. The signal that emerges from A1 is thus an equalised version of the input which acquires both general treble lift and specific boost around 7 kHz as the pan control is rotated in either direction away from the centre position. The degree of equalisation is adjustable by means of PR1 and PR2.

After the signal emerges from A1, it is fed to the relatively conventional panning control formed by Rp and $VR1_B$. The impedances are so arranged that the output signal is -3 dB in either channel at the centre position relative to extreme pan. A complication to this pan circuit are the capacitors strapped across each half of the pan control (Cp). These cause a degree of general high-frequency cut which acts mostly around the centre position and helps counteract the effect of the impedance at the slider of $VR1_A$ which still causes some equalisation, even at its centre position.

The signal then passes to A4 and A5, and associated components. This is an amplifier with a DC common mode gain of 1 and a DC differential gain of 2. The action of the 22 nF capacitors around the top leg of the feedback resistors is to limit the high-frequency differential mode gain to 1 while maintaining low-frequency differential gain as close to 2.

So, as the pan control is rotated to an extreme position, the differential mode amplifier gradually feeds low-frequency anti-phase information into the 'quiet' channel. At high frequency the differential mode amplifier has no effect but, instead, passes a high-frequency equalised version of the signal (with the few, salient auricle-coloration simulating high-frequency boost frequencies) to the left and right output. Figures 11.23 to 11.26 are oscillograms of typical output waveforms for a 400 Hz square-wave stimulus.

The subsidiary circuit (formed around A6) follows the circuitry described above on the rear-pan channels. This was largely designed to accomplish a phase inversion on one channel at low frequencies. This

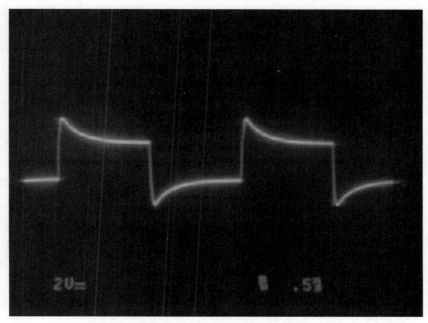

Figure 11.23 *OM, either channel, centre position*

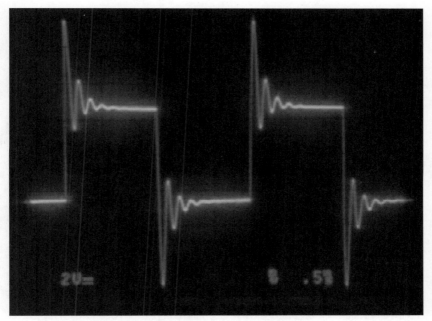

Figure 11.24 *OM, live channel, extreme pan*

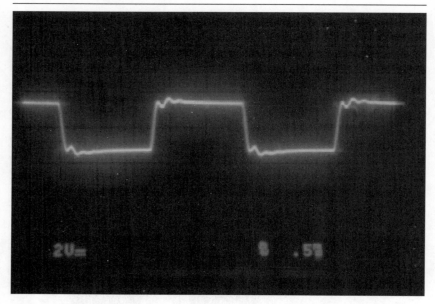

Figure 11.25   *OM, dead channel, extreme pan (note opposite phase to waveform in Figure 11.24)*

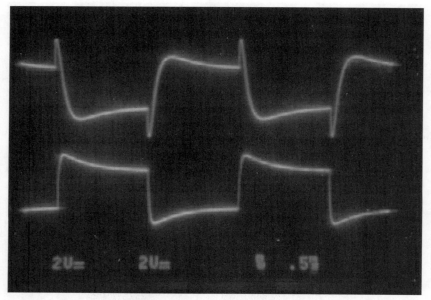

Figure 11.26   *OM, both channels, centre rear (note majority of low-frequency information is out of phase)*

technique facilitated a degree of mono compatibility. In effect, rear pan in the OM system was accomplished using a simple interchannel ratio technique on a signal which was largely composed of out-of-phase information at low frequencies. By adjusting the potentiometers which controlled high-frequency and high-Q boost appropriately it was further possible to create an appropriate attenuation of high frequencies in the rear arc – which is what is required from the psychoacoustic data. How this technique works is still not entirely known, its discovery being largely serendipitous. What is known is that head related movements reinforce a rear image generated this way and this is one of the most robust features of the OM system.

References

Bergault, D.R. (1994) *3-D sound for virtual reality and multimedia*. Academic Press Inc.
Blauert, J. (1983) *Spatial Hearing*. MIT Press.
Brice, R. (1997) *Multimedia and Virtual Reality Engineering*. Newnes.
Edeko, F.O. (1988) Improving stereophonic image sharpness. *Electronics and Wireless World*, Vol. 94, No. 1623 (Jan.).
Malham, D.G. (1995) *Basic Ambisonics*. York University Web pages.
Nisbett, A. (1979) *The Techniques of the Sound Studio*. Focal Press.

Notes

1 For the engineer developing spatial sound systems such a clear, concise theory has many attractions. However, experiments devised to investigate the localisation of transient sounds (as opposed to pure tones) appear to indicate to most psychologists that the situation is a little more complicated than Lord Rayleigh had supposed. However, it may be demonstrated that their evidence supports Duplex theory as an explanation for the perception of the localisation (or lateralisation) of transients (Brice 1997).

2 My original appreciation of this simple fact arose from a thought experiment designed to investigate the logical validity of the, oft cited, anecdotal preference among the golden-eared cognoscenti that vinyl records (and analogue recording equipment) can still offer a 'lifelike' quality unmatched by the technically superior digital compact disc. I reasoned that if I measured the differences between the signals produced from the pick-up cartridge and a CD player the signals would differ in several ways:

(i) The vinyl replay would have higher distortion and lower signal to noise ratio than its digital counterpart. There seemed little point in further investigating this aspect of recording and replay performance because so much work has been done on the subjective effects of dynamic range and linearity all suggesting, with very little room for error, that increased dynamic range and improving linearity correlate positively with improved fidelity and subjective preference.

(ii) The vinyl replay would have a frequency response which was limited with respect to the digital replay. This may play a part in preference for vinyl replay. It has been known since the 1940s that 'the general public' overwhelmingly prefer restricted bandwidth circuits for monaural sound reproduction.

(iii) Interchannel crosstalk would be much higher in the case of analogue recording and replay equipment. Furthermore, because crosstalk is usually the result of a negative reactance (either electrical or mechanical) this crosstalk would tend to increase proportionately with increasing signal frequency.

While attempting to imagine the subjective effect of this aperiodic crosstalk, it suddenly occurred to me that this mechanism would cause a progressive narrowing of the stereo image with respect to frequency which would (if it happened at the right rate) map the high-frequency intensity derived stereo image on top of the low-frequency, interaural delay derived image, thereby achieving the same effect as achieved by Edeko's loudspeakers and Blumlein's Shuffler circuit.

3 The term quadraphonic is due only to the fact that the systems employed four loudspeakers; by implication this would make stereo – biphonic!

Fact Sheet #11: An improved stereo microphone technique

- Blumlein's stereo
- A modern practical implementation

Blumlein's stereo

What makes Blumlein's 1933 'stereo' patent (REF 1) so important is his originality in realising the principle (explained in Chapter 11) that interchannel intensity differences alone produce both high-frequency interaural intensity differences and low-frequency inter-aural phase differences when listening with loudspeakers. Intriguingly, Blumlein regarded the principle of pan-potted stereo as trivial – it seems, even in 1933, the principle of positioning a pre-recorded single mono sound-signal by means of intensity control was well known. The technological problem Blumlein set out to solve was how to 'capture' the sound field; so that directional information was encoded solely as intensity difference.

Blumlein noted that a crossed-pair of velocity microphones mounted at 45 degrees to the centre of the stereo image has the technological advantage that a pure intensity-derived stereo signal may be obtained from such a configuration without the use of electrical matrixing. His instinct proved right because this has become one of the standard arrangements for the acquisition of intensity coded stereophony, to such an extent that this configuration has become associated exclusively with his name, often being referred to as, the 'Blumlein-pair', an eponymous, and somewhat incorrect label! In fact, the greater part of Blumlein's patent is concerned with a primitive 'dummy-head' (quasi-binaural) stereophonic microphone arrangement in which,

'two pressure microphones a1 and a2 [are] mounted on opposite sides of a block of wood or baffle b which serves to provide the high frequency intensity differences at the microphones in the same way as the human head operates upon the ears' (Figure F11.1).

Blumlein noted that, when listened to with headphones, the direct output from the microphones produced an excellent stereo effect but, when replayed through loudspeakers, the stereo effect was very disappointing. The transformation Blumlein required was the translation of low-frequency, inter-microphone phase differences into inter-channel intensity differences. He proposed the following technique:

'The outputs from the two microphones are taken to suitably arranged network circuits which convert the two primary channels into two secondary channels which may be called the summation and difference channels arranged so that the current flowing in the summation channel will represent the mean of the currents flowing in the two original channels, while

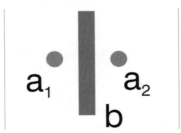

Figure F11.1   *Blumlein's microphone arrangement*

the current flowing into the difference channel will represent half the difference of the currents in the original channels. . . . Assuming the original currents differ in phase only, the current in the difference channel will be PI/2 different in phase from the current in the summation channel. This difference current is passed through two resistances in series between which is a condenser which forms a shunt arm. The voltage across this condenser will be in phase with that in the summation channel. By passing the current in the summation channel through a plain resistive attenuation network comprised of resistances a voltage is obtained which remains in phase with the voltage across the condenser in the difference channel. The voltages are then combined and re-separated by [another] sum and difference process. . . so as to produce two final channels. The voltage in the first final channel will be the sum of these voltages and the second final channel will be the difference between these voltages. Since these voltages were in phase the two final channels will be in phase but will differ in magnitude-only.

Blumlein's comments on the perpendicularly of the sum and difference vectors are far from obvious. However, consider Figure F11.2.

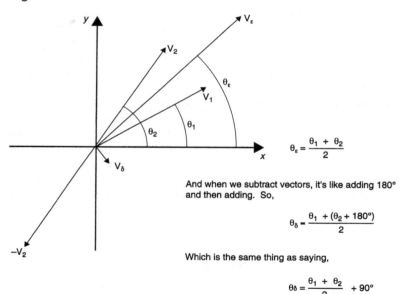

$$\theta_\varepsilon = \frac{\theta_1 + \theta_2}{2}$$

And when we subtract vectors, it's like adding 180° and then adding. So,

$$\theta_\delta = \frac{\theta_1 + (\theta_2 + 180°)}{2}$$

Which is the same thing as saying,

$$\theta_\delta = \frac{\theta_1 + \theta_2}{2} + 90°$$

Figure F11.2 *Provided the magnitude of two vectors remains identical, the sum vector and the difference vector are always perpendicular, as shown*

A modern practical implementation

The circuit described below is designed so that maximum stereo obliquity is achieved when the inter-microphone delay is 500 μs. Other calibrations are possible *mutatis mutandis*. Table F11.1 below tabulates the phase-angle which 500 μs represents at various frequencies.

Table F11.1

30 Hz	5.4 degrees
60 Hz	10.8 degrees
300 Hz	54 degrees
1 kHz	180 degrees

Consider the 30 Hz case. The circuit operates by first deriving the sum and difference of the phasor (vector) quantities derived from the primary left and right channels, i.e.

Let $V_1 = (0, 1)$ and

$V_2 = (\sin 5.4 \text{ degrees}, \cos 5.4 \text{ degrees}) = (0.1, 0.996)$

$V_{sum} = V_1 + V_2 = (0.1, 1.996)$, which has a magnitude $= 2$.

$V_{diff} = V_1 - V_2 = (-0.1, 0.004)$ which has a magnitude $= 0.1$.

So, at 30 Hz, the difference channel is 20 times (26 dB) smaller than the signal in the sum channel.
    Now consider the situation at 300 Hz, where

$V_2 = (\sin 54 \text{ degrees}, \cos 54 \text{ degrees}) = (0.81, 0.59)$

$V_{sum} = (0.81, 1.59)$, magnitude $= 1.78$

$V_{diff} = (-0.81, -0.41)$, magnitude $= 0.9$.

at 300 Hz the signal is approximately 2 times smaller (6 dB) compared with the signal in the sum channel.
    Now 300 Hz is nearly three octaves away from 30 Hz and the gain is 20 dB different demonstrating that the signal in the difference channel rises by 6 dB/octave. This confirms Blumlein's statement that, 'for a given obliquity of sound the phase difference is approximately proportional to frequency, representing a fixed time delay between sound arriving at the two ears.'

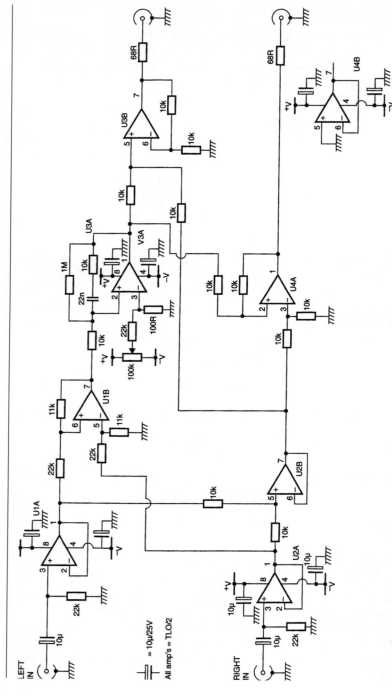

Figure F11.3 *Circuit schematic for the binaural to summation stereophony transcoder*

Looking now at the circuit diagram for the binaural to summation stereophony transcoder illustrated in Figure F11.3, consider the role of the integrator circuit implemented around U3a. The role of this circuit is both, to rotate the difference phasor by 90 degrees (and thus align it with the axis of the phasor in the sum channel) and to provide the gain/frequency characteristic to compensate for the rising characteristic of the signal in the difference channel. This could be achieved with a simple integrator. However, at intermediate and high frequencies (>1000 Hz), it is necessary to return the circuit to a straightforward matrix arrangement which transmits the high frequency differences obtained due to the baffling effect of the block of wood directly into the stereo channels. This is implemented by returning the gain and phase characteristic of the integrator-amplifier to 0 dB and 0 degrees phase-shift at high frequencies. This is the function of the 10 k resistor in series with the 22 nF integrator capacitor. (The actual circuit returns to 180 degrees phase shift at high frequencies – i.e. not 0 degrees; this is a detail which is compensated for in the following sum and difference arrangement.)

Clearly all the above calculations could be made for other microphone spacings. For instance, consider the situation in which two spaced omni's (6 ft apart) are used as a stereo pick-up arrangement. With this geometry, 30 Hz would produce nearly 22 degrees of phase shift between the two microphones for a 30 degree obliquity. This would require,

Magnitude of sum phasor = 1.97

Magnitude of difference phasor = 0.39

that is, an integrator with a gain of 5. The gain at high frequency would once again need to fall to unity. At first this seems impossible because it requires the stand-off resistor in the feedback limb to remain 10 k as drawn in the figure above. However, consideration reveals that the transition region must begin at commensurately lower frequencies for a widely spaced microphone system (since phase ambiguities of >180 degrees will arise at lower frequencies) so that all that needs to be scaled is the capacitor, revealing that there is a continuum of possibilities of different microphone spacings and translation circuit values.

References

1. Blumlein, A. British Patent 394,325 June 14th 1933

12
Let's Stick Together – Recording consoles

Introduction

This chapter is about recording consoles, the very heart of a recording studio. Like our own heart, whose action is felt everywhere in our own bodies, consideration of a recording console involves wide-ranging considerations of other elements within the studio system. These, too, are covered in this chapter.

In pop and rock music – as well as in most jazz recordings – each instrument is almost always recorded onto one track of multi-track tape and the result of the 'mix' of all the instruments combined together electrically inside the audio mixer and recorded onto a two-track, (stereo) master tape for production and archiving purposes. Similarly in the case of sound reinforcement for rock and pop music and jazz concerts, each individual musical contributor is separately miked and the ensemble sound mixed electrically. It is the job of the recording or balance engineer to control this process. This involves many aesthetic judgements in the process of recording the individual tracks (tracking) and mixing down the final result. However there exist relatively few parameters under her/his control. Over and above the office of correctly setting the input gain control so as to ensure best signal to noise ratio and control of channel equalisation, her/his main duty is to judge and adjust each channel gain fader and therefore each contributor's level within the mix. A further duty, when performing a stereo mix, is the construction of a stereo picture or image by controlling the relative contribution each input channel makes to the two, stereo mix amplifiers. In the cases of both multi-track mixing and multi-microphone mixing, the apparent position of each instrumentalist within the stereo picture (image) is controlled by a special stereophonic panoramic potentiometer or pan pot for short.

Standard levels and level meters

Suppose I asked you to put together a device comprising component parts I had previous organised from different sources. And suppose I had paid very little attention to whether each of the component parts would fit together (perhaps one part might be imperial and another metric). You would become frustrated pretty quickly because the task would be impossible. So it would be too for the audio mixer, if the signals it received were not, to some degree at least, standardised. The rational behind these standards and the tools used in achieving this degree of standardisation are the subjects of the first few sections of this chapter.

The adoption of standardised studio levels (and of their associated line-up tones) ensures the interconnectability of different equipment from different manufacturers and ensures that tapes made in one studio are suitable for replay and/or rework in another. Unfortunately, these 'standards' have evolved over many years and some organisations have made different decisions which, in turn, have reflected upon their choice of operating level. National and industrial frontiers exist too, so that the subject of maximum and alignment signal levels is fraught with complication.

Fundamentally there exist only two absolute levels in any electronic system, maximum level and noise floor. These are both illustrated in Figure 12.1. Any signal which is lower than the noise floor will disappear as it is swamped by noise and signal which is larger than maximum level will be distorted. All well recorded signals have to sit comfortably between the 'devil' of distortion and the 'deep blue sea' of noise. Actually, that's the fundamental job of any recording engineer!

In principle, maximum level would make a good line-up level. Unfortunately, it would also reproduce over loudspeakers as a very loud noise indeed and would therefore, likely as not, 'fray the nerves' of those people working day after day in recording studios! Instead a lower level is

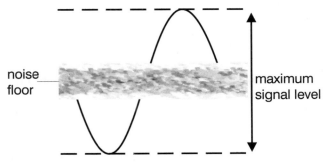

Figure 12.1 *System dynamic range*

used for line-up which actually has no physical justification at all. Instead it is cleverly designed to relate maximum signal level to the perceptual mechanism of human hearing and to human sight as we shall see. Why sight? Because it really isn't practical to monitor the loudness of an audio signal by sound alone. Apart from anything else, human beings are very bad at making this type of subjective judgement. Instead, from the very earliest days of sound engineering, visual indicators have been used to indicate audio level, thereby relieving the operator from making subjective auditory decisions. There exist two important and distinct reasons to monitor audio level.

The first is to optimise the drive, the gain or sensitivity of a particular audio circuit, so that the signal passing through it is at a level whereby it enjoys the full dynamic range available from the circuit. If a signal travels through a circuit at too low a level it unnecessarily picks up noise in the process. If it is too high, it may be distorted or 'clipped' as the stage is unable to provide the necessary voltage swing, as shown in Figure 12.1.

The second role for audio metering exists in, for instance, a radio or television continuity studio where various audio sources are brought together for mixing and switching. Listeners are justifiably entitled to expect a reasonably consistent level when listening to a radio (or television) station and do not expect one programme to be unbearably loud (or soft) in relation to the last. In this case, audio metering is used to judge the apparent loudness of a signal and thereby make the appropriate judgements as to whether the next contribution should be reduced (or increased) in level compared with the present signal.

The two operational requirements described above demand different criteria of the meter itself. This pressure has led to the evolution of two types of signal monitoring meter, the volume unit (VU) meter and the peak programme meter (PPM).

The VU meter

A standard VU meter is illustrated in Figure 12.2a. The VU is a unit intended to express the level of a complex wave in terms of decibels above or below a reference volume, it implies a complex wave – a programme waveform with high peaks. A 0 VU reference level therefore refers to a complex-wave power-reading on a standard VU meter. A circuit for driving a moving coil VU meter is given in Figure 12.2b. Notice that the rectifiers and meter are fed from the collector of TR_a which is a current source in parallel with Re. Because Re is a high value in comparison with the emitter load of TR_a the voltage gain during the part of the input cycle when the rectifier diodes are not in conduction is very large. This alleviates most of the problem of the Si diodes' offset voltage. From the circuit it is clear that a VU meter is an indicator of the average power of a waveform; it therefore accurately

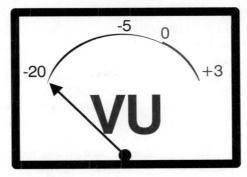

Figure 12.2a *VU meter*

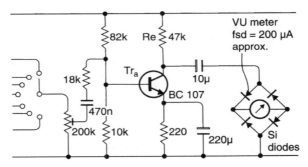

Figure 12.2b *VU meter circuit*

represents the apparent loudness of a signal because the ear too mathematically integrates audio waveforms with respect to time. But, because of this, the VU is not a peak-reading instrument. A failure to appreciate this, and on a practical level this means allowing the meter needle to swing into the red section on transients, means the mixer is operating with inadequate system headroom. This characteristic has led the VU to be regarded with suspicion in some organisations.

To some extent this is unjustified, because the VU may be used to monitor peak levels provided the action of the device is properly understood. The usual convention is to assume that the peaks of the complex wave will be 10 to 14 dB higher than the peak value of a sine wave adjusted to give the same reference reading on the VU meter. In other words, if a music or speech signal is adjusted to give a reading of 0 VU on a VU meter, the system must have at least 14 dB headroom – over the level of a sine wave adjusted to give the same reading – if the system is not to clip the programme audio signal. In operation the meter needles should only very occasionally swing above the 0 VU reference level on complex programme.

The PPM meter

Whereas the VU meter reflects the perceptual mechanism of the human hearing system, and thereby indicates the loudness of a signal, the PPM is designed to indicate the value of peaks of an audio waveform. It has its own powerful champions, notably the BBC and other European broadcasting institutions. The PPM is suited to applications in which the balance engineer is setting levels to optimise a signal level to suit the dynamic range available from a transmission (or recording) channel. Hence its adoption by broadcasters who are under statutory regulation to control the depth of their modulation and therefore fastidiously to control their maximum signal peaks. In this type of application, the balance engineer does not need to know the 'loudness' of the signal, but rather needs to know the maximum excursion (the peak value) of the signal.

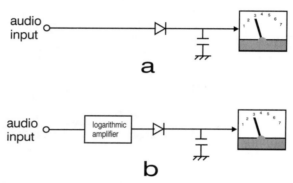

Figure 12.3 *Peak reading meters*

It is actually not difficult to achieve a peak reading instrument. The normal approach being a meter driven by a buffered version of a voltage stored on a capacitor, itself supplied by a rectified version of the signal to be measured (see Figure 12.3(a)). In fact, the main limitation of this approach lies with the ballistics of the meter itself which, unless standardised, leads to different readings. The PPM standard demands a defined and consistent physical response time of the meter movement. Unfortunately, the simple arrangement is actually unsuitable as a volume monitor due to the highly variable nature of the peak to average ratio of real-world audio waveforms, a ratio known as crest factor. This enormous ratio causes the meter needle to flail about to such an extent that it is difficult to interpret anything meaningful at all! For this reason, to the simple arrangement illustrated in Figure 12.3(a), a logarithmic amplifier is appended as shown at (b). This effectively compresses the dynamic range of the signal prior to its display; a modification which (together with a

controlled decay time constant) greatly enhances the PPM's readability –
albeit at the expense of considerable complexity.

The peak programme meter of the type used by the BBC is illustrated in
Figure 12.4. Notice the scale marked 1 to 7, each increment representing
4 dB (except between 1 and 2 which represents 6 dB). This constant
deflection per decade is realised by the logarithmic amplifier. Line-up tone
is set to PPM4 and signals are balanced so that peaks reach PPM6, that is

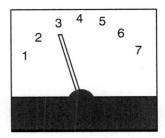

Figure 12.4   *BBC style PPM*

8 dB above reference level. (BBC practice is that peak studio level is 8 dB
above alignment level.) BBC research has shown that the true peaks are
actually about 3 dB higher than those indicated on a BBC PPM and that
operator errors cause the signal to swing occasionally 3 dB above indicated
permitted maximum, i.e. a total of 14 dB above alignment level.

PPM dynamic performance

PPM

In BS55428 Part 9, the PPM is stated as requiring, 'an integration time of
12 ms and a decay time of 2.8 s for a decay from 7 to 1 on the scale'. This
isn't entirely straightforward to understand. However, an earlier standard
(British Standard BS4297:1968) defined the rise time of a BBC style PPM
in terms of reading relative to 5 kHz tone-burst durations such that, for a
steady tone adjusted to read scale 6, bursts of various values should be
within the limits given below:

Burst duration	*Meter reading (relative to 6)*
Continuous	0 dB
100 ms	0 ± 0.5 dB
10 ms	−2.5 ± 0.5 dB
5 ms	−4.0 ± 0.5 dB
1.5 ms	−9.0 ± 1.0 dB

This definition has the merit of being testable.

VU meter

The VU meter is essentially a milliammeter with a 200 mA FSD fed from a full-wave rectifier installed within the case with a series resistor chosen such that the application of a sine-wave of 1.228 V RMS, (i.e. 4 dB above that required to give 1 mW in 600 R) causes a deflection of 0 VU. Technically, this makes a VU an RMS reading volt meter. Of course, for a sine-wave the relationship between peak and RMS value is known (3 dB or $1/\sqrt{2}$), but there exists no simple relationship between RMS and pk for real-world audio signals.

In frequency response terms, the response of the VU is essentially flat (0.2 dB limits) between 35 Hz and 10 kHz. The dynamic characteristics are such that when a sudden sine-wave type signal is applied, sufficient to give a deflection at the 0 VU point, the pointer shall reach the required value within 0.3 second and shall not overshoot by more than 1.5% (0.15 dB).

Opto-electronic level indication

Electronic level indicators range from professional bargraph displays which are designed to mimic VU or PPM alignments and ballistics, through the various peak-reading displays common on consumer and prosumer goods (often bewilderingly calibrated), to simple peak-indicating LEDs. The latter, can actually work surprisingly well – and actually facilitate a degree of precision alignment which belies their extreme simplicity.

In fact, the difference between monitoring using VUs and PPMs is not as clear cut as stated. Really both meters reflect a difference in emphasis: the VU meter indicates loudness – leaving the operator to allow for peaks based on the known, probabilistic nature of real audio signals. The PPM, on the other hand, indicates peak – leaving it to the operator to base decisions of apparent level on the known stochastic nature of audio waveforms. However, the latter presents a complication because, although the PPM may be used to judge level, it does take experience. This is because the crest factor of some types of programme material differs markedly from others especially when different levels of compression are used between different contributions. To allow for this, institutions which employ the PPM apply ad hoc rules to ensure continuity of level between contributions and/or programme segments. For instance, it is BBC practice to balance different programme material to peak at different levels on a standard PPM.

Despite its powerful European opponents, a standard VU meter combined with a peak-sensing LED is very hard to beat as a monitoring device, because it both indicates volume and, by default, average crest factor. Any waveforms which have unusually high peak to average ratio are indicated by the illumination of the peak LED. PPMs unfortunately do not indicate loudness and their widespread adoption in broadcast accounts for the many uncomfortable level mismatches between different contributions, especially between programmes and adverts.

Polar CRT displays

One very fast indicator of changing electrical signals is a cathode ray tube (CRT – see Chapter 15). With this in mind there has, in recent years, been a movement to use CRTs as a form of fast audio monitoring, not as in an oscilloscope, with an internal timebase, but as a polar, or XY display. The two-dimensional polar display has a particular advantage over a classic, one-dimensional device like a VU or PPM in that it can be made to display left and right signals at the same time. This is particularly useful because, in so doing, it permits the engineer to view the degree to which the left and right signals are correlated; which is to say the degree to which a stereo signal contains in-phase, mono components and the degree to which it contains out-of-phase or stereo components.

In the polar display, the Y plates inside the oscilloscope are driven with a signal which is the sum of the left and right input signal (suitably amplified). The X plates are driven with a signal derived from the stereo difference signal (R – L), as shown in Figure 12.5. Note that the left signal will create a single moving line along the diagonal L axis as shown. The right signal clearly does the same thing along the R axis. A mono (L = R) signal will create a single vertical line, and an out-of-phase mono signal will produce a horizontal line. A stereo signal produces a woolly ball

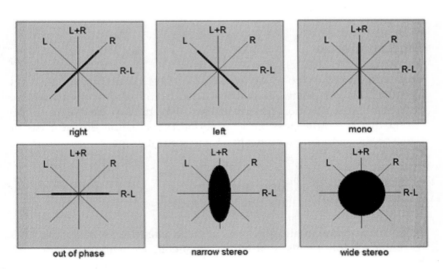

Figure 12.5 *Audio polar displays*

centred on the origin; its vertical extent governed by the degree of L/R correlation and its horizontal extent governed by L/R de-correlation. And herein lies the polar display's particular power, that it can be used to assess the character of a stereo signal, alerting the engineer to possible transmission or recording problems, as illustrated in Figure 12.5.

One disadvantage of the polar display methodology is that, in the absence of modulation, the cathode ray will remain undeviated and bright spot will appear at the centre of the display, gradually burning a hole on the phosphor! To avoid this, commercial polar displays incorporate cathode modulation (k mod) so that, if the signal goes below a certain value, the cathode is biased until the anode current cuts-off, extinguishing the beam.

Standard operating levels and line-up tones

Irrespective of the type of meter employed, it should be pretty obvious that a meter is entirely useless unless it is calibrated in relation to a particular signal level (think about it if rulers had different centimetres marked on them!).

Three important line-up levels exist (see Fig. 12.6):

PPM4 = 0 dBu = 0.775 V RMS, used by UK broadcasters.
0 VU = +4 dBu = 1.23 V RMS, used in commercial music sector.
0 VU = –10 dBV = 316 mV RMS, used in consumer and 'prosumer'
equipment.

Digital line-up

The question of how to relate 0 VU and PPM 4 to digital maximum level of 0 dBFS (for 0 dB relative to Full Scale) has been the topic of hot debate. Fortunately, the situation has crystallised over the last few years to the extent that it is now possible to describe the situation on the basis of widespread implementation in USA and European broadcasters. Essentially,

0 VU = +4 dBu = –20 dBFS (SMPTE RP155)
PPM4 = 0 dBu = –18 dBFS (EBU R64-1992)

Sadly, these recommendations are not consistent. And while the EBU recommendation seems a little pessimistic in allowing an extra 4 dB headroom above their own worst-case scenario, the SMPTE suggestion looks positively gloomy in allowing 20 dB above alignment level. Although

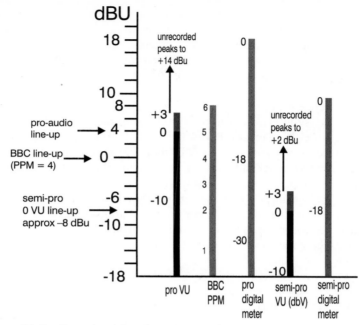

Figure 12.6 *Standard levels compared*

this probably reflects the widespread – though technically incorrect – methodology, when monitoring with VUs, of setting levels so that peaks often drive the meter well into the red section.

Sound mixer architecture and circuit blocks

The largest, most expensive piece of capital electronic equipment in any professional studio is the main studio mixer. So much so that in publicity shots of a recording studio this component is always seen to dominate the proceedings! Perhaps that's why there seem to be so many terms for it: mixer, mixing desk, or just desk, audio console, console and simply 'the board', to name but a few. To many people outside of the recording industry, the audio mixer represents the very essence of sound recording. This is partially correct, for it is at the console that are made many of the important artistic decisions that go towards mixing a live band or a record. However, in engineering terms, this impression is misleading. The audio console is a complicated piece of equipment but, in its electronic essence, its duties are relatively simple. In other words, the designer of an audio mixer is more concerned with the optimisation of relatively simple circuits – which may then be repeated many, many times – than she is with the design of clever or imaginative signal processing. But before we investigate the individual circuit elements of an audio mixer, it is important to understand the way in which these blocks fit together. This is usually termed system architecture.

System architecture

There is no simple description of audio console system architecture. That's because there exist different types of consoles for different duties and because every manufacturer (and there are very many of them) each has their own ideas about how best to configure the necessary features in a manner which is both operationally versatile, ergonomic and which maintains the 'cleanest' signal path from input to output. However, just as houses all come in different shapes and sizes and yet all are built relying upon the same underlying assumptions and principles, most audio mixers share certain system topologies.

Input strip

The most conspicuous 'building block' in an audio console, and the most obviously striking at first glance, is the channel input strip. Each mixer channel has one of these and they tend to account for the majority of the panel space in a large console. A typical input strip for a small recording console is illustrated in Figure 12.9. When you consider that a large console may have 24, 32 or perhaps 48 channels – each a copy of the strip illustrated in Figure 12.9 – it is not surprising that large commercial studio consoles look so imposing. But always remember, however 'frightening' a console looks, it is usually only the sheer bulk which gives this impression. Most of the panel is repetition and once one channel strip is understood, so are all the others!

Much harder to fathom, when faced with an unfamiliar console, is the bus and routing arrangements which feed the group modules, the monitor and master modules. These 'hidden' features relate to the manner in which each input module may be assigned to the various summing amplifiers within the console. And herein lies the most important thing to realise about a professional audio console; that it is many mixers within one console. First let's look at the groups.

Groups

Consider mixing a live rock band. Assume it is a quintet; a singing bass player, one guitarist, a keyboard player, a saxophonist and a drummer. The inputs to the mixer might look something like this:

Channel 1   Vocal mic
Channel 2   Bass amp mic

Channel 3	Lead guitar amp mic
Channel 4	Backing mic (for keyboardist)
Channel 5	Backing mic (for second guitarist)
Channel 6	Sax mic
Channel 7	Hi-hat mic
Channel 8	Snare mic
Channel 9	Bass drum mic
Channel 10 & 11	Drum overheads
Channel 12 & 13	Stereo line piano input
Channel 14 & 15	Sound module line input

Clearly inputs 7 to 11 are all concerned with the drums. Once these channels have been set so as to give a good balance between each drum, it is obviously convenient to group these faders together so that the drums can be adjusted relative to the rest of the instruments. This is the role of the separate summing amplifiers in a live console, to group various instruments together. That's why these smaller 'mixers within mixers' are called groups. These group signals are themselves fed to the main stereo mixing amplifier, the master section. Mixer architecture signal flow is, therefore, channelled to groups and groups to master output mixer as shown in Figure 12.7. A block schematic of a simplified live-music mixer is given in Figure 12.8 in which this topology is evident.

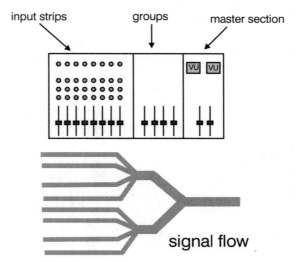

Figure 12.7 *Signal flow*

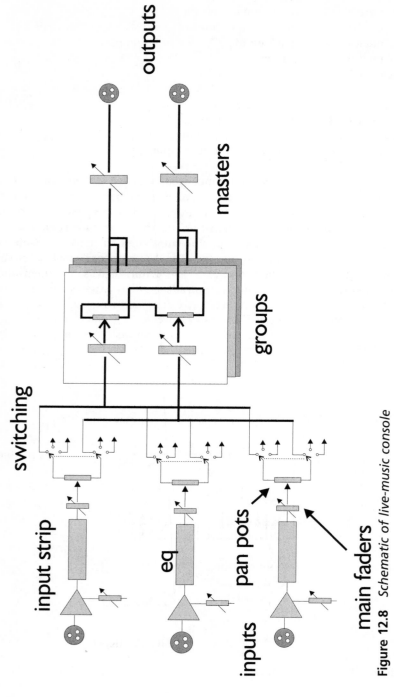

inputs

input strip

eq

pan pots

main faders

switching

groups

masters

outputs

Figure 12.8 *Schematic of live-music console*

Pan control

As we saw in the last chapter, stereophonic reproduction from loudspeakers requires that stereo information is carried by interchannel intensity differences alone – there being no requirement for interchannel delay differences. The pan control progressively attenuates one channel while progressively strengthening the other as the knob is rotated, the input being shared equally between both channels when the knob is in its centre (12 o'clock) position. In the sound mixer shown in Figure 12.8, notice that the channel pan control operates in a rather more flexible manner, as a control which may be used to 'steer' the input signal between either of the pairs of buses selected by the routing switches. The flexibility of this approach becomes evident when a console is used in a multi-track recording session.

Effect sends and returns

Not all the functions required by the balance engineer can be incorporated within the audio mixer. To facilitate the interconnection with outboard equipment, most audio mixers have dedicated mix amplifiers and signal injection points called effect sends and returns which make incorporation of other equipment within the signal flow as straightforward as possible.

The groups revisited

In a recording situation, the mixing groups may well ultimately be used in the same manner as described in relation to the live console, to group sections of the arrangement so as to make the task of mixing more manageable. But the groups are used in a totally different way during recording or tracking. During this phase, the groups are used to route signals to the tracks of the multi-track tape machine. From an electronic point of view, the essential difference here is that, in a recording situation, the group outputs are directly utilised as signals and a recording mixer must provide access to these signals. Usually a multi-track machine is wired so that each group output feeds a separate track of the multi-track tape recorder. Often there are not enough groups to do this, in which case, each group feeds a number of tape machine inputs, usually either adjacent tracks or in 'groups of groups', so that, for instance, groups 1 to 8 will feed inputs 1 to 8 and 9 to 16 and so on.

The recording console

So far, this is relatively straightforward. But a major complication arises during the tracking of multi-track recordings because not only must signals be routed to the tape recorder via the groups, tape returns must be routed back to the mixer to guide the musicians as to what to play next. And this must happen at the same time! In fact, it is just possible for a good sound engineer, using 'crafty' routing, to cope with this using a straightforward live mixing desk. But it is very difficult. What is really required is a separate mixer to deal with the gradually increasing numbers of tape replay returns, thereby keeping the main mixer free for recording duties. Various mixer designers have solved this problem in different ways. Older consoles (particularly of English origin) have tended to provide an entirely separate mixer (usually to the right of the group and main faders) devoted to dealing with the return signals. Such a mixer architecture is known as a split console. The alternative approach – which

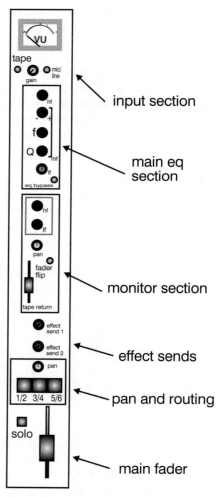

Figure 12.9 *Input strip of in-line console*

is now very widespread – is known as the in-line console; so named because the tape-return controls are embedded within each channel strip, in line with the main fader. This is the type of console which is considered in detail below. From an electronic point of view there exists very little difference between these approaches, the difference is more one of operational philosophy and ergonomics.

Both the split and in-line console are yet another example of a 'mixer within a mixer'. In effect, in the in-line console, the tape returns feed an entirely separate mixer so that each tape return signal travels via a separate fader (sometimes linear, sometimes rotary) and pan control before being summed on an ancillary stereo mix bus known as the monitor bus. The channel input strip of an in-line console is illustrated in Figure 12.9. The monitor bus is exceptionally important in a recording console because it is the output of this stereo mix amplifier that supplies the signal that is fed to the control room power amplifier during the tracking phase of the recording process. (Usually control room outputs are explicitly provided on the rear panels of a recording console for this purpose.) The architecture is illustrated in Figure 12.10. During mixdown, the engineer will want to operate using the main faders and pan controls, because these are the most operationally convenient controls, being closest to the mixer edge nearest the operator. To this end, the in-line console includes the ability to switch the tape returns back through the main input strip signal path, an operation known as 'flipping' the faders. The circuitry for this is illustrated in Figure 12.11.

Talkback

Talkback exists so that people in the control room are able to communicate with performers in the studio. So as to avoid sound 'spill' from loudspeaker into open microphones (as well as to minimise the risk of howl-round), performers inside the studio invariably wear headphones and therefore need a devoted signal which may be amplified and fed to special headphone amplifiers. In the majority of instances this signal is identical to the signal required in the control room during recording (i.e. the monitor bus). In addition a microphone amplifier is usually provided within the desk which is summed with the monitor bus signal and fed to the studio headphone amplifiers. This microphone amplifier is usually energised by a momentary switch to allow the producer or engineer to communicate with the singer or instrumentalist but which cannot, thereby, be left open thus distracting the singer or allowing them to hear a comment in the control room which may do nothing for their ego!

Equalisers

For a variety of reasons, the signals arriving at the recording console may require spectral modification. Sometimes this is due to the effect of

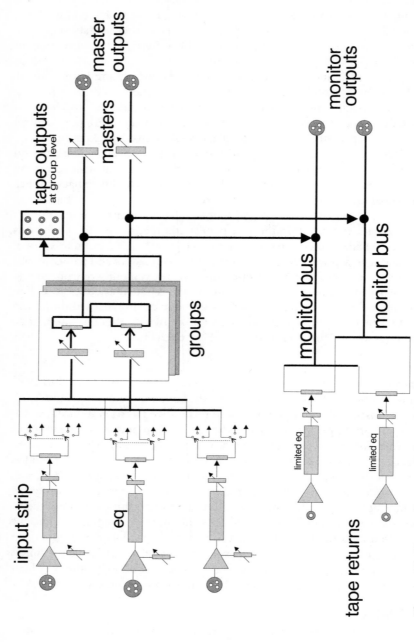

Figure 12.10 *In-line console architecture*

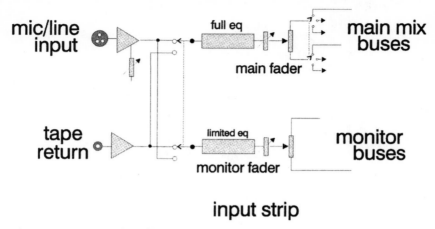

Figure 12.11 *Fader 'flip' mechanism*

inappropriate microphone choice or of incorrect microphone position. Sometimes it is due to an unfortunate instrumental tone (perhaps an unpleasant resonance). Most often, the equaliser (or simply EQ) is used in a creative fashion and to enhance or subdue a band (or bands) of frequencies so as to blend an instrument into the overall mix, or boost a particular element so that its contribution is more incisive.

It is this creative element in the employment of equalisation which has created the situation that exists today, that the quality of the EQ is often a determining factor in a recording engineer's choice of one console over another. The engineering challenges of flexible routing, low interchannel crosstalk, low noise and good headroom having been solved by most good manufacturers, the unique quality of each sound desk often resides in the equaliser design. Unfortunately this state of affairs introduces a subjective (even individualistic) element into the subject of equalisation which renders it very difficult to cover comprehensively. Sometimes it seems that every circuit designer, sound engineer and producer each has his, or her, idea as to what comprises an acceptable, an average and an excellent equaliser! A simple equaliser section – and each control's effect – is illustrated in Figure 12.12.

Audio mixer circuitry

Now we understand the basic architecture of the mixer, it is time to look at each part again and understand the function of the electrical circuits in each stage in detail.

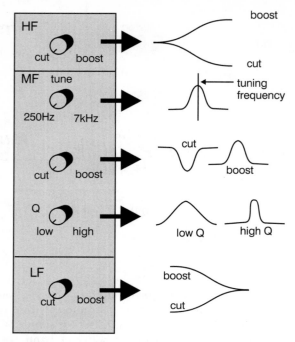

Figure 12.12   *Equaliser controls*

The input strip was illustrated in Figure 12.9. Notice that below the VU meter, the topmost control is the channel gain trim. This is usually switchable between two regimes: a high gain configuration for microphones and a lower gain line level configuration. This control is set with reference to its associated VU meter. Below the channel gain are the equalisation controls, the operation of these controls is described in detail below.

Microphone pre-amplifiers

Despite the voltage amplification provided by a transformer or pre-amplifier, the signal leaving a microphone is still, at most, only a few millivolts. This is much too low a level to be suitable for combining with the outputs of other microphones inside an audio mixer. So, the first stage of any audio mixer is a so-called microphone pre-amplifier which boosts the signal entering the mixer from the microphone to a suitable operating level. The example below is taken from a design (Brice 1990) which was for a portable, battery driven mixer – hence the decision to use discrete transistors, rather than use current-thirsty op-amps. Each of the input stage microphone amplifiers is formed of a transistor 'ring of three'. The

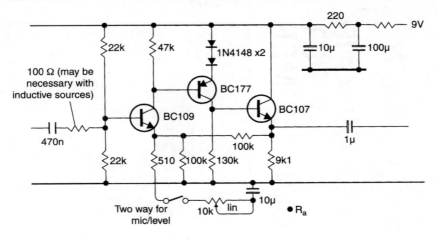

Figure 12.13 *Microphone pre-amplifier*

design requirement is for good headroom and a very low noise figure. The final design is illustrated in Figure 12.13.[1] The current consumption for each microphone pre-amplifier is less than 1 mA.

For more demanding studio applications a more complex microphone pre-amplifier is demanded. First, the stage illustrated in Figure 12.13 is only suitable for unbalanced microphones and the majority of high quality microphones utilise balanced connection, since they are especially susceptible to hum pick-up if this type of circuit is not employed. Second, high quality microphones are nearly always capacitor types and therefore require a polarising voltage to be supplied for the plates and for powering the internal electronics. This supply (known as phantom power as mentioned above) is supplied via the microphone's own signal leads and must therefore be isolated from the microphone pre-amplifier. This is one area where audio transformers (Figure 12.14) are still used, and an example of a microphone input stage using a transformer is illustrated, although there exist plenty of practical examples where transformerless stages are used for reasons of cost.

Notice that the phantom power is provided via resistors which supply both phases on the primary side of the transformer, the current return being via the cable screen. The transformer itself provides some voltage gain at the expense of presenting a much higher output impedance to the following amplifier. However, the amplifier has an extremely high input impedance (especially so when the negative feedback is applied) so this is really of no consequence. In Figure 12.14, amplifier gain is made variable so as to permit the use of a wide range of microphone sensitivities and applications. An ancillary circuit is also provided which

Data Sheet

jensen transformers
INCORPORATED

JE-110K-HPC
MICROPHONE INPUT TRANSFORMER

The JE-110K-HPC is a printed circuit type 150/10K with a winding similar to the JE-115K-E. The multiple interleaved layer winding exhibits very low leakage inductance requiring no series RC network across the 100K ohm secondary load resistor when used with an amplifier incorporating 2 μS phase lead compensation in the feedback circuit. Since the PC bobbin contains a smaller stack of laminations than the wire lead JE-115K-E, the JE-110K-HPC uses more total turns of smaller wire. The result is higher maximum level capability at low frequencies and the distortion is the lowest of all types in this size (0.11% @ 20Hz), but the higher series losses increase the noise by 0.9dB, compared to the wire lead version JE-115K-E.

The pin pattern is compatible with the JE-6110K-APC.

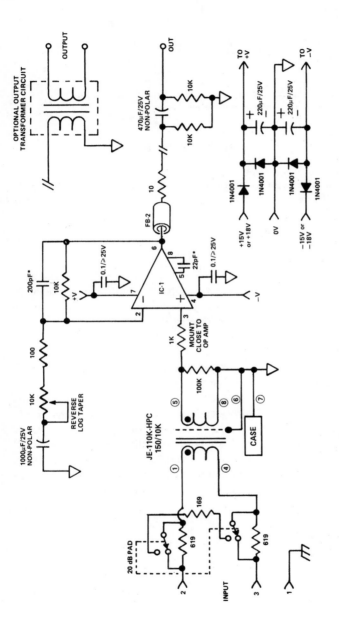

SCHEMATIC DIAGRAM OF TYPICAL MICROPHONE PREAMPLIFIER UTILIZING JE-110K-HPC

NOTES:

1. IC-1 = integrated circuit opamp such as MA-332 or NE-5534.
2. Gain Range: +24dB ➔ +58dB.
3. Keep traces short between transformer and opamp.
4. All resistors = 1%, metal film.
5. 200pF cap in feedback = 2 μsec compensation.
6. FB-2 = ferrite bead available from Jensen.
7. Capacitors marked * = polystyrene or polypropylene.

Figure 12.14 *Microphone transformer*

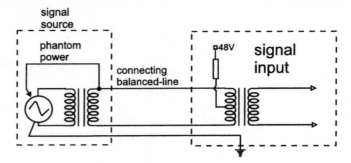

Figure 12.15   *Transformer-based input stage*

enables the signal entering the circuit to be attenuated by 20 dB, this is to allow for very high output microphones without overloading the electronics.

Insert points

After the channel input amplifier, insert points are usually included. These allow external (outboard) equipment to be patched into the signal path. Notice that this provision is usually via jack connectors which are normalised when no plug is inserted (Figure 12.16).

Equalisers and tone-controls

At the very least, a modern recording console provides a basic tone-control function on each channel input, like that shown in Figure 12.17. This circuit is a version of the classic circuit due to Baxandall. But this type of circuit only provides fairly crude adjustment of bass and treble ranges, as illustrated in Figure 12.12. This type of response (for fairly obvious reasons) is often termed a 'shelving' equaliser. So, the Baxandall shelving EQ is invariably appended with a mid-frequency equaliser which is tuneable over a range of frequencies; thereby enabling the sound engineer to adjust the middle band of frequencies in relation to the whole

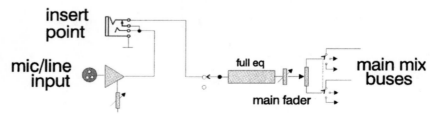

Figure 12.16   *Insert points*

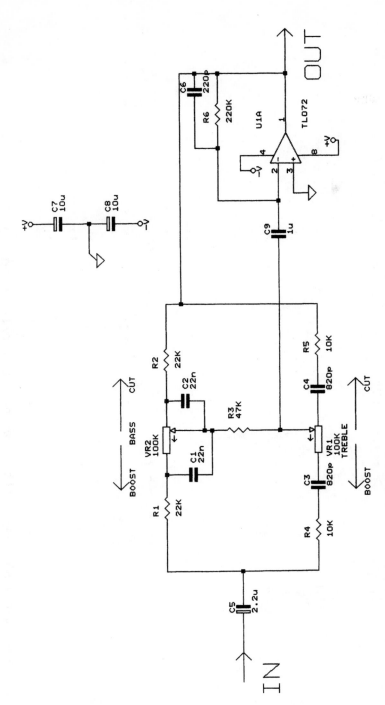

Figure 12.17 *Simple tone control circuit*

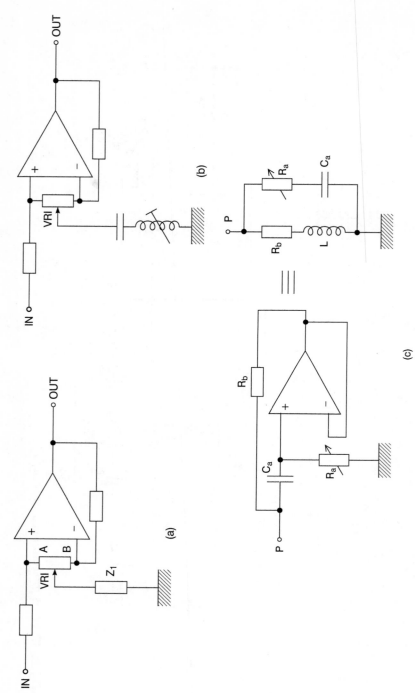

Figure 12.18a–c *Derivation of mid-band parametric EQ circuit – see text*

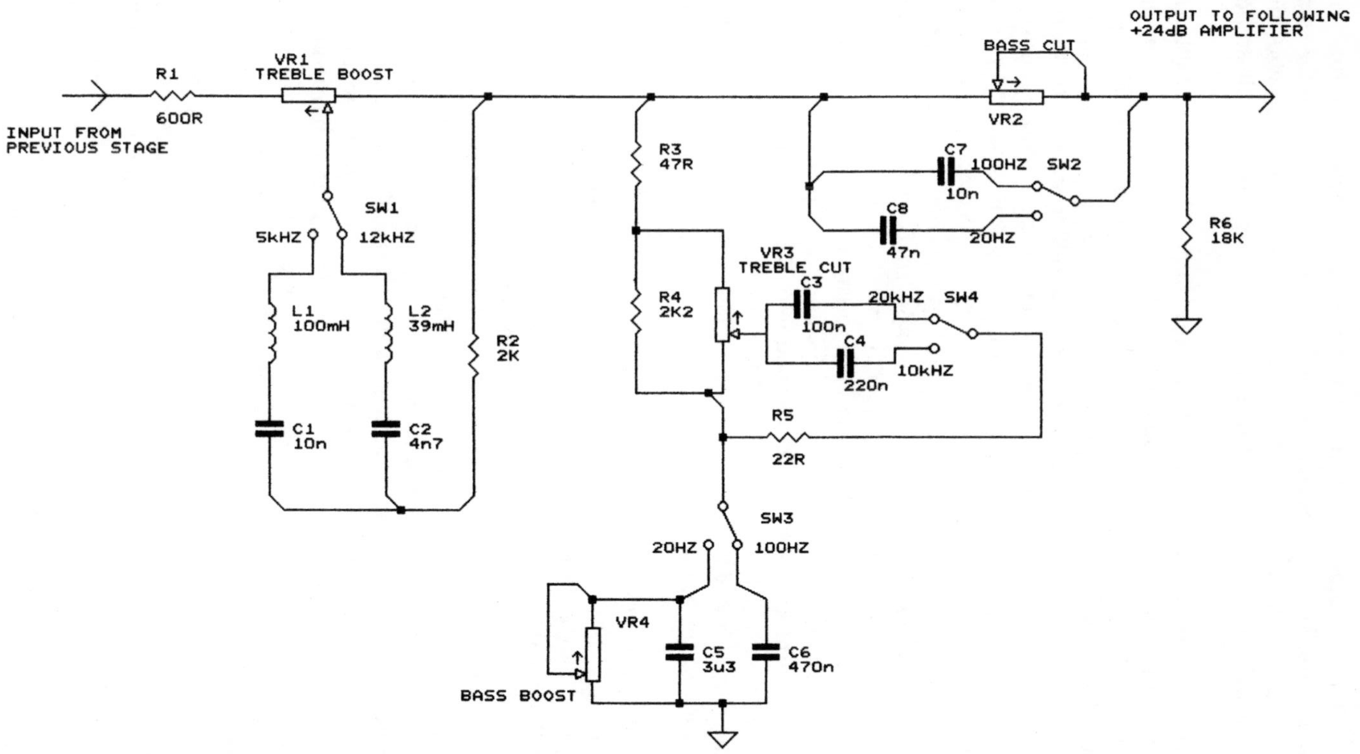

Figure 12.18d *Older, passive equaliser circuit*

spectrum (see Figure 12.12 also). A signal manipulation of this sort requires a tuned circuit which is combined within an amplifier stage which may be adjusted over a very wide range of attenuation and gain (perhaps as much as ±20 dB). How such a circuit is derived is illustrated in Figure 12.18.

Imagine Z1 is a small resistance. When the slider of VR1 is at position A, the input is heavily attenuated and the lower feedback limb of the op-amp is at its greatest value (i.e. the gain of the stage is low). Now imagine moving the slider of VR1 position B. The situation is reversed; the input is much less attenuated and the gain of the amplifier stage is high. This circuit therefore acts as a gain control because point A is arranged to be at the extreme anti-clockwise position of the control. As a tone control this is obviously useless. But, in the second part of Figure 12.18 (b), Z1 is replaced with a tuneable circuit, formed by a variable inductor and a capacitor. This tuned circuit has a high impedance – and therefore little effect – except at its resonant frequency; whereupon it acquires a very low dynamic impedance. With the appropriate choice of inductor and capacitor values, the mid-frequency equaliser can be made to operate over the central frequency range. The action of VR1 is to introduce a bell-shape EQ response (as illustrated in Figure 12.12) which may be used to attenuate or enhance a particular range of frequencies; as determined by the setting of the variable inductor.

Inductor-gyrators

As shown, the frequency adaptive component is designated as a variable inductor. Unfortunately, these components do not readily exist at audio frequencies and to construct components of this type expressly for audio-frequency equalisation would be very expensive. For this reason the variable inductors in most commercial equalisers are formed by gyrators; circuits which emulate the behaviour of inductive components by means of active circuits which comprise resistors, capacitors and op-amps. An inductor-gyrator circuit is illustrated in Figure 12.18c. This is known as the 'bootstrap' gyrator and its equivalent circuit is also included within the figure. Note that this type of gyrator circuit (and indeed most others) presents a reasonable approximation to an inductor which is grounded at one end. Floating inductor-gyrator circuits do exist but are rarely seen.

Operation of the bootstrap gyrator circuit can be difficult to visualise: but think about the two frequency extremes. At low frequencies C_a will not pass any signal to the input of the op-amp. The impedance presented at point P, will therefore be the output impedance of the op-amp (very low) in series with R_b which is usually designed to be in the region of a few hundred ohms. Just like an inductor, at low frequencies the reactance is low. Now consider the high-frequency case. At HF, Ca will pass signal, so that the input to the op-amp will be substantially that presented at point P. Because the op-amp is a follower; the output will be a low-impedance copy of its input. By this means, resistor R_b will thereby have little or no potential across it – because the signal at both its ends is the same. Consequently no signal current will pass through it. In engineering slang, the low-value resistor R_b is said to have been 'bootstrapped' by the action of the op-amp and therefore appears to have a much higher resistance than it actually has. Once again, just like a real inductor, the value of reactance at high-frequencies at point P is high. The inductor-gyrator circuit is made variable by the action of R_a, which is formed by a variable resistor component. This alters the break-point of the RC circuit C_a/R_a, and thereby, the value of the 'virtual' inductor.

In recent years, the fascination in 'retro' equipment has brought about a resurgence of interest in fixed inductor-capacitor type equalisers. Mostly outboard units, these are often passive EQ circuits (often of great complexity, as illustrated in Figure 12.18d), followed by a valve line-amplifier to make up the signal gain lost in the EQ. In a classic LC type equaliser, the variable-frequency selection is achieved with switched inductors and capacitors.

'Q'

Often it is useful to control the Q of the resonant circuit too so that a very broad, or a very narrow, range of frequencies can be effected as appropriate. Hence the inclusion of the Q control as shown in Figure 12.12. This effect is often achieved by means of a series variable resistor in series with the inductor-capacitor (or gyrator-capacitor) frequency-determining circuit.

Effect send and return

The effect send control feeds a separate mix amplifier. The output of this submix circuit is made available to the user via an effect send output on the back of the mixer. An effect return path is usually provided too, so that the submix (suitably 'effected' – usually with digital reverberation) can be reinjected into the amplifier at line level directly into the main mix amplifiers.

Faders and pan controls

Beneath the effect send control is the pan control and the main fader. Each channel fader is used to control the contribution of each channel to the overall mix as described above. A design for a fader and panpot is illustrated in Figure 12.19. The only disadvantage of the circuit is that it introduces loss. However, its use is confined to a part of the circuit where the signal is at a high level and the Johnson noise generated in the network is very low since all the resistances are low. The circuit takes as its starting point that a semi-log audio-taper fader can be obtained by

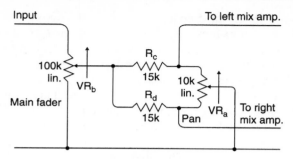

Figure 12.19 *Fader and pan pot circuit*

using a linear potentiometer when its slider tap and 'earthy' end is shunted with a resistor one-tenth of the value of the total potentiometer resistance. Resistors R_c and R_d and VR_a, the pan pot itself, form this one-tenth value network. Because the slider of the pan pot is grounded and the output signal is obtained from either end of the pan pot it is clear that in either extreme position of VR_a, all the signal will appear on one channel and none on the other. It is then only a matter of deciding whether the control law of the pan pot is of the correct type. The calculation of the control law obtained from the circuit is complicated because the signal level fed to left and right channels is not proportional to the resistive value of each part of VR_a. This is because the total value of the network R_c, R_d and VR_a, although reasonably constant, is not the same irrespective of the setting of VR_a and so the resistive attenuator comprising the top part of VR_b and its lower part, shunted by the pan pot network, is not constant as VRa is adjusted. Furthermore as VR_a is varied, so the output resistance of the network changes and, since this network feeds a virtual-earth summing amplifier, this effect too has an influence on the signal fed to the output because the voltage to current converting resistor feeding the virtual-earth node changes value. The control law of the final circuit is non-linear: the sum of left and right, when the source is positioned centrally, adding to more than the signal appearing in either channel when the source is positioned at an extreme pan position. This control law is very usable with a source seeming to retain equal prominence as it is 'swept across' the stereo stage.

Mix amplifiers

The mix amplifier is the core of the audio mixer. It is here that the various audio signals are combined together with as little interaction as possible. The adoption of the virtual earth mixing amplifier is universal. An example of a practical stereo mix amplifier (ibid.) is shown in Figure 12.20. Here, the summing op-amp is based on a conventional transistor

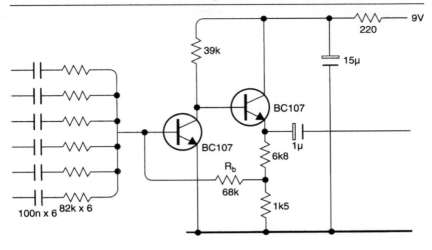

Figure 12.20   *Mix amplifier*

pair circuit. The only difficult decision in this area is the choice of the value for R_b. It is this value combined with the input resistors that determines the total contribution each input may make to the final output.

Line-level stages

Line-level audio stages are relatively straightforward. Signals are at a high level, so noise issues are rarely encountered. The significant design parameters are: linearity, headroom and stability. Another issue of some importance is signal balance; at least in professional line-level stages which are always balanced. Of these, output-stage stability is the one most often ignored by novice designers.

A high degree of linearity is achieved in modern line-level audio stages by utilising high open-loop gain op-amps and very large feedback factors. The only issue remaining, once the choice of op-amp has been made, is available headroom. This is almost always determined by choice of power supply rails. Taking a typical audio op-amp (TL072, for example), maximum output swing is usually limited to within a few volts of either rail. So, furnished with 12 V positive and negative supplies, an op-amp could be expected to swing 18 V pk-pk. This is equivalent to 6.3 V RMS, or 16 dBV, easily adequate then for any circuit intended to operate at 0 VU = −10 dBV. In a professional line-output circuit, like that shown in Figure 12.21, this voltage swing is effectively doubled because the signal appears 'across' the two opposite phases. The total swing is therefore 12.6 V RMS which is equivalent to 24 dBu. For equipment intended to operate at 0 VU = +4 dBu, such a circuit offers 20 dB headroom which is adequate.

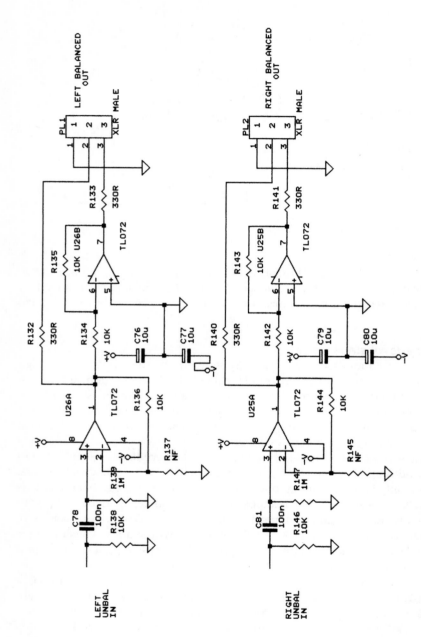

Figure 12.21 *Line output circuit*

However, ±12 V supplies really represent the lowest choice of rail volts for professional equipment and some designers prefer to use 15 V supplies for this very reason.

Looking again at the output stage circuit illustrated in Figure 12.21, notice the inclusion of the small value resistors in the output circuit of the op-amps. These effectively add some real part to the impedance 'seen' by the op-amp when it is required to drive a long run of audio cables. At audio frequencies, the equipment interconnection cable 'looks' to the output stage as a straightforward – but relatively large – capacitance. And a large negative reactance is, almost always, an excellent way to destabilise output circuits. Output 'padding' resistors, like R132 and R133, help a great deal in securing a stable performance into real loads.

Line-level input stages present different problems. Notice that the function performed by the circuit in Figure 12.21, over and above its duty to drive the output cables, is to derive an equal and opposite signal so as to provide a balanced audio output from a single-ended, unbalanced input signal. This is a common feature of professional audio equipment because, although balanced signals are the norm outside the equipment, internally most signals are treated as single ended. The reasons for this are obvious; without this simplification all the circuitry within – for example – a console would be twice as complex and twice as expensive.

The line-level input stage on professional equipment therefore has to perform a complementary function to the output stage, to derive a single-ended signal from the balanced signal presented to the equipment. Conceptually, the simplest circuit is a transformer – like that shown in Figure 12.22. In many ways this is an excellent solution for the following reasons: it provides electrical isolation, it has low noise and distortion and it provides good headroom, provided the core doesn't saturate. But, most important of all, it possesses excellent common-mode rejection or CMR. That means that any signal which is common (i.e. in phase) on both signal phases is rejected and does not get passed on to following equipment. By contriving the two signal conductors within the signal cable to occupy – as nearly as possible – the same place, by twisting them together, any possible interference signal is induced equally in both phases. Such a signal thereafter cancels in the transformer stage because a common signal cannot cause a current to flow in the primary circuit and cannot, therefore, cause one to flow in the secondary circuit. This is illustrated in Figure 12.22 as well.

Another advantage of a balanced signal interface is that the signal circuit does not include ground. It thereby confers immunity to ground-sourced noise signals. On a practical level it also means that different equipment chassis can be earthed, for safety reasons, without incurring the penalty of multiple signal return paths and the inevitable 'hum-loops' this creates. However, transformers are not suitable in many applications for a number

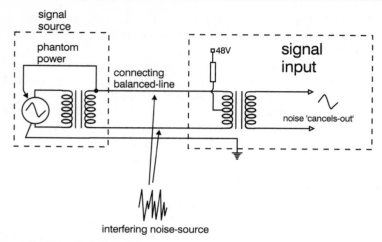

Figure 12.22   *Balanced input circuit – and CM rejection*

of reasons. First, they are very expensive. Second, they are heavy, bulky and tend to be microphonic (i.e. they have a propensity to transduce mechanical vibration into electrical energy!) so electronically balanced input stages are widely employed instead. These aim to confer all the advantages of a transformer, cheaply, quietly and on a small scale. To some degree, an electronic stage can never offer the same degree of CMR, as well as the complete galvanic isolation offered by a transformer.

Mixer automation

Mixer automation consists (at its most basic level) of computer control over the individual channel faders during a mixdown. Even the most dextrous, and clear thinking, balance engineer obviously has problems when controlling perhaps as many as 24 or even 48 channel faders at once. For mixer automation to work, several things must happen. First, the controlling computer must know precisely which point in the song or piece has been reached, in order that it can implement the appropriate fader movements. Second, the controlling computer must have, at its behest, hardware which is able to control the audio level on each mixer channel, swiftly and noiselessly. This last requirement is fulfilled a number of ways, but most often a voltage controlled amplifier (VCA) is used.

A third requirement of a fader automation system is that the faders must be 'readable' by the controlling computer so that the required fader movements can be implemented by the human operator and memorised by the computer for subsequent recall.

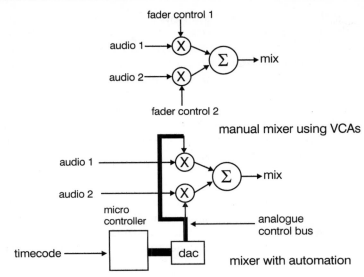

Figure 12.23 *Fader automation system*

A complete fader automation system is shown in schematic form in Figure 12.23. Notice that the fader does not pass the audio signal at all. Instead the fader simply acts as a potentiometer driven by a stabilised supply. The slider potential now acts as a control voltage which could, in theory, be fed directly to the voltage controlled amplifier VCA1. But this would be to miss the point. By digitising the control voltage, and making this value available to the microprocessor bus, the fader 'position' can be stored for later recall. When this happens, the voltage (at the potentiometer slider) is recreated by means of a DAC and this is applied to the VCA, thereby reproducing the operator's original intentions.

One disadvantage of this type of system is the lack of operator feedback once the fader operation is overridden by the action of the VCA: importantly, when in recall mode, the faders fail, by virtue of their physical position, to tell the operator (at a glance) the condition of any of the channels and their relative levels. Some automation systems attempt to emulate this important visual feedback by creating an iconic representation of the mixer on the computer screen. Some even allow these virtual faders to be moved, on screen, by dragging them with a mouse. Another more drastic solution, and one which has many adherents on sound quality grounds alone, is to use motorised faders as part of the control system. In this case the faders act electrically as they do in a non-automated mixer; carrying the audio signal itself. The control system loop is restricted to reading and 'recreating' operator fader physical movements. Apart from providing unrivalled operator feedback (and the quite thrilling spectacle of banks of faders moving as if under the aegis of ghostly hands!) the advantage of this type of automation system is the lack of VCAs in the signal path. VCA circuits are necessarily complicated and their operation is beset with various deficiencies, mostly in the areas of insufficient dynamic range and control signal breakthrough. These considerations have kept motorised faders as favourites among the best mixer manufacturers, this despite their obvious complexity and cost.

Timecode

Timecode is the means by which an automation system is kept in step with the music recorded onto tape. Normally, a track of the multi-track tape is set aside from audio use and devoted to recording a pseudo audio signal comprised of a serial digital code. This code is described in detail in Chapter 15.

Digital consoles

Introduction to digital signal processing (DSP)

Digital signal processing involves the manipulation of real-world signals (for instance, audio signals, video signals, medical or geophysical data signals etc.) within a digital computer. Why might we want to do this? Because, these signals, once converted into digital form (by means of an analogue to digital converter – see Chapter 10), may be manipulated using mathematical techniques to enhance, change or display the data in a particular way. For instance, the computer might use height or depth data from a geophysical survey to produce a coloured contour map or the computer might use a series of two-dimensional medical images to build up a three-dimensional virtual visualisation of diseased tissue or bone. Another application, this time an audio one, might be to remove noise from a music signal by carefully measuring the spectrum of the interfering noise signal during a moment of silence (for instance during the run-in groove of a record) and then subtracting this spectrum from the entire signal, thereby removing only the noise – and not the music – from a noisy record.

DSP systems have been in existence for many years but, in these older systems, the computer might take many times longer than the duration of the signal acquisition time to process the information. For instance, in the case of the noise reduction example, it might take many hours to process a short musical track. This leads to an important distinction which must be made in the design, specification and understanding of DSP systems;

that of non-real time (where the processing time exceeds the acquisition or presentation time) and real-time systems which complete all the required mathematical operations so fast, that the observer is unaware of any delay in the process. When we talk about DSP in digital audio it is always important to distinguish between real-time and non-real-time DSP. Audio outboard equipment which utilises DSP techniques is, invariably, real-time and has dedicated DSP chips designed to complete data manipulation fast. Non-real time DSP is found in audio processing on a PC or Apple Mac where some complex audio tasks may take many times the length of the music sample to complete.

Digital manipulation

So, what kind of digital manipulations might we expect? Let's think of the functions which we might expect to perform within a digital sound mixer. First, there is addition. Clearly, at a fundamental level, that is what a mixer is – an 'adder' of signals. Second, we know that we want to be able to control the gain of each signal before it is mixed. So multiplication must be needed too. So far, the performance of the digital signal processing 'block' is analogous with its analogue counterpart. The simplest form of digital audio mixer is illustrated in Figure 12.24. In this case, two digital audio signals are each multiplied by coefficients (k1 and k2) derived from the position of a pair of fader controls; one fader assigned to either signal. The signals issuing from these multiplication stages are subsequently added together in a summing stage. All audio mixers possess this essential architecture, although it may be supplemented many times over.

But, in fact, the two functions of addition and multiplication, plus the ability to delay signals easily within digital systems, allow us to perform all the functions required within a digital sound mixer, even the equalisation functions. That's because equalisation is a form of signal filtering on successive audio samples which is simply another form of mathematical

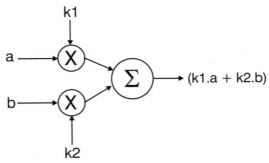

Figure 12.24 *Simple digital audio mixer*

manipulation, even though it is not usually regarded as such in analogue circuitry.

Digital filtering

The simplest form of analogue low-pass filter is shown in Figure 12.25. Its effect on a fast rise-time signal wavefront (an 'edge') is also illustrated. Notice that the resulting signal has its 'edges' slowed down in relation to the incoming signal. Its frequency response is also illustrated, with its turnover frequency. Unfortunately, in digital circuits there are no such

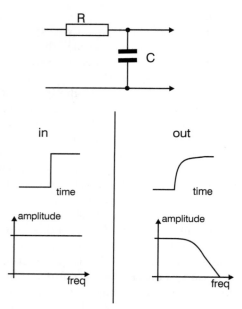

Figure 12.25   *RC low-pass filter*

things as capacitors or inductors which may be used to change the frequency response of a circuit. However, if you remember, we've come across situations before in sections as diverse as microphones to flanging, phasing and chorus wherein a frequency response was altered by the interaction of signals delayed with respect to one another. This principle is the basis behind all digital filtering and may be extended to include several stages of delay as shown in Figure 12.26. By utilising a combination of adder and variable multiplication factors (between the addition function and the signal taps) it is possible to achieve a very flexible method of signal filtering in which the shape of the filter curve

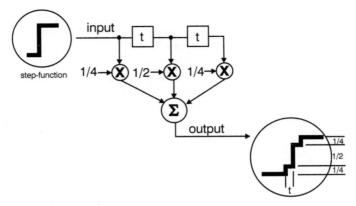

Figure 12.26   *Principle of digital filter*

may be varied over a very wide range of shapes and characteristics. While such a technique is possible in analogue circuitry, note that the 'circuit' (shown in Figure 12.26) is actually not a real circuit at all, but a notional block diagram. It is in the realm of digital signal processing that such a filtering technique really comes into its own: the DSP programmer has only to translate these processes into microprocessor type code to be run on a microcontroller IC which is specifically designed for audio applications – a so-called DSP IC. Herein lies the greatest benefit of digital signal processing; that, by simply reprogramming the coefficients in the multiplier stages, a completely different filter may be obtained. Not only that, but if this is done in real time too, the filter can be made adaptive, adjusting to demands of the particular moment in a manner which might be useful for signal compression or noise reduction.

Digital mixer architecture

Because of the incredible flexibility and 'programmability' of digital signal processing based mixers, architecture is much harder (less necessary!) to define. This, alone, is a great advantage. For instance, digital processing too has blurred the traditional distinction between split and in-line consoles because, with the aid of configurable signal paths and motorised faders, the same small group of faders can be used to 'flip' between the role of the recording mixer and that of the playback mixer.

Reference
Brice, R. (1990) Audio mixer design. *Electronics and Wireless World*, July.

Note

1 The amplifier shown has an input noise density of 3 nV per root hertz
and a calculated input noise current density of 0.3 pA per root hertz
ignoring flicker noise. The frequency response and the phase response
remain much the same regardless of gain setting. This seems to go
against the intractable laws of gain-bandwidth product: as we increase
the gain we must expect the frequency response to decrease and vice
versa. In fact, the ring-of-three circuit is an early form of 'current-mode-
feedback' amplifier which is currently very popular in video applications. The explanation for this lies in the variable gain-setting resistor
Ra. This not only determines the closed-loop gain by controlling the
proportion of the output voltage fed back to the inverting port but it
also forms the dominating part of the emitter load of the first transistor
and consequently the gain of the first stage. As the value of Ra
decreases, so the feedback diminishes and the closed-loop gain rises. At
the same time the open-loop gain of the circuit rises because TR1's
emitter load falls in value. Consequently the performance of the circuit
in respect of phase and frequency response, and consequently stability,
remains consistent regardless of gain setting.

Fact Sheet #12: Digital signal processing

- Architecture of DSP devices
- Convolution
- Impulse response
- FIR and IIR digital filters
- Design of digital filters
- Frequency response
- Derivation of band-pass and high-pass filters
- Digital frequency domain analysis – the z-transform
- Problems with digital signal processing

Architecture of DSP devices

The first computers, including those developed at Harvard
University, had separate memory space for program and data;
this topology being known as Harvard architecture. In fact the
realisation, by John von Neumann – the Hungarian-born
mathematician – that program instructions and data were only
numbers and could share the same 'address-space' was a great
breakthrough at the time and was sufficiently radical that this
architecture is often named after its inventor. The advantage of

the von Neumann approach was great simplification but at the expense of speed because the computer can only access either an instruction or data in any one processing clock cycle. The fact that virtually all computers follow this latter approach illustrates that this limitation is of little consequence in the world of general computing.

However, the speed limitation 'bottleneck', inevitable in the von Neumann machine, can prove to be a limitation in specialist computing applications like digital audio signal processing. As we have seen, in the case of digital filters, digital signal processing contains many, many multiply and add type instructions of the form,

$$A = B.C + D$$

Unfortunately, a von Neumann machine is really pretty inefficient at this type of calculation so the Harvard architecture lives on in many DSP chips meaning that a multiply and add operation can be performed in one clock cycle; this composite operation being termed a Multiply ACcumulate (MAC) function. A further distinction pertains to the incorporation within the DSP IC of special registers which facilitate the managing of circular buffers such as those discussed in Chapter 6, for the implementation of reverb, phasing, chorus and flanging effects.

The remaining differences between a DSP device and a general purpose digital microcomputer chip relate to the provision of convenient interfaces thereby allowing direct connection of ADCs, DACs and digital transmitter and receiver ICs.

Convolution

In the simple three-stage digital filter considered above at the end of Chapter 12, we imagined the step function being multiplied by a quarter, then by a half and finally by a quarter again: and, at each stage, the result was added-up to give the final output. This actually rather simple process is given a frightening name in digital signal processing theory, where it is called convolution.

Discrete convolution is a process which provides a single output sequence from two input sequences. In the example above, a time-domain sequence – the step function – was convolved with the filter response yielding a filtered output sequence. In textbooks convolution is often denoted by the character '*'. So if we call the input sequence $h(k)$ and the input sequence $x(k)$, the filtered output would be defined as,

$$y(n) = h(k) * x(k)$$

Impulse response

A very special result is obtained if a unique input sequence is convolved with the filter coefficients. This special result is known as the filter's impulse response and the derivation and design of different impulse responses is central to digital filter theory. The special input sequence we use to discover a filter's impulse response is known as the 'impulse input'. (The filter's impulse response being its response to this impulse input.) This input sequence is defined to be always zero, except for one single sample which takes the value 1 (i.e. the full-scale value). We might define, for practical purposes, a series of samples like this,

0, 0, 0, 0, 0, 1, 0, 0, 0, 0

Now imagine these samples being latched through the three-stage digital filter shown above. The output sequence will be:

0, 0, 0, 0, 0, 1/4, 1/2, 1/4, 0, 0, 0, 0

Obviously the zeros don't really matter, what's important is the central section: 1/4, 1/2, 1/4. This pattern is the filter's impulse response.

FIR and IIR digital filters

Notice that the impulse response of the filter above is finite: in fact, it only has three terms. So important is the impulse response in filter theory that this type of filter is actually defined by this characteristic of its behaviour and is named a Finite Impulse Response (FIR) filter. Importantly, note that the impulse response of an FIR filter is identical to its coefficients.

Now look at the digital filter in Figure F12.1. This derives its result from both the incoming sequence and from a sequence which is fed-back from the output. Now if we perform a similar

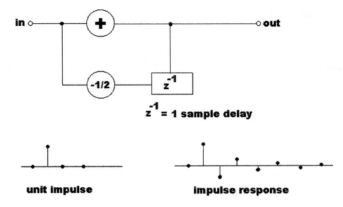

Figure F12.1 *Infinite impulse response (IIR) filter*

thought-experiment to the convolution example above and imagine the resulting impulse-response from a filter of this type, it results in an output sequence like that illustrated in the figure: that's to say, an infinitely decaying series of values. Once again, so primordial is this characteristic, that this category of filter is termed an Infinite Impulse Response (IIR) filter.

IIR filters have both disadvantages and advantages over the FIR type. Firstly, they are very much more complicated to design; because their impulse response is not simply reflected by the tap coefficients, as in the FIR. Secondly, it is in the nature of any feedback system (like an analogue amplifier), that some conditions may cause the filter to become unstable if it is has not been thoroughly designed, simulated and tested. Furthermore, the inherent infinite response may cause distortion and/or rounding problems as calculations on smaller and smaller values of data are performed. Indeed, it's possible to draw a parallel between IIR filters and analogue filter circuits: they share the disadvantages of complexity of design and possible instability and distortion, but they also share the great benefit that they are efficient. An IIR configuration can be made to implement complex filter functions with only a few stages, whereas the equivalent FIR filter would require many hundreds of taps with all the drawbacks of cost and signal delay that this implies. (Sometimes FIR and IIR filters are referred to as 'Recursive' and 'Non-Recursive' respectively; these terms directly reflecting the filter architecture.)

Design of digital filters

Digital filters are nearly always designed from a knowledge of the required impulse response. IIR and FIR filters are both designed in this way; although the design of IIR filters is complicated because the coefficients do not represent the impulse response directly. Instead, IIR design involves various mathematical methods which are used to analyse and derive the appropriate impulse response from the limited number of taps. This makes the design of IIR filters from first principles rather complicated and maths-heavy! Fortunately, FIRs are easier to understand and a brief description gives a good deal of insight into the design principles of all digital filters.

We already noted (in Chapter 12) that the response type of the 1/4, 1/2, 1/4 filter was a low-pass; remember it 'sloweddown' the fast rising edge of the step waveform. If we look at the general form of this impulse response, we will see that this is a very rough approximation to the behaviour of an ideal low-pass filter which we already met in Chapter 10, in relation to reconstruction filters. There we saw that the (sin x)/x function defines the behaviour of an ideal, low-pass filter and the derivation of this function is given in Figure F12.2. Sometimes termed a sinc function, it has the characteristic that it is infinite; gradually decaying with ever smaller oscillations about zero.

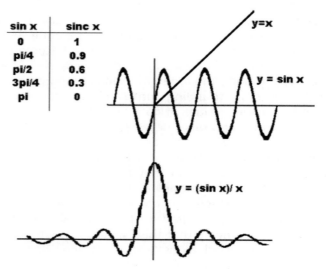

Figure F12.2 *Derivation of sin x/x function*

This illustrates that the perfect low-pass FIR filter would require an infinite response, an infinite number of taps and the signal would take an infinitely long time to pass through it! Fortunately for us, we do not need such perfection.

However the 1/4, 1/2, 1/4 filter is really a very rough approximation indeed. So let's now imagine a better estimate to the true sinc function and design a relatively simple filter using a 7-tap FIR circuit. I have derived the values for this filter in Figure F12.2. This suggests a circuit with the following tap values;

0.3, 0.6, 0.9, 1, 0.9, 0.6, 0.3

The only problem these values present is that they total to a value greater than 1. If the input was the unity step-function input, the output would take on a final amplitude of 4.6. This might overload the digital system, so we normalise the values so that the filter's response at DC (zero frequency) is unity. This leads to the following, scaled values;

0.07, 0.12, 0.2, 0.22, 0.2, 0.12, 0.07

The time-domain response of such an FIR filter to a step and impulse response is illustrated in Figure F12.3. The improvement over the three-tap filter is already obvious.

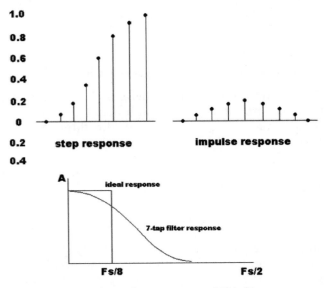

Figure F12.3 *Finite impulse response (FIR) filter*

Frequency response

But how does this response relate to the frequency domain response? For it is usually with a desired frequency response requirement that filter design begins. This important question is really asking, how can we express something in the time domain in terms of the frequency domain? It should be no surprise by now that such a manipulation involves the Fourier transform. Normal text-book design methods involve defining a desired frequency response and computing (via the Fourier transform) the required impulse response; thereby defining the tap coefficients of the FIR filter.

However, this is a little labour intensive and is not at all intuitive, so here's a little rule of thumb which helps when you're thinking about digital filters. If you count the number of sample-periods in the main lobe of the sinc curve and give this the value, n. Then, very roughly, the cut-off frequency of the low-pass filter will be the sampling frequency divided by n,

$$F_c = F_s / n$$

So, for our 7-term, FIR filter above, $n = 8$ and F_c is roughly $F_s/8$. In audio terms, if the sample rate is 48 kHz, the filter will show a shallow roll-off with the turn-over at about 6 kHz. The frequency response of this filter (and an ideal response) is shown in Figure F12.3. In order to approach the ideal response, a filter of more than 30 taps would be required.

Derivation of band-pass and high-pass filters

All digital filters start life as low-pass filters and are then transformed into band-pass or high-pass types. A high-pass is derived by multiplying each term in the filter by alternating values of +1 and –1. So, our low-pass filter,

0.07, 0.12, 0.2, 0.22, 0.2, 0.12, 0.07 is transformed into a high-pass like this,

+0.07, –0.12, +0.2, –0.22, +0.2, –0.12, +0.07

The impulse response and the frequency of this filter is illustrated in Figure F12.4. If you add up these high-pass filter terms, you'll notice they come nearly to zero. This demonstrates that the high-pass filter has practically no overall gain at

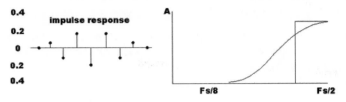

Figure F12.4 *Digital high-pass filter*

DC, as you'd expect. Notice too how the impulse response looks 'right', in other words, as you'd anticipate from an analogue type.

A band-pass filter is derived by multiplying the low-pass coefficients with samples of a sine-wave at the centre frequency of the band-pass. Let's take our band-pass to be centred on the frequency of $F_s/4$. Samples of a sine-wave at this frequency will be at the 0 degree point, the 90 degree point, the 180 degree point and the 270 degree point and so on. In other words,

0, 1, 0, –1, 0, 1, 0, –1

If we multiply the low-pass coefficients by this sequence we get the following,

0, 0.12, 0, –0.22, 0, +0.12, 0

The impulse response of this circuit is illustrated in Figure F12.5. This looks intuitively right too, because the output can be seen to 'ring' at $F_s/4$, which is what you'd expect from a resonant filter. The derived frequency response is also shown in the diagram.

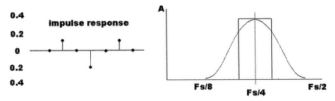

Figure F12.5 *Digital band-pass filter*

Digital frequency domain analysis – the z-transform

The z-transform of a digital signal is identical to the Fourier transform except for a change in the lower summation limit. In fact, you can think of 'z' as a frequency variable which can take on real and imaginary (i.e. complex) values. When the z-transform is used to describe a digital signal, or a digital process (like a digital filter) the result is always a rational function of the frequency variable z. That's to say, the z-transform can always be written in the form,

$$X(z) = N(z)/D(z) = K(z - z_1)/(z - p_1)$$

Where the z's are known as 'zeros' and the p's are known as 'poles'.

A very useful representation of the z-transform is obtained by plotting these poles and zeros on an Argand diagram; the resulting two-space representation being termed the 'z-plane'. When the poles and zeros are plotted in this way, they give us a very quick way of visualising the characteristics of a signal or digital signal process.

Problems with digital signal processing

As we have already seen (Chapter 10), sampled systems exhibit aliasing effects if frequencies above the Nyquist limit are included within the input signal. This effect is usually no problem because the input signal can be filtered so as to remove any offending frequencies before sampling takes

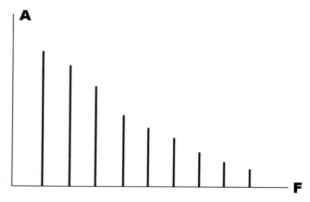

Figure F12.6   *Generation of harmonics due to non-linearity*

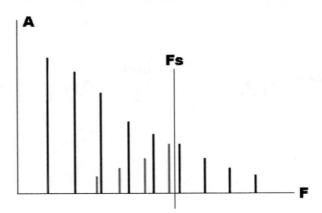

Figure F12.7 *Aliasing of harmonic structure in digital, non-linear processing*

place. However, consider the situation in which a band-limited signal is subjected to a non-linear process once in the digital domain. This process might be as simple as a 'fuzz' type overload effect; created with a plug-in processor. This entirely digital process generates a new large range of harmonic frequencies (just like its analogue counterpart), as shown in Figure F12.6. The problem arises that many of these new harmonic frequencies are actually above the half-sampling frequency limit and get folded back into the pass-band; creating a rough quality to the sound and a sonic signature quite unlike the analogue 'equivalent' (see Figure F12.7). This effect may account for the imperfect quality of many digital 'copies' of classic analogue equipment.

13
Unchained Melody – Amplifiers

Two different philosophies

Two different philosophies underlie amplifier design. These are the philosophies we identified in Chapter 1. One is appropriate for musical applications, the other for high quality monitoring and reproduction. The problem is, many entirely specious and time-wasting debates centre on a failure to make an adequate distinction between the two.

A 'straight wire with gain'

For high quality monitoring purposes only one philosophy is appropriate; that the practical amplifier should approach – as nearly as possible – the perfect amplifier. Something Peter Walker of the Acoustical Manufacturing Company Ltd, erstwhile manufacturers of the Quad mark, expressed as a 'straight wire with gain!' Like the perfect op-amp, the perfect audio amplifier should possess: a very high input impedance, so as to avoid loading upstream equipment; a very low output impedance, so as to damp loudspeaker oscillations – see Chapter 4; a perfectly linear transfer characteristic; and a 'ruler-flat' frequency response. Moreover it should possess a power output suitable for reproducing a realistic dynamic range; at least equivalent to that possessed by 16-bit digital audio. Despite a widely held belief to the contrary, most modern semiconductor amplifiers embody all the above characteristics save, perhaps, the last. There exists a body of evidence that it is very difficult indeed, if not completely impossible, to detect by listening alone any differences between modern semiconductor power amplifiers despite enormous variations in technology and price.

Cool!

The alternative philosophy, adopted by the manufacturers of instruments and instrumental amplifiers, is to design and build an amplifier for a particular 'sound'. This is entirely justifiable. It may involve, for instance, the perpetuation of valve technology in order to obtain a deliberately distorting transfer characteristic or 'flattering' tonal feature. (Chapter 4 discussed some ways in which valve power amplifiers can ingratiate themselves by virtue of certain technical attributes.) The 'hi-fi fraternity' has done a great deal to confuse these two motives claiming greater accuracy and objectivity for products preferred on entirely subjective grounds.

Power amplifiers

It is the job of the power amplifier to transform the electrical signals at an operating level of a few volts pk-pk across $600\,\Omega$ to a signal of perhaps a few hundred volts pk-pk across as small an impedance as possible – certainly less than $1\,\Omega$ – which is typical of the signal required to drive a moving coil loudspeaker. These figures illustrate the very considerable power gain performed by the final stage of the audio amplification process and explain the term power amplifier. Power amplifiers may be divided into two main classes: Class A and Class B. A third type of power amplifier, which doesn't really belong to either class, is known as a *current dumping* amplifier. All three are described below.

It is possible to have two types of amplifiers because of the demonstrable property of all musical sounds: all involve a physical mechanism which vibrates minutely back and forth when it is blown, struck or excited via a bow. When a loudspeaker produces the sounds of each of these instruments, it too must vibrate back and forth exactly (or as nearly exactly as possible) following the contours of the original sound. Similarly, the current at the output terminals of the amplifier must vibrate back and forth in order to drive the loudspeaker.

When a designer chooses between a Class-A and a Class-B amplifier design he (or she) is choosing between how the amplifier deals with the equal and opposite nature of musical signals. Put simply, the Class-A amplifier has a single circuit devoted to producing both the positive half and the negative half of the musical sounds. A Class-B amplifier has two circuits, one devoted to handling the positive going halves of each cycle, the other devoted to handling the negative portions of the signal. Each type of amplifier has its merits and demerits. The Class-A amplifier, because it has a single output circuit has the advantage of simplicity. But it pays for its straightforwardness with a lack of efficiency. The Class-A amplifier works as hard when it is reproducing silence as when it reproduces the most awe-inspiring orchestral crescendo. And, as in most of nature, a lack of efficiency leads to heat being generated. Indeed the one sure cachet of a Class-A amplifier is heat!

Class A

Thermionic valves are ideally suited to Class-A amplification because they do not mind running hot and the simple circuitry is appropriate to a technology where the amplifying devices themselves remain relatively expensive and bulky. Transistors, on the other hand, are less well suited and dissipation in the output transistors of a Class-A audio power amplifier must be very carefully controlled if they are not to self-destruct due to thermal runaway. Interestingly, Class B was never considered for high quality amplification until the advent of transistor power amplifiers where the threat of thermal runaway dictated this more complex topology.

Class B

Class-B amplifiers are more efficient because they contain two circuits, one devoted to handing the positive half and the other devoted to handling the negative half of each signal cycle. Virtually all but a very few semiconductor power amplifiers employ Class-B circuits. The important feature of this type of circuit is that while one circuit is on (say the positive half-cycle circuit) the other is off. Each has no job to do while the other is operating. So half the amplifier always remains idle, consuming no power. And, when the amplifier produces silence, both halves can effectively be off! Now, it would be a silly exaggeration to say that most music is composed of silence, but it is certainly true (as anyone who owns an amplifier with a power-meter will tell you) that for the majority of the time, the output power produced by an amplifier is very small indeed. It is in this way that the Class-B amplifier scores over its rival, the Class A. Whereas the Class-A type dissipates power all the time waiting to produce a fortissimo chord, the Class-B amplifier only dissipates heat power as – and when – it is called upon to produce electrical power and that is only significant for a very small proportion of the time. There is a catch of course. As anyone who has ever broken something intricate into two pieces knows, once broken it is very hard to put the object back together again without revealing a flaw where the two halves meet. So it is with Class-B amplifiers. Our electrical schizophrenic must switch personalities completely and instantaneously if it is to work correctly. Mr Hyde must fall asleep precisely as Dr Jekyll wakes. One circuit must not linger, dozing, as the other comes awake. Neither must it fall asleep before the other is fully conscious. If these conditions are not fulfilled, the Class-B amplifier produces a very unpleasant form of distortion known, not surprisingly, as crossover distortion.

Amplifiers for high quality monitoring

Audio amplifiers are best understood by 'breaking down' the design into separate stages. We shall see that all audio amplifiers (even valve amplifiers) can be regarded as being formed of three distinct stages: an input stage (often combined with the mechanism by which loop feedback is applied); a voltage amplification stage; and an output stage. For reasons other than sheer perversity it is practical to examine a power amplifier 'in reverse' i.e. looking at the output stage first, followed by the voltage amplification stage, followed by the input stage. This is because the loudspeaker requirement defines the important parameters of the output stage; the output stage defines the voltage amplification stage and so on.

Valve amplifiers

Valve plate impedances are very large (as we saw in Chapter 4) so a plate circuit is unsuitable for driving a low impedance load like a loudspeaker. Instead a transformer is almost universally employed to modify the impedance seen by the valve by the square of the turns ratio of the transformer. In its simplest form the valve amplifier is constructed with a single valve as shown in Figure 13.1. This scheme has a number of significant disadvantages, the most serious of which is the requirement for the stage quiescent current to flow in the transformer. This results in a constant magnetising current which, in turn, demands the use of a large transformer capable of tolerating the necessarily high flux density. A far better arrangement is illustrated in Figure 13.2 which is the type of circuit almost comprehensively employed in high quality valve amplifiers. This so-called 'push-pull' arrangement has the advantage that the quiescent current splits between the two valve circuits (which work in anti-phase) and therefore cancels the magnetic effect within the core of the transformer.

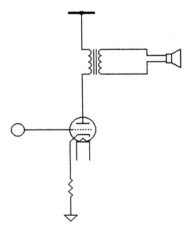

Figure 13.1 *Single-ended valve output stage*

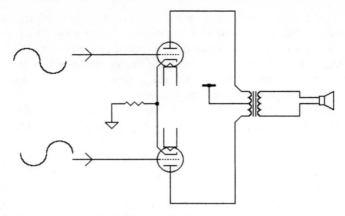

Figure 13.2   *Push-pull, balanced valve output stage*

A complication ensues from the adoption of a push-pull output stage because of the requirement to feed each of the output valves with a phase-opposed signal. This means the voltage amplification stage is necessarily combined with a stage which is known as a phase splitter. Such a circuit is illustrated in Figure 13.3. This is probably the simplest form of phase splitter but it suffers from the drawback that it has no voltage gain. Far better is the circuit illustrated in Figure 13.4 which combined the phase-splitting function with voltage amplification.

The input stage of a typical valve amplifier combines voltage amplification and the injection point for the loop feedback. A typical

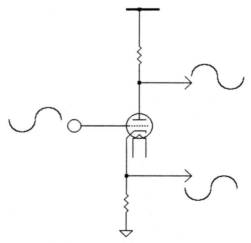

Figure 13.3   *Split-load phase-splitter circuit*

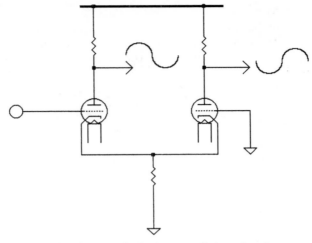

Figure 13.4 *Cathode-coupled phase-splitter circuit*

circuit is given in Figure 13.5. This illustrates a traditional triode amplification stage where the input signal is applied on the grid and the output is taken from the anode circuit. But note the addition of the feedback network in the cathode circuit of the valve. The valve thus 'sees' the difference in voltage between the grid input signal and a resistively divided version of the output signal at its cathode. It thereby forms the subtraction circuit necessary in any feedback amplifier.

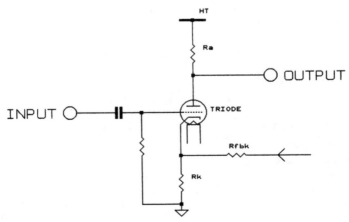

Figure 13.5 *Triode input stage*

Output transformers

You may already have guessed that the transformer which converts the impedance of the loudspeaker into something which the output valves can drive is no ordinary component. The design of output transformers

Technical Data Sheet, No. I

Quality Range of Output Transformers PPO/0, PPO/1, PPO/2

General.
This range of push-pull output transformers is intended for use in equipment reproducing the full audio-frequency range with the lowest distortion.

Power Rating.
12 watts for 0.5% harmonic distortion at 50 cps.

Anode to Anode Load.
PPO/0   5,000 to 6,000 ohms
PPO/1   8,000 ohms   } for loading triodes, tetrodes or pentodes
PPO/2   10,000 ohms

Permissible d.c. Unbalance.
10%.

Secondary Load.
All secondary windings are brought out in two equal sections which can be connected in series to match a 15 ohms load or in parallel for a 3.75 ohms load.

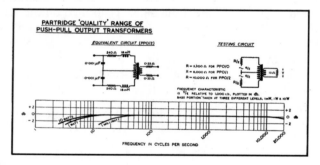

Frequency Characteristics.
The diagram shows the frequency characteristics taken under matched impedance conditions, e.g., 10,000 ohms source working into the primary winding in the case of the PPO/2.

These characteristics will be modified under actual conditions. With triodes the bass will be better than as shown while the treble will be not quite so good. With tetrodes or pentodes however the bass cut off will be shifted up in frequency by approximately one octave, and the treble will be correspondingly extended.

Leakage Inductances, Winding Self Capacity, d.c. Resistance, etc.
Figures are given on the diagram and they vary between types in direct proportion to the anode to anode load impedance.

Primary Induction at 50 c.p.s.
Primary inductance at very low levels of flux density is 45 H. rising to about 600 H. at 12 watts. These figures are for the PPO/2, those for the other types being directly proportional to the appropriate anode to anode load.

PARTRIDGE TRANSFORMERS LTD.,
LONDON.

Figure 13.6   *High quality output transformer*

(for so are these components named) is an art and a science all of its own. Essentially, two requirements conflict: that of a large primary inductance – which ensures a good low-frequency performance – and that of a small primary capacitance which ensures a good high-frequency performance (and a good stability margin). Unfortunately, the first requires many primary turns and this destroys the second. Practical transformers are constructed with many primary turns separated into sections so as to reduce the effect of primary shunt capacitance to the minimum. The fact remains however, that a high quality valve output transformer is a complex and costly component. The specification of a high quality valve-amplifier output transformer is given in Figure 13.6.

Triode or pentode

The circuit for the Williamson amplifier is given in Figure 13.7. The Williamson was a notable amplifier on performance grounds. In many ways, the design set the performance levels for high quality amplification for a generation – perhaps for all time. Since 1947, amplifiers may have become cheaper, cooler, lighter and more reliable (and power output levels are available now which would have been very difficult with valves) but the essential levels of electronic performance in terms of linearity, stability and frequency response have not changed since Williamson's day. An interesting feature of the Williamson is the way in which the output valves have been arranged to operate as triodes. The KT88 valves employed are actually beam-tetrodes. However, these valves, and indeed any beam-tetrode or pentode valve, may be made to act as a triode by connecting ('strapping') the screen grid to the anode as shown in Figure 13.8. Triodes, despite their advantages, are relatively inefficient. For instance, a comparison of a valve amplifier using two EL34 valves operated as triodes or pentodes reveals the figures in Table 13.1.

Ultra-linear and cathode-coupled amplifiers

An interesting approach to this problem was taken by Blumlein who discovered a tetrode or pentode could be made to operate in an intermediate position somewhere between triode and pentode and enjoy the benefits of both regimes. Figure 13.8 illustrates the general principle. If the screen grid is connected as shown at (a), the valve operates as a triode. If it is connected as shown at (b), it operates as a pentode. But if it is connected as shown at (c), the triode is said to operate in an ultra-linear mode, a circuit which enjoys the efficiency of the pentode with the distortion characteristics of the triode. Many classic high quality valve power amplifier designs employed this circuit arrangement, the British manufacturer Leak being especially partial. The Leak 25 Plus amplifier is illustrated in Figure 13.9.

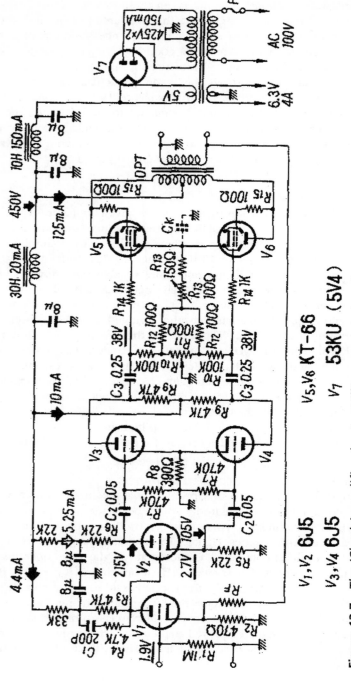

Figure 13.7 The 'Classic' amplifier due to Williamson

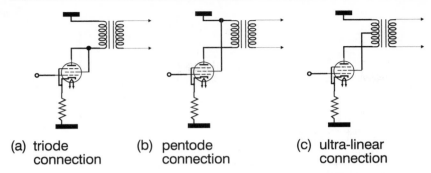

(a) triode
    connection

(b) pentode
    connection

(c) ultra-linear
    connection

Figure 13.8 *Comparison of triode, pentode and ultra-linear circuits*

Table 13.1

	Triodes (in push-pull)	*Pentodes (in push-pull)*
HT volts	430 V	450 V
Anode volts	400 V	
Rg2	100R (each valve)	1k (common to both)
Rk (per valve)	440R	465R
Ra – a	5k	6.5k
Drive volts	48 V	54 V
Power output	19 W	40 W
Distortion	1.8%	5.1%

Figure 13.9 *Leak 25 Plus amplifier*

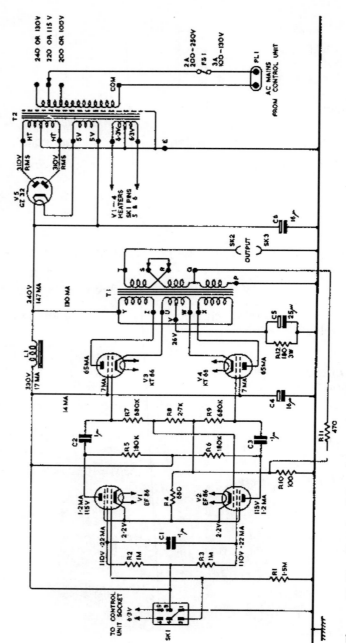

Figure 13.10 *Circuit of Quad 2 monitoring amplifier*

An alternative arrangement, which used a similar technique to combine part of the cathode circuit in the output transformer, was developed by Quad and the circuit of the classic Quad 2 is given in Figure 13.10.

Semiconductor amplifiers

Once again we will examine the semiconductor amplifier in reverse, looking at the output stage first. A loudspeaker requires a low impedance drive, so the simplest form of transistor amplifier is an emitter follower. Unfortunately such a circuit has a very low efficiency so this scheme is usually not employed; however, a circuit which substitutes the emitter load with an extra complementary emitter follower forms the basis of virtually all semiconductor audio amplifiers, and this circuit is illustrated in Figure 13.11. Sometimes (especially in amplifiers of earlier design) truly

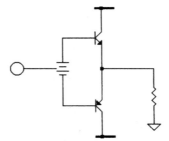

Figure 13.11 *Complementary emitter-follower output circuit*

matched, complementary bipolar transistors were not available so output transistors of the same polarity are combined with complementary ancillary transistors in the form of complementary Darlington-Esaki pairs. Such a topology is known as a quasi-complementary design and this is illustrated in Figure 13.12.

Note that the semiconductor output circuit possesses a host of advantages over its valve equivalent. First, and most significant of all, it does not require an output transformer. Second, it only requires an input signal of one polarity. It may therefore be fed by a single-ended stage. One slight drawback exists in the fact that the complementary emitter output stage has no voltage gain, therefore all the voltage gain has to be supplied by the voltage amplification stage. Also note the bias voltage generator between the two transistor bases. The adjustment of this voltage controls the conduction angle of the output stage. If this voltage is set so that both output transistors remain in conduction for both halves of the output waveform, the stage operates in Class A (as explained above). If, however, the bias voltage is set to zero, there exists a part of the conduction cycle

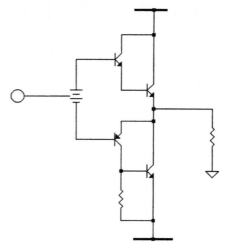

Figure 13.12 *Quasi-complementary output stage*

where neither transistor will carry current. This is not a usual condition for an audio amplifier except one or two special types – one of which is described below. Usually this bias voltage is set so that one transistor stops conducting just as the other begins to conduct. In practice the control is set so as to reduce high-frequency crossover distortion.

An important consideration arises if bias is set so that this output stage works as a Class-B amplifier because, unlike in a Class-A amplifier, the input impedance of a transistor pair changes radically depending on the emitter current through the device. Because this is always changing when set in a Class-B condition, this type of circuit creates a great deal of distortion if it is fed from a voltage source. Instead it must be fed with a current source (i.e. a very high output impedance). Fortunately such a requirement sits happily with a requirement for large voltage gain.

The voltage gain of a common emitter amplification stage, like that shown in Figure 13.13, is given by:

$$g_m \cdot R_c$$

Clearly, the higher the collector load, the higher the stage gain, but there exists a practical maximum value of collector resistor for a given quiescent current and voltage rail. Happily, there exist several schemes for increasing the stage output impedance and voltage gain and these are frequently employed in practical audio amplifiers. The first consists of the application of positive feedback from the other side of the output devices, as shown in Figure 13.14. This maintains a constant voltage across Rc and

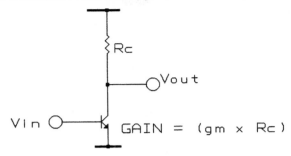

Figure 13.13 *Simple common-emitter amplifier*

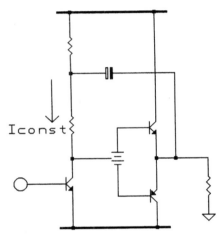

Figure 13.14 *Bootstrapped voltage-gain stage*

this increases, enormously, the impedance at the collector of Tr1. (Because, if the current hardly appears to change for a given change in voltage at the collector, it is the same effect as connecting a very high resistance between the collector and a very high voltage rail.) An alternative arrangement involves the adoption of an active constant-current load, like that shown in Figure 13.15. This circuit, or a variant of it, is very common in modern audio amplifiers. Note also in this diagram how a simple potentiometer may be used to generate the bias voltage between the two output device bases.

The input stage of a modern audio amplifier is invariably formed by a long-tailed pair, reminiscent of the phase splitter in the valve amplifier. However, its adoption here is for different reasons. Notably it provides a convenient point for the injection of loop feedback as is shown in Figure

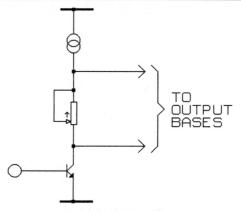

Figure 13.15   *Voltage-gain stage with active load*

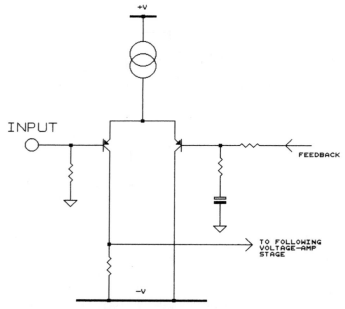

Figure 13.16a   *Long-tailed pair input-stage*

13.16a in which the input signal is seen applied to Tr1, the feedback to Tr2 and the output signal taken from Tr1 collector. A full circuit (due to Linsley Hood) of a typical, class B power amplifier is given in Figure 13.16b.

Current dumping

Current dumping is the term used to describe the amplifier topology invented and originally exploited by the Acoustical Manufacturing

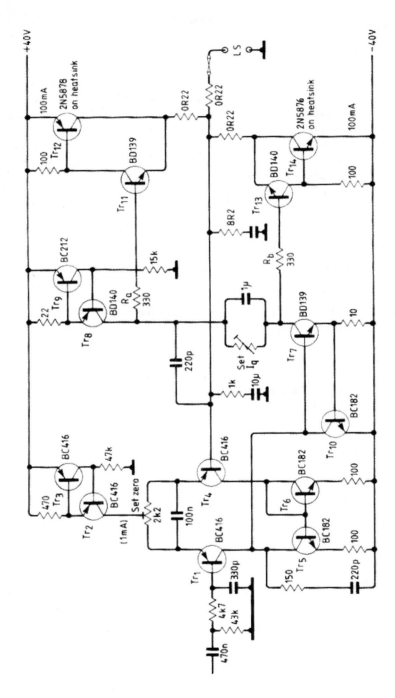

Figure 13.16b *Typical class B, transistor power amplifier circuit (due to Linsley Hood, Electronics and Wireless World Nov 1987)*

Company in the Quad 405, and pertains to the relatively straightforward task undertaken by the output transistors in an amplifier so named. A simple power amplifier with a voltage amplification stage (A1) and a complementary bipolar transistor output stage is illustrated in Figure 13.17a. In many ways this looks like any other complementary power amplifier except that there is no static bias applied to the output transistors (compare with Figure 13.11). Negative feedback encloses the whole and attempts to maintain linearity despite the large transfer-characteristic discontinuity introduced by the unbiased output stage. Provided the gain/bandwidth of A1 is large enough, the output signal is remarkably undistorted, especially at low frequencies. Essentially this arrangement would be entirely acceptable were it possible to construct the amplifier A1 perfectly, so that it 'slewed' infinitely quickly across the crossover 'dead band'. The manner in which A1 behaves is illustrated by the sketch of the signal waveform at the bases of the two transistors. Of course, it isn't possible to construct a perfect amplifier for A1 and, in practice, as frequency increases, crossover distortion starts to make itself heard. The essence of the idea behind current dumping is illustrated by the inclusion of Rd (shown with dotted connections in Figure 13.17a). Rd feeds current directly to the load during the proportion of the output cycle when both the output transistors are off. In effect, Rd reduces the 'gap' A1 is required to slew across and permits a practical amplifier to be used instead of a mythological, perfect amplifier. Stripped of their duty during the essential and fragile crossover region, the output transistors are only called upon to deliver drive into the load at powers above about 100 mW while A1 does all the clever bits in between – hence the term 'current dumping' to describe the uncomplicated job they perform. In

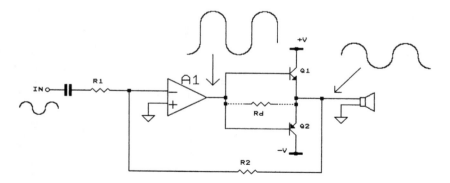

PRIMITIVE CURRENT–DUMPING AMPLIFIER

Figure 13.17a *Simplified current-dumping stage*

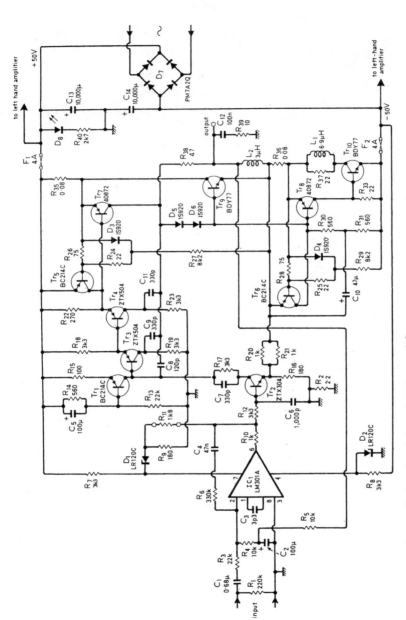

Figure 13.17b *Full circuit of Quad 405 amplifier*

Figure 13.18   *Quad 405 amplifier*

Quad's final circuit, the resistor is augmented by a capacitor and inductor which act in a reactance-bridge arrangement to enhance the effect of Rd across the whole audio bandwidth (see Figure 13.17b). The Quad 405 amplifier is illustrated in Figure 13.18.

MOSFETs

MOSFETs represent an alternative to bipolar junction transistors (BJTs) in commercial power amplifier design and there is much debate surrounding modern solid-state power amplifiers about the use of MOSFETs. Much of the dialectic revolves around the inherent linearity of BJTs vs. FETs. Specifically, that the open-loop transfer characteristic of the crossover region in a MOSFET amplifier is less 'tidy' than the corresponding region in a BJT amplifier when the latter is adjusted for optimum bias. However, more significant, when considering the design options, is relative price; with a MOSFET output device costing pounds rather than the pence a BJT commands – even in relatively large quantities. Furthermore, MOSFETs have the disadvantage that they have a finite – and relatively large – on resistance which is wasteful of power and necessitates the use of paralleled devices in order to drive low impedance loudspeakers. These two considerations affect 'watts per dollar' calculations considerably and would seem to account for the relative dearth of commercial MOSFET amplifiers rather more than considerations of inherent device linearity. However, MOSFETs do possess a number of favourable characteristics, most notably an inherent protection against thermal runaway and secondary breakdown (which in turn permits simpler overcurrent protection circuitry) and a transition frequency which remains the same irrespective of drain current. These alone may justify the MOSFET a place in power amplifier designs except where low cost is an overwhelming

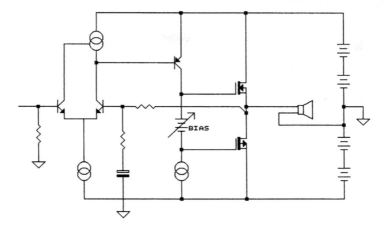

Figure 13.19 *MOSFET amplifier*

factor in the design criteria. The circuit for a typical Class-B power amplifier circuit employing MOSFETs is given in Figure 13.19.

Pre-amplifiers

Pre-amplifiers do not feature much in professional audio, their role is reserved for domestic applications where the pre-amplifiers is the control unit preceding the power amplifier. In professional applications all the duties performed by the pre-amplifiers are undertaken by the mixer. Sometimes the pre-amplifier is separate from the power amplifier but this is fairly rare. With modern, high-level sources such as tape and CD, the pre-amplifiers can be as simple as a selector switch and dual-ganged volume control – a combination somewhat risibly termed a passive pre-amplifier. Alternatively it may contain tone controls, high-pass and low-pass filters and other signal processing functions like channel summing for mono switching, mute, dim and so on. An important feature of older pre-amplifiers was the amplification and equalisation circuitry necessary for the replay of vinyl records. Due to their fragile structure, recording on records was pre-equalised, whereby the bass was cut and the treble boosted during the cutting of the master disc. The replay electronics was therefore required to present a complimentary characteristic and to boost the signal to a useful level before being applied to the power amplifier. This equalisation characteristic was defined by the Recording Industries Association of America and the replay circuit universally dubbed the RIAA pre-amplifiers. A circuit of a simple pre-amplifier for domestic applications

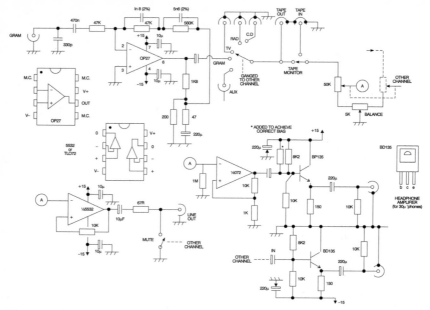

Figure 13.20 *Signal pre-amplifier*

is illustrated in Figure 13.20. Note the RIAA pre-amplifiers constructed around the OP–27 low noise op-amp.

Instrument amplifiers

One of the unifying themes in this book is the distinction between predominantly linear recording and reproducing technology and the non-linear technology of creative music making. So it should be no surprise to find out that many instrument amplifiers are less 'well behaved' than their studio counterparts. That is, an instrumental amplifier may pay less attention to linearity and stability than a monitoring amplifier. However, this distinction requires some careful qualification. With so many purely electronic instruments forming the backbone of today's musical production sound sources, non-linear amplification is not appropriate. When used with these instruments, the job of the amplifier is simply a matter of 'voicing' the instrument's inherent sound. And in practical terms this is often simply part of the PA (in live applications) or supplied via the sound desk and monitoring amplifier (in the case of studio recording). The term instrumental amplifiers is thus rather specific and refers (here at least) to amplifiers which are a 'part' of the instrumental sound. In other words, they 'colour' the sound in their own right. The prime example of this is guitar amplifiers.

Originally the role of the amplifier was not intended to be a feature of the sound; indeed many of the amplifiers were intended to be 'distortionless'. However, guitarists pushed the equipment to its limits in search of expressive potential and thereby uncovered limits of the performance envelope unforeseen by the equipment's designers. Ironically, often it is precisely in these limits that the greatest potential for artistic utterance was found, thereby establishing the sonic signature of a particular performance limitation as a *de facto* standard for acolytes and imitators alike. In turn, manufacturers have been forced to continue to build equipment which is deliberately designed to expose a design limitation or else find a way of simulating the effect with more modern equipment. Perhaps the strongest evidence of this is in the perpetuation of the valve amplifier.

Relieved of a duty to be accurate, instrumental amplification is very difficult to analyse objectively! However, a few observations may be made with some certainty. First, most amplification is not usually designed with a deliberately modified frequency response. This is more usually a function of the designer's choice of loudspeaker and housing. Amplifiers (both low level and power level) are more usually engineered for their distortion characteristics. (As mentioned before, valve amplification is almost certainly preferred for its longer transition band from 'non-distorting' to 'distorting' regimes. This gives the instrumentalist a wider and more controllable tonal and expressive palette.) This characteristic is enhanced by very limited amounts of negative feedback.

Some research has been done to connect various transfer curve characteristics with subjective perceptions. Once again the ear proves to be a remarkably discerning apparatus. So that, in spite of the fact that all distortion mechanisms perform roughly the same 'function', each commercially available amplifier has its own distinctive sound and loyal adherents – some units having acquired an almost cult status. While many of these differences might be difficult, if not impossible, to analyse, a number of distinguishing characteristics are obvious enough. First, the forward gain of the amplifier has an effect on the rate of discontinuity between the linear and non-linear portions of the transfer characteristic. A unit with a low forward gain and a small degree of, or no, negative feedback will show a sluggish transition into the overload region. Such a unit will produce a distortion on a sine-wave input like that illustrated in Figure 13.21(b). Whereas a unit with a high forward gain and a good deal of negative feedback, and thus a faster transition into the non-linear region, will produce an output waveform more like that shown in Figure 13.21(c). Second – and this is probably the most important distinguishing feature of all – the character of the distorted sound is dependent to a large

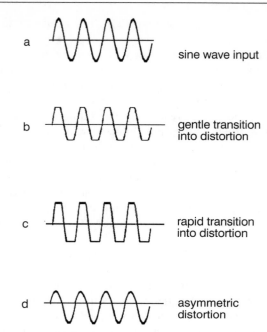

a   sine wave input

b   gentle transition
           into distortion

c   rapid transition
           into distortion

d   asymmetric
           distortion

Figure 13.21   *Comparison of circuit distortion characteristics*

measure on the degree of asymmetry imparted to the output waveform by the overdriven amplifier stage. An asymmetrically distorted waveform (like that shown in Figure 13.21 (d)) has a far higher proportion of even harmonics than the waveform shown in Figure 13.21(c), which has a high proportion of odd harmonics. As stated in Chapter 2, even harmonics tend to be musically related to the original input tone, whereas high order, odd harmonics are musically unrelated to the original guitar signal. This suggests that an amplifier producing a symmetrical overload characteristic will tend to sound 'harsher' than a unit yielding asymmetrical distortion, and subjectively this is indeed the case. The necessarily low gain and asymmetric characteristic of valve amplifiers is almost certainly why they are preferred to their elaborate semiconductor-amplifier counterparts, which, due to a high level of inherent linearity and very high forward gain, tend to elicit a rasping, strident tone when in overload. Unfortunately, the designer has no option when faced with the design of a solid-state amplifier. Being of essentially Class-B design, these amplifiers cannot function without large amounts of negative feedback and the designer of such an amplifier is forced to adopt upstream electronics to try to emulate the gradual distortion characteristics of a valve amplifier! With the advent of digital electronics, this philosophy has blossomed, as the next section illustrates.

DSP and amplification

An alternative approach to instrumental amplification has emerged in the last few years with the availability of medium-priced DSP technology. This is illustrated in Figure 13.22 where a modern, high quality 'characterless'

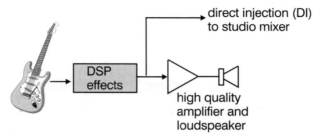

Figure 13.22 *Contemporary approach to instrumental amplification*

amplifier is shown driving a high-quality loudspeaker via a DSP engine, thereby simulating the virtual chain of a subjectively chosen amplifier and loudspeaker design. The advantage being that, by reprogramming the DSP, any amplifier and loudspeaker combination can be chosen – literally – at the flick of a switch.

Fact Sheet #13: New techniques

- Introduction
- Convolutional reverberation
- Pitch-correction techniques

Introduction

This fact sheet deals with a few of the 'cutting-edge' signal processing techniques which have appeared in the last few years.

Convolutional reverberation

As we have seen, despite practical complication, reverberation is simply the repetition of a sound event many times over; as (in physical terms) the original sound source 'bounces' around the original performance environment. It is also well known that particular performance spaces are more attractive than others: great concert halls, opera houses and certain recording studios being particularly famous for their reverberation characteristic.

Unfortunately not all vocal performances can be captured in the hallowed halls of La Scala, Milan. So how might we capture the particular characteristic of an individual performance space? This is the aim of convolutional reverb, a technique realised in the DRE S777 product from Sony (Figure F13.1).

Figure F13.1 *Sony convolutional reverb unit*

There exists the interesting possibility that, if a very short pulse (known as an impulse function – see Fact Sheet 12) is played over loudspeakers in an acoustic environment: and the resulting reflection pattern recorded with microphones and stored in a digital filter, that any signal – passed through the resulting filter – would assume the reverberation of the original performance space. Truly one could sing in La Scala! This is exactly the principle behind the Sony DRE S777, and it is capable of producing a most realistic and flattering reverberation. On a practical matter, it is noteworthy that the amount of data collected in this technique is enormous and different reverberation 'settings' (or more properly, programs) are loaded into the unit via CD-ROMs.

One interesting aside, concerning Sony's technique in capturing the original reverberation pattern of an acoustic space, concerns their use of a frequency sweep; rather than a single pulse. As we saw in Fact Sheet 3, a short pulse contains all frequencies but – inevitably, because it only lasts a very short time – at rather low power. Sony therefore used a frequency sweep to record the reverb characteristic and then back-calculated the effect, as if the original excitation had been a short pulse. This renders a much higher signal to noise ratio. This procedure is well known from radar where the repetitive frequency sweeps (rather than pulses) are described, graphically, as 'chirps'.

Pitch-correction

Poor or faulty intonation is the bane of the recording engineer's life. Vocalists or instrumentalists with insufficient practice or experience (or talent!) often require many, laborious takes to assemble a decent performance. Furthermore, the act of recording itself puts much greater demands on a musician in terms of intonation for, whilst the odd slightly mistuned note would go un-noticed in a concert, repeated listening exposes each and every blemish with every subsequent audition. In order to correct for faulty intonation, several digital signal processing algorithms have been developed which aim to re-tune incoming signals to a preset scale of frequency values; a sort of pitch quantisation.

One of the most successful of these is Auto-Tune due to Antares Systems of Applegate, California. Auto-Tune started life as an audio workstation plug-in. Developed in 1997, it was originally developed for ProTools but is now available in a number of different forms, including; Digital Performer, Logic Audio and CuBase for Mac and Cakewalk.

Depending on the version, Antares' Auto-Tune costs a few hundred pounds as a plug-in. The Auto-Tune algorithm may be used in two modes; graphical and automatic. The automatic screen is shown in Figure F13.2 and, as you can see, it's

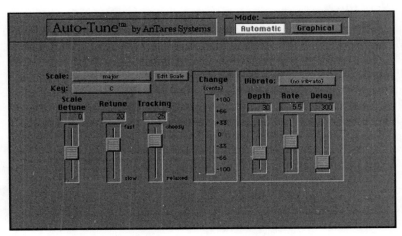

Figure F13.2   *'AnTares Auto-Tune' software*

simplicity itself. In this mode, the software instantaneously detects the pitch of the input, identifies the closest pitch in a user specified scale and corrects the input pitch to match the scale pitch. Major, minor, chromatic, and 26 historical and microtonal scales are available. The slider marked 'Retune, slow-fast' affects the speed with which the re-tune algoritm is applied. Too fast and the pitch correction will strip vibrato. Conversely, a vibrato can be introduced with any depth, rate, and delay; as controlled by the sliders on the right. The 'Graphical Mode' displays the detected pitch and allows you to draw in the desired pitch on a graph. This mode gives complete control over adding or repairing scooping pitches and large vibratos.

Algorithms like Auto-Tune only operate upon highly periodic input waveforms; such as those produced by a solo voice or instrument. Correction may only be applied to signals of this type and which have sufficient dynamic range and are not mixed with other sounds or noise. Nonetheless, these limitations not withstanding, the quality of the output is spectacular, unsteady vocals and solo instruments have their pitch corrected and the resulting audio contains virtually inaudible by-products of this remarkable process.

Antares have recently introduced a real-time hardware version of Auto-Tune, in the form of the ATR-1. This has hardware controls which mimic the sliders in the screen-shot. Overall processing delay is a remarkably short 4 ms.

VariPhrase

VariPhrase is a term invented by the Roland corporation, to describe their VP-9000 processor (Figure F13.3). This clever, eclectic blend of processing techniques combines sampling, pitch shifting, time stretching and formant filtering to greatly

Figure F13.3 *Rolands VP-9000 'VariPhrase' processor*

enhance the scope of conventional sampling techniques. The VP-9000 can instantly match loops from different sources to the same key and tempo and bend notes in real-time without changing the phrase length.

The effect of formants limiting the pitch range over which a simple sample may be employed is discussed above (in Chapter 7). However, imagine being able to pre-process the sample in order to analyse and record its formant structure. This formant structure could then be removed from the sounds (in a filter), the new sound could be mapped over a much larger diapason and the formant filters could be re-applied as a final process. This is the process applied in the VariPhrase processor and greatly augments the range over which vocal or instrumental samples may be used. This is especially useful if the original instrument is not available for multi-sampling; the process by which several samples are employed – each used over a limited gamut of keys.

A similar, time-based phenomenon occurs due to the effect of a sample being used well outside its original pitch – this effect being particularly noticeable on long loops and rhythm loops; that the duration (and therefore tempo) changes. This drawback too is dealt with in the VariPhase processor by employing a sophisticated time-stretch algorithm which is intelligent applied in real time on the basis of a pre-processing analysis of the original sample.

14
Shout – Loudspeakers

Moving-coil loudspeaker theory

Loudspeakers convert electrical energy back into acoustic energy. Normally, they perform this in an arrangement whereby the oscillating electric current from the power amplifier flows in a coil. This coil is wound upon a former at the apex of a paper cone which is itself free to oscillate backwards and forwards because it is supported at its edges and apex by an elastic suspension. The addition of a large magnet in close proximity to the coil completes the assembly. The oscillating magnetic field which is generated as the electric current moves to and fro in the coil interacts with the field generated by the permanent magnet and creates alternating attractive and repulsive forces which propel the loudspeaker cone in and out, thereby creating sound. The mechanism is illustrated in Figure 14.1. Every loudspeaker is enclosed within a box. The reason for this is simple.

Enclosures

Suppose for a moment that the loudspeaker is reproducing the sound of a kick-drum and that the signal from the amplifier has caused the cone of the loudspeaker to move forward. The forwards motion produces a pressure wave in front of the loudspeaker, and it is this pressure wave that our ears must detect if we are to experience the sound of the original drum. Unfortunately the same forwards motion that produces a pressure to the front of the loudspeaker will produce a small vacuum behind it. If the pressure wave is to travel to our ears so that we can hear the kick-drum, it must be prevented from moving around the edge of the loudspeaker to neutralise the vacuum behind. Without the baffling effect of a box this is exactly what is allowed to happen with all low-frequency

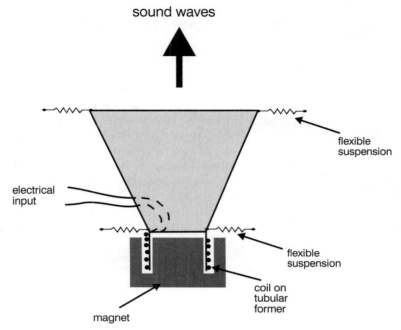

Figure 14.1 *Moving coil loudspeaker*

sounds. (If you've ever heard the sound of a loudspeaker without a box you will know this is true because the sound is almost entirely lacking in low frequencies and it sounds 'thin' or 'tinny'.)

Sealed-box

To obtain a good bass response from any loudspeaker, it is necessary to isolate the radiation from the rear of the cone from the radiation from the front and the easiest way to do that is to put it in a box. Technically this is known as the infinite baffle or, more simply, a sealed-box enclosure. Many hundreds of successful designs from the bookshelf size to the gargantuan have exploited this straightforward principle. A pair of fully sealed monitoring loudspeakers are illustrated in Figure 14.2. Notice that, in this design, the cone material is spun aluminium and not the, more usual, plastic or paper. The fully sealed loudspeaker enclosure, despite its popularity and widespread adoption, is not without its engineering problems. This is because trapped inside the apparently empty loud-speaker enclosure is air. And, though it is a truism to say so – because it is invisible – there's more to air than meets the eye.

When we try to walk in a swimming pool, it is very obvious to us the resistance the water offers to our motion. On the other hand, when we

Figure 14.2 *Infinite-baffle monitoring loudspeakers*

move through air we do so relatively unimpeded, so it is easy to forget that it is there at all. But air is always there, gravity holds it to our planet and therefore it has both weight and density. (It is, after all, dense enough to hold jumbo jets aloft!) The loudspeaker designer must always remember that air has mass and appreciate too, an enclosed volume of air – like that trapped inside a loudspeaker box – possesses compliance or 'springiness'. If it is hard to believe that the invisible and intangible ocean which we unconsciously navigate everyday has any properties at all, think back to the days of pumping up bicycle tyres and what happened when you put your finger over the outlet of the bicycle pump. You'll remember that the plunger acted against the invisible spring of the air trapped inside the sealed pump.

Like air, the loudspeaker unit itself possesses mass and compliance. In acoustic terms, the mass of the loudspeaker is not the mass of the heavy magnet and frame assembly but the mass of the paper, aluminium or plastic cone. The compliance is provided by the surround and support spider at the apex of the cone. As any bungy-jumper will tell you, it is the property of any compliance and mass that they will oscillate at their natural period or resonant frequency. The support compliance and cone's mass determine one of the fundamental limitations to any loudspeaker's performance – the bass resonant frequency (see Figure 14.3). Near the bass resonant frequency the acoustic output becomes distorted in a number of ways. First, because the loudspeaker 'wants' to oscillate at its

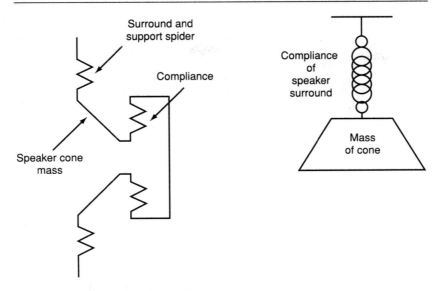

Figure 14.3 *Mechanism of bass resonance*

natural frequency, there is a very large increase in its acoustic output when it tries to reproduce notes in the vicinity of its resonance (see Chapter 3). Second, because the loudspeaker cone is very free to move at these frequencies, the cone's movement becomes non-linear, giving rise to harmonic or intermodulation distortion. Third, the loudspeaker becomes so easy to drive as far as the amplifier is concerned that it may 'look' to the amplifier as if there is no loudspeaker connected at all. The amplifier thus loses control over the loudspeaker's movement.

Bass-reflex enclosure

The addition of an enclosing box does nothing to improve the performance limitations imposed by the loudspeaker's natural resonance. Very often, the situation is actually worsened because the springiness of the enclosed air inside the box augments the loudspeaker's own compliance and raises the bass resonant frequency of the combined loudspeaker and enclosure, further limiting its useful bass response. The good the enclosure has done in increasing the bass output, it has partially undone by raising the bass resonant frequency (see Figure 14.4). The use of a port or tunnel in a bass-reflex design can improve all three areas of a loudspeaker's performance in the region of the bass resonant frequency. It can reduce the hump in the acoustic output, it can improve the load presented to the amplifier, and hence enhance the control exercised by a

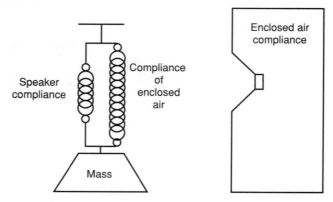

Figure 14.4   *The effect of a sealed box on resonance*

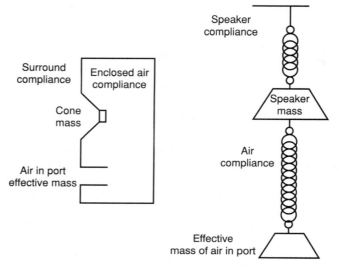

Figure 14.5   *Mechanism of bass-reflex loudspeaker*

good amplifier, and it can reduce distortion. A simple bass-reflex loudspeaker is illustrated in Figure 14.5

You might think from a superficial inspection of the bass-reflex loudspeaker that the port, which after all communicates the rear of cone loudspeaker radiation to the front, would forgo the benefits of the enclosed cabinet. You might reflect, why doesn't a pressure wave from the front of the loudspeaker simply disappear down the port to neutralise the vacuum created within the box? At very low frequencies this is indeed what does happen and it is for this reason that the very low-frequency response of a bass-reflex loudspeaker enclosure is less good than that of a sealed box. (This fact alone has kept the sealed box a favourite among some designers.) However, it is at the resonant frequency of the loudspeaker that the bass-reflex design really comes into its own. By calculating the compliance of the air trapped inside the enclosure and the mass of air resting within the port, the designer of the bass-reflex loudspeaker arranges for this mass and compliance to resonate near, or at, the same frequency as the bass resonance of the loudspeaker unit alone.

Try to picture in your mind's eye that, at this frequency, the mass of air bounces upon the compliance of the air within the box. Mechanically the arrangement is similar to a mass and spring suspended from another mass and spring. The really wonderful property possessed by the bass-reflex loudspeaker is that at the bass resonant frequency the air in the port moves out as the loudspeaker cone moves out and moves in as the cone moves in. It thus selectively stiffens the combined compliance of the surround and support spider and the air enclosed within the box. It thus restricts the movement of the cone at, and near, the bass resonant frequency lowering the load presented to the amplifier, restraining cone movement and thus reducing distortion as are shown in Figure 14.5.

Theile–Small parameters

The exact choice of resonant frequency, as well as the ratio of mass/ compliance required to achieve this resonance, is critical in achieving the desired LF frequency response from a bass-reflex design. In turn, these parameters cannot be chosen without knowing important details about the low-frequency loudspeaker unit itself. The whole subject was analysed in an important paper by Theile (1961), who identified the necessary parameters required to design a bass-reflex cabinet. Until recently, the determination of these parameters required precision electronic and mechanical test equipment. But Theile's work (and that of another researcher, Small) is now so well recognised that the modern enclosure designer is aided by the widespread acceptance of these Theile–Small parameters, which are usually published in the manufacturer's data for the loudspeaker unit.

Crossover networks

Clearly one of the most important design criteria for a good loudspeaker is an even frequency response. Unfortunately the physical requirements for a loudspeaker unit with good high-frequency (HF) response and a good low-frequency (LF) and middle-frequency (MF) response conflict. For instance good, even high-frequency reproduction demands a light diaphragm which would be destroyed in trying to reproduce low frequencies. The only option in fulfilling excellent reproduction at frequency extremes is to divide the electrical input between two or more loudspeaker units, each tailored to reproducing a particular range of frequencies. Adequate performance is usually obtainable from just two units. A design of this type being known as two way. Dividing the electrical input is not done proportionately but is divided by frequency range with one, larger loudspeaker handling the low and middle frequencies (perhaps up to about 3 kHz) and a smaller high-frequency unit handling the frequencies above. The concept thus being to design a complementary filter network. This electrical filter network is called a crossover circuit. A simple passive crossover network is illustrated in Figure 14.6; passive because all the components contained within the network are passive components. The increasing reactance with frequency of the inductor attenuates high frequencies and prevents them reaching the low-frequency loudspeaker unit. The decreasing reactance with frequency of the capacitor prevents low frequencies from reaching and damaging the delicate high-frequency unit.

Practical crossover networks are often substantially more complicated than the circuit shown in Figure 14.6. One such example is given in Figure

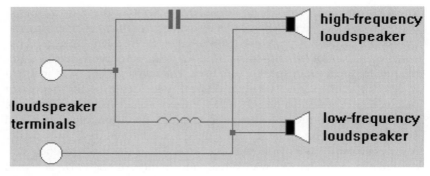

Figure 14.6   *Simple passive crossover network*

14.7 and the loudspeakers for which this circuit was designed are illustrated in Figure 14.8. Here, the crossover divides the sound signals three ways (known as a three-way design) into a low-frequency unit, a mid-frequency unit and a high-frequency unit. Notice also that the crossover filters are of a higher order, thereby increasing the selectivity over which each unit operates. Many books include design equations for the design of crossover networks. These should be treated with extreme caution. Most, if not all, assume a filter design based on a terminating

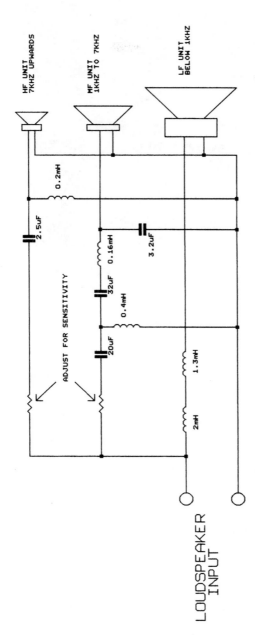

Figure 14.7 *Practical crossover circuit*

Figure 14.8   *Three-way, bass reflex monitor design*

impedance which is purely resistive. This, unfortunately, is nonsense because the one thing that can be said, without reservation, about a loudspeaker is that it does not present a purely resistive impedance – especially outside its intended working region. Failure to take account of this can result in very anomalous responses from the filters as well as presenting the amplifier with an 'unpleasant' load.[1] To design an effective crossover, first you must know the impedance characteristics of each driver, and then design the filter around those impedances.

Active crossovers

Despite their ubiquity, passive *LC* crossovers are wasteful of heat (because all reactive components aren't ideal) and wasteful of resources because the inductive components particularly are bulky and expensive. These reasons, coupled with the not inconsiderable difficulties in designing passive networks which match the capricious impedances of real-world loudspeaker units, have led some designers to abandon passive crossover networks in favour of electronic filtering earlier in the reproduction chain; for instance, before the power amplifiers. Of course this spells the necessity for more power amplifiers; twice the number if the design is a two-way design and three times (i.e. a total of six amplifiers for stereo) if the design is three way! Nevertheless, the engineering superiority of this approach is obvious and modern semiconductor power amplifiers are so efficient and cheap that the economic reasons for not adopting this approach are slight. The case for active crossovers becomes stronger still with the advent of digital audio and digital filtering which allow the possibility of very carefully controlled filtering in the digital domain prior to conversion to analogue and subsequent amplification. These filters can even be made to compensate for shortcomings in the design of the loudspeaker units themselves. Figure 14.9 illustrates a schematic of an

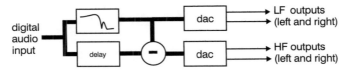

Figure 14.9   *Digital active crossover circuit*

active, digital crossover network prior to DACs and power amplifiers. Notice how a truly complementary response may be achieved by subtracting the low-pass filtered digital signal from the original digital signal; the latter being suitably delayed to allow for the propagation delay through the low-pass digital filter.

Slots

The adoption of a slot is not common in loudspeaker designs and yet its espousal can confer advantages to the acoustic radiation pattern of the loudspeaker (Harwood 1968). Figure 14.10 illustrates a design of my own which incorporated a slot on from of the LF/MF unit as shown. Figure 14.11 illustrates the action of a slot, effectively diffracting the sound from the loudspeaker. The diffraction from the slot becomes more pronounced as slot width/wavelength (a/l) tend to zero. However, as Harwood noted:

It has been shown in quantitative tests, that for small values of (a/l) the predictions do not match the experimental evidence. In fact what appears to happen is that for values up to $a/l = 0.7$ the slit is indeed working but, because of this sound energy flows along the front of the cabinet and is re-radiated at its edge.

Figure 14.10 *Monitors incorporating acoustic 'slot'*

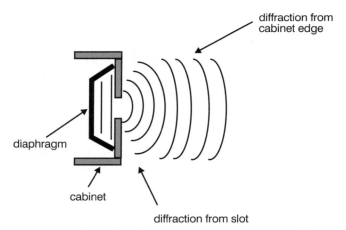

Figure 14.11 *Action of slot and cabinet on sound diffraction*

This effect is illustrated in Figure 14.11 too. This consideration alone has caused designers to try to keep the geometry of loudspeaker cabinets as narrow as possible. Where this is not possible, an alternative approach is to raise the mid-frequency and tweeter unit onto a raised lozenge as was done in the loudspeakers illustrated in Figure 14.8.

Leslie loudspeaker

The Leslie Speaker System, named after its inventor, Don Leslie, operates by rotating a directional sound source around a fixed pivot point as illustrated in Figure 14.12. This produces an amplitude modulation feature known as tremolo or vibrato. In addition, due to the Doppler effect, 'The Leslie' causes a sound source apparently to raise in pitch as it approaches and fall as it recedes. This produces a set of FM sidebands of complex tonality. It is this effect which gives the Leslie its distinctive sound. It is, after a fashion, the acoustical analogy of the electronic chorus (see Chapter 6) and the 'Leslie sound' may be approximated by an electronic vibrato plus chorus effect.

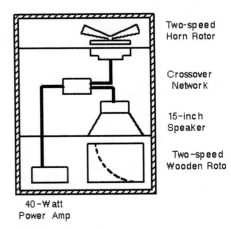

Figure 14.12 *Leslie Speaker System*

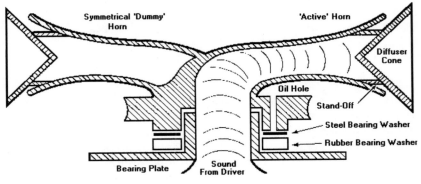

Figure 14.13 *Leslie horn construction*

The classic Leslie loudspeakers, (models 145, 147 and 122) all share the same basic construction (see Figure 14.12), comprising a 40 watt valve amplifier, a passive 12 dB per octave crossover circuit (at 800 Hz) and a rotating treble horn. The treble loudspeaker unit faces upwards into the treble horn assembly which is actually only open on one side – the other side acting solely as a counterweight. This horn arrangement is rotated by means of an AC motor. In addition, there is a rotating ported drum beneath the bass speaker, into which the bass speaker fires. Both rotating speakers may be spun quickly and slowly. Louvres located in the top and bottom compartments let out treble and bass sound respectively as illustrated in Figure 14.14.

Figure 14.14 *Leslie cabinet*

The Leslie 145 loudspeaker was marketed as an accessory to the Hammond organ but found much wider application during the psychedelic 1960s where the swirling, phasey sound the unit generates suited exactly the composers' desire for a mysterious, swimming sound palate. Examples include *Pet Sounds* by the Beach Boys and *Tomorrow Never Knows* by the Beatles – where it is used on vocals! Probably the most famous example of the Leslie (in combination with the B3 Hammond) is Procul Harem's

Whiter Shade of Pale. Listen also to the work of Billy Preston. Preston makes use of a widely adopted technique of speeding the rotor to the higher speed in the course of a solo to increase the dramatic intensity.

Horn loudspeakers

Ordinary open-cone, direct-radiator types of loudspeaker suffer from very poor efficiency. Practical efficiencies in the transduction of electrical power to acoustic power are in the region of 1%! This is not a great disadvantage as might be thought, because the acoustical powers involved in producing really very loud noises are still small – especially in a confined environment like a domestic living room, where the combination of a powerful amplifier and inefficient loudspeaker is preferable to a small amplifier and efficient loudspeaker. Nevertheless in applications such as public address and sound systems for outdoor rock events, greater efficiency from the loudspeakers is required, lest the amplifiers become Leviathans! Poor efficiency at low frequencies is aided by the addition of a ported enclosure as described earlier because this utilises the radiation from both the front and the rear of the port and also controls the impedance of the loudspeaker seen by the amplifier, rendering more of the load resistive and thereby allowing power to be delivered into it from the amplifier.

In the mid-range and at high frequencies, the major limiting factor in the efficiency of loudspeakers is due to the discrepancy between the density of air and that of the lightest practical material from which the diaphragm may be formed. Paper, for instance, which is very often employed as a material for loudspeaker diaphragms, is about 700 times more dense than air. This means the amplifier is principally employed moving the cone mass rather than the mass of air it is intended to excite! The answer is to make the air

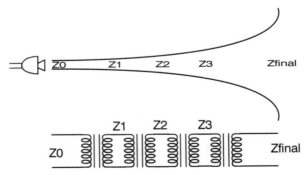

Figure 14.15 *Principle of horn loudspeaker*

load 'look' different to the loudspeaker, to transform the air load's impedance. This is precisely what the addition of a horn does as illustrated in Figure 14.15. The horn acts as a kind of acoustical gearbox, coupling the loudspeaker more efficiently to the air load.

Electrostatic loudspeakers

When a capacitance of capacitance C has a voltage applied (V), the charge (Q) stored on the capacitor is equal to C times V. Or:

$$Q = CV$$

In every capacitor in this condition, there exists an attractive force between the plates of the capacitor which is related to the size (value) of the capacitor and the applied voltage (V) times the charge (Q). Like this:

Attractive force $= k.QV/2$ . . . (where k is a constant)

If one of the plates of the capacitor is allowed to move and an AC voltage source is connected to the capacitor, sound will be created. But the arrangement is not yet suitable as a loudspeaker for the emitted sound will be at twice the applied frequency, because the force of attraction results from a voltage of either polarity. This drawback may be overcome by arranging for a permanent high-voltage DC bias to be present, to ensure the signal voltage never reverses polarity. Graced with this addition, this device acts as a primitive loudspeaker, its attraction lying in its extreme mechanical simplicity.

However, because $Q = CV$, the force on the plates is related to the square of the applied voltage, which implies that the loudspeaker will generate large amounts of second-harmonic distortion unless a means is employed to keep the charge Q constant irrespective of applied voltage. This is accomplished by supplying the bias voltage through a very high value resistor, so that the charge on the capacitor can only leak away very slowly. A further drawback involves the unidirectional bias voltage, which requires that the permanent electrical attractive force be balanced by a similarly permanent mechanical restraining force to stop the moving plate from falling on the static plate. This problem may be overcome by arranging the loudspeaker in a balanced fashion, as illustrated in Figure 14.16. The modulating signal is fed in phase opposition to the two fixed plates and the static force, of one fixed plate on the moving plate (the diaphragm), is balanced by the equal and opposite force of the other plate. This is the mechanical arrangement of all commercial electrostatic loudspeakers, note that the static plates must be manufactured so as to be as acoustically

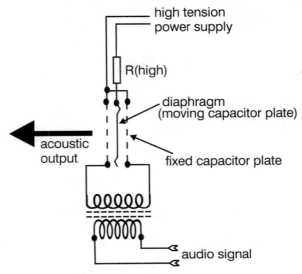

Figure 14.16   *Electrostatic loudspeaker*

transparent as possible. Due to their linear relationship of signal voltage to force and their extreme mechanical elegance, electrostatic loudspeakers are capable of a performance which outshines many (if not all) moving coil designs. Their drawbacks lie in their impedance, which is almost a pure reactance, which limits their efficiency and can upset driving amplifiers which are not unconditionally load stable and in their mechanical fragility which limits their usefulness to domestic applications and to classical music monitoring.

References

Harwood, H.D. (1968) The New BBC Monitoring Loudspeaker. *Wireless World*, April.
Theile, A.N. (1961) Loudspeakers in Vented Boxes. *Proceedings of the IRE Australia*, August.

Note

1 I have been called in more than once to 'rescue' a commercial design where precisely this kind of mistake had been made by the original designer.

Fact Sheet #14: A top-down, non-linear model of auditory localisation

- The Francinstien stereo image enhancement system and the OM 3D sound-processor were both developed commercially by Perfect Pitch Music Ltd. In designing these systems, I developed a top-down model of the human auditory localisation system to underpin my experiments. That model is presented here.[1]

An engineer attempting to solve the various issues relating to the design of a communications system must first possess an understanding of the requirement. In this, a working model of the target system is of great value. Figure F14.1 illustrates the working model of the human auditory spatial perception mechanism. It highlights the data highways to which consideration has been given in the development of the stereophonic systems described in the preceeding chapters.

The auricles are modelled as linear, time-invariant (LTI) systems, this is almost undoubtedly true. The function of the auricle – to re-map spatial attributes as changes in the spectral balance of the acoustic signal arriving at the entrance to the auditory canal is both linear (in that the output of this system contains only the frequencies which are present at its input and no others) and time-invariant (in that the properties of the system do not change with time). The same cannot be said about the modelling of the middle-ear or the electro-mechanical Fourier analyser of the cochlea and basilar membrane. The latter has been assumed also to be an LTI system performing a real-time, continuous, analogue computation and positional digital coding of the Fourier transform of the signal arriving at the auditory canal. There is evidence that the operation of the cochlea is neither linear (due to non-linear vibration of the basilar membrane) or time-invariant (evidenced by the detection of cochlea echoes and that the frequency discrimination of cadaver ears is less good than the frequency discrimination of a living ear). Nonetheless for the purposes of the discussion below this model is assumed.

The qualitative importance of the position and role of the cochlea is its formulation of auditory 'data' in terms of a frequency-domain representation. The Spatial Audition CPU (SACPU) following the two cochlea is thus operating on the magnitude and phase spectra of the signals arriving at the ears. Inside the SACPU (once again treated as a number of LTI systems)

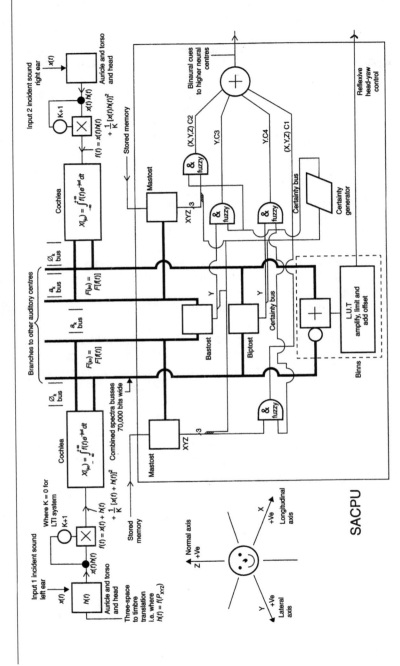

Figure F14.1 *Auditory localisation model – see text*

each ear's digitally coded signal passes to a number of different analysis sub-systems; two Monaural Amplitude Spectrum to Three-space translators (MASTOSTs), one Binaural Amplitude Spectrum to Two-Space Translator (BASTOST) and one Binaural Phase spectrum to Two Space Translator (BIPTOST). Notice, that the model does not specifically incorporate a discrete h.f. envelope generation and phase detection mechanism which would account particularly for Henning's findings mentioned in Chapter 11. Instead this effect is explained by the incorporation of a non-linearity-stage into the model at the position of the middle-ear. The introduction of non-linearity at this stage accounts for both the middle-ear AGC type effects and the non-linearity of the cochlea transduction stage which introduces components in the frequency and phase spectra which could then be processed by the proposed LTI SACPU to generate the perceptions which Henning discovered. Reflexive head control is initiated by means of a feedback system incorporating the BINNS (Binaural Null-Search) sub-system. This reflexive control system attempts to place a sound to coincide with the best image obtainable on the fovea.

The four dimensions of space into which each sub-system codes its output are: x, position on the longitudinal axis; y, position on the lateral axis; z, position on the normal axis and C which is a unipolar scalar metric of certainty. The head is taken as the origin of the co-ordinate system and the polarity conventions are as shown in Figure F14.1.

Notes

1 This model was first presented in an appendix to my book *'Multimedia and Virtual Reality Engineering'* (Newnes, 1997).

15
Synchronicity – Video and synchronisation

Introduction

Audio, for all its artistic power and technological intricacy, is just one element in today's multimedia entertainment phalanx. Today's recording engineers are more likely to find themselves working in a MIDI studio locked to SMPTE timecode than they are to be rigging microphones. Today's composer has a better chance of making his/her living by proactively seeking work for television and film (for which he/she will require an understanding of the medium) than to wait for a commission for a symphony from a rich patron! This chapter contains a description of the principles and concepts which form the technical foundations of an understanding of analogue and digital television.

Persistence of vision

The human eye exhibits an important property which has great relevance to the film and video industries. This property is known as the persistence of vision. When an image is impressed upon the eye, an instantaneous cessation of the stimulus does not result in a similarly instantaneous cessation of signals within the optic nerve and visual processing centres. Instead, an exponential 'lag' takes place with a relatively long time required for total decay. The cinema has exploited this effect for over 100 years. Due to the persistence of vision, if the eye is presented with a succession of still images at a sufficiently rapid rate, each frame differing only in the positions moving within a fixed frame of reference, the impression is gained of a moving image. In a film projector each still frame of film is drawn into position in front of an intense light source while the source of light is shut off by means of a rotating shutter. Once the film frame has stabilised, the light is allowed through – by opening the shutter

– and the image on the frame is projected upon a screen by way of an arrangement of lenses. Experiments soon established that a presentation rate of about 12 still frames per second was sufficiently rapid to give a good impression of continuously flowing movement but interrupting the light source at this rate caused unbearable flicker. This flicker phenomenon was also discovered to be related to level of illumination; the brighter the light being repetitively interrupted, the worse the flicker. Abetted by the low light output from early projectors, this led to the first film frame-rate standard of 16 frames per second (fps). A standard well above that required simply to give the impression of movement and sufficiently rapid to ensure flicker was reduced to a tolerable level when used with early projection lamps. As these lamps improved flicker became more of a problem until an ingenious alteration to the projector fixed the problem. The solution involved a modification to the rotating shutter so that, once the film frame was drawn into position, the shutter opened, then closed, then opened again, before closing a second time for the next film frame to be drawn into position. In other words, the light interruption frequency was raised to twice that of the frame rate. When the film frame rate was eventually raised to the 24 fps standard which is still in force to this day, the light interruption frequency was raised to 48 times per second, a rate which enables high levels of illumination to be employed without causing flicker.

Cathode ray tube and raster scanning

To every engineer, the cathode ray tube (CRT) will be familiar enough from the oscilloscope. The evacuated glass envelope contains an electrode assembly and its terminations at its base whose purpose is to shoot a beam of electrons at the luminescent screen at the other end of the tube. This luminescent screen fluoresces to produce light whenever electrons hit it. In an oscilloscope the deflection of this beam is effected by means of electric fields – a so-called electrostatic tube. In television the electron beam (or beams in the case of colour) is deflected by means of magnetic fields caused by currents flowing in deflection coils wound around the neck of the tube where the base section meets the flare. Such a tube is known as an electromagnetic type.

Just like an oscilloscope, without any scanning currents the television tube produces a small spot of light in the middle of the screen. This spot of light can be made to move anywhere on the screen very quickly with the application of the appropriate current in the deflection coils. The brightness of the spot can be controlled with equal rapidity by altering the rate at which electrons are emitted from the cathode of the electron gun assembly. This is usually effected by controlling the potential between the grid and the cathode electrodes of the gun. Just as in an electron tube or valve, as the grid electrode is made more negative in relation to the cathode, the flow of electrons to the anode is decreased. In the case of the CRT the anode is formed by a metal coating on the inside of the tube flare. A decrease in grid voltage – and thus anode current – results in a darkening of the spot of light. Correspondingly, an increase in grid voltage results in a brightening of the scanning spot.

In television, the bright spot is set up to move steadily across the screen from left to right (as seen from the front of the tube). When it has completed this journey it flies back very quickly to trace another path across the screen just below the previous trajectory. (The analogy with the movement of the eyes as they 'scan' text during reading can't have escaped you!) If this process is made to happen sufficiently quickly, the eye's persistence of vision, combined with an afterglow effect in the tube phosphor, conspire to fool the eye, so that it does not perceive the moving spot but instead sees a set of parallel lines drawn on the screen. If the number of lines is increased, the eye ceases to see these as separate too – at least from a distance – and instead perceives an illuminated rectangle of light on the tube face. This is known as a raster. In the broadcast television system employed in Europe this raster is scanned twice in $\frac{1}{25}$ of a second. One set of 312.5 lines is scanned in the first $\frac{1}{50}$ of a second and a second interlaced set – which is not superimposed but is staggered in the gaps in the preceding trace – is scanned in the second $\frac{1}{50}$. The total number of lines is thus 625. In North America, a total of 525 lines (in two interlaced passes of 262.5) are scanned in $\frac{1}{30}$ of a second.

This may seem like a complicated way of doing things and the adoption of interlace has caused television engineers many problems over the years. Interlace was adopted in order to accomplish a 2 to 1 reduction in the bandwidth required for television pictures with very little noticeable loss of quality. It is thus a form of perceptual coding – what we would call today a data compression technique. Where bandwidth is not so important – as in computer displays – non-interlaced scanning is employed. Note also that interlace is, in some respects, the corollary of the double exposure system used in the cinema to raise the flicker frequency to double the frame rate.

Television signal

The television signal must do two things, the first is obvious, the second less so. First, it must control the instantaneous brightness of the spot on the face of the cathode ray tube in order that the brightness changes which constitute the information of the picture may be conveyed. Second, it must control the raster scanning, so that the beam travels across the tube face in synchronism with the tube within the transmitting camera. Otherwise information from the top left-hand side of the televised scene will not appear in the top left-hand side of the screen and so on! In the analogue television signal this distinction between picture information and scan synchronising information (known in the trade as sync-pulse information) is divided by a voltage level known as black level. All information above black level relates to picture information, all information below relates to sync information. By this clever means, all synchronising information is 'below' black level. The electron beam therefore remains cut-off – and the screen remains dark – during the sync information. In digital television the distinction between data relating to picture modulation and sync is established by a unique codeword preamble which identifies the following byte as a sync byte.

Horizontal and vertical sync

The analogy between the eye's movement across the page during reading and the movement of the scan spot in scanning a tube face has already been made. Of course the scan spot doesn't move onto another page like the eyes do once they have reached the bottom of the page. But it does have to fly back to start all over again once it has completed one whole set of lines from the top to the bottom of the raster. The spot thus flies back in two possible ways: a horizontal retrace, between lines, and a vertical retrace, once it has completed one whole set of lines and is required to start all over again on another set. Obviously to stay in synchronism with the transmitting camera the television receiver must be instructed to perform both horizontal retrace and vertical retrace at the appropriate times – and furthermore not to confuse one instruction for the other!

It is for this reason that there exist two types of sync information known reasonably enough as horizontal and vertical. Inside the television monitor these are treated separately and respectively initiate and terminate the horizontal and vertical scan generator circuits. These circuits are similar – at least in principle – to the ramp or sawtooth generator circuits discussed in Chapter 5. As the current gradually increases in both the horizontal and vertical scan coils, the spot is made to move from left to right and top to bottom, the current in the top to bottom circuit growing 312.5 times more slowly than in the horizontal deflection coils so that 312.5 lines are drawn in the time it takes the vertical deflection circuit to draw the beam across the vertical extent of the tube face.

The complete television signal is illustrated in Figures 15.1 and 15.2 which display the signal using two different timebases. Notice the amplitude level which distinguishes the watershed between picture information and sync information. Known as black level this voltage is set

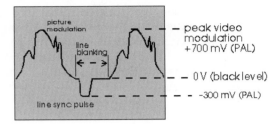

Figure 15.1 *Television signal (viewed at line rate)*

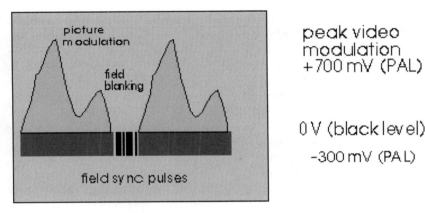

Figure 15.2 *Television signal (viewed at field rate)*

to a standard 0 V. Peak white information is defined not to go beyond a level of 0.7 V above this reference level. Sync information, the line or horizontal sync, 4.7 microsecond pulse is visible in the figure and should extend 0.3 V below the black reference level. Note also that the picture information falls to the black level before and after the sync pulse. This interval is necessary because the electron beam cannot instantaneously retrace to the left-hand side of the screen to restart another trace. It takes a little time – about 12 microseconds. This period, which includes the duration of the 4.7 microsecond line-sync pulse during which time the beam current is controlled 'blacker than black', is known as the line-blanking period. A similar, much longer, period exists to allow the scan spot to return to the top of the screen once a whole vertical scan has been accomplished, this interval being known as the field blanking or vertical interval.

Looking now at Figure 15.2, a whole 625 lines are shown, in two fields of 312.5 lines. Notice the wider sync pulses which appear between each field. In order that a monitor may distinguish between horizontal and vertical sync, the duration of the line-sync pulses are extended during the vertical interval (the gap in the picture information allowing for the field retrace) and a charge-pump circuit combined with a comparator is able to detect these longer pulses as different from the shorter line-sync pulses. This information is sent to the vertical scan generator to control the synchronism of the vertical scan.

Colour perception

Sir Isaac Newton discovered that sunlight passing through a prism breaks into the band of multicoloured light which we now call a spectrum. We perceive seven distinct bands in the spectrum – red, orange, yellow, green, blue, indigo, violet. We see these bands distinctly because each represents a particular band of wavelengths. The objects we perceive as coloured are perceived thus because they too reflect a particular range of wavelengths. For instance, a daffodil looks yellow because it reflects predominantly wavelengths in the region 570 nm. We can experience wavelengths of different colour because the cone cells, in the retina at the back of the eye, contain three photosensitive chemicals each of which is sensitive in three broad areas of the light spectrum. It is easiest to think of this in terms of three separate but overlapping photochemical processes: a low-frequency (long-wavelength) RED process, a medium-frequency GREEN process and a high-frequency BLUE process. (Electronic engineers might prefer to think of this as three, shallow-slope band-pass filters!) When light of a particular frequency falls on the retina, the action of the light reacts selectively with this frequency-discriminating mechanism. When we perceive a red object we are experiencing a high level of activity in our long wavelength (low-frequency) process and low levels in our other two. A blue object stimulates the short wavelength or high-frequency process and so on. When we perceive an object with an intermediate colour, say the yellow of the egg yoke, we experience a mixture of two chemical processes caused by the overlapping nature of each of the frequency-selective mechanisms. In this case, the yellow light from the egg causes stimulation in both the long wavelength RED process and the medium-wavelength GREEN process (Figure 15.3). Because human beings possess three separate colour vision processes we are classified as trichromats. People afflicted with colour blindness usually lack one of the three chemical responses in the normal eye; they are known as dichromats although a few rare individuals are true monochromats. What has not yet been discovered, among people or other animals, is a more-than-three colour perception system. This is fortunate for the engineers who developed colour television!

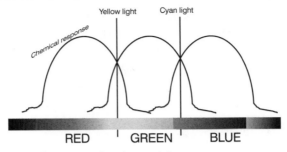

Figure 15.3 *Mechanism of colour vision*

The fact that our cone cells only contain three chemicals is the reason that we may be fooled into experiencing the whole gamut of colours with the combination of only three, so called, primary colours. The television primaries of red, green and blue were chosen because each stimulates only one of the photosensitive chemicals found in the cone cells. The great television swindle is that we can, for instance, be duped into believing we are seeing yellow by activating both the red and green tube elements simultaneously – just as would a pure yellow source. Similarly we may be hoodwinked into seeing light blue cyan with the simultaneous activation of green and blue. We can also be made to experience paradoxical colours like magenta by combining red and blue, a feat that no pure light source could ever do! This last fact demonstrates that our colour perception system effectively 'wraps around', mapping the linear spectrum of electromagnetic frequencies into a colour circle, or a colour space. And it is in this way that we usually view the science of colour perception: we can regard all visual sense as taking place within a colour three space. A television studio vectorscope allows us to view colour three space end on, so it looks like a hexagon – Figure 15.4(a). Note that

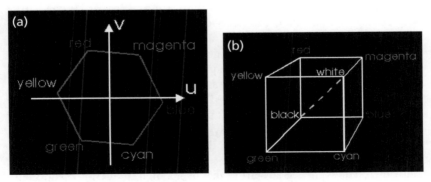

Figure 15.4 *(a) Studio vectorscope display (b) Colour 3 space*

each colour appears at a different angle, like the numbers on a clock face. Hue is the term used in image processing and television to describe a colour's precise location on this locus. Saturation is the term used to describe the amount a pure colour is 'diluted' by white light. The dotted axis shown in Figure 15.4(b). is the axis of pure luminance. The more a particular shade moves towards this axis from a position on the boundary of the cube, the more a colour is said to be desaturated.

Colour television

From the discussions of the trichromatic response of the eye and of the persistence of vision, it should be apparent that a colour scene may be rendered by the quick successive presentation of the red, green and blue components of a colour picture. Provided these images are displayed frequently enough, the impression of a full colour scene is indeed gained. Identical reasoning led to the development of the first colour television demonstrations by Baird in 1928 and the first public colour television transmissions in America by CBS in 1951. Known as a field-sequential system, in essence the apparatus consisted of a high field-rate mono-chrome television system with optical red, green and blue filters presented in front of the camera lens and the receiver screen which, when synchronised together, produced a colour picture. Such an electromechanical system was not only unreliable and cumbersome but also required three times the bandwidth of a monochrome system (because three fields had to be reproduced in the period previously taken by one). In fact, even with the high field rate adopted by CBS, the system suffered from colour flicker on saturated colours and was soon abandoned after transmissions started. Undeterred, the engineers took the next most obvious logical step for producing coloured images. They argued that rather than to present sequential fields of primary colours, they would present sequential dots of each primary. Such a (dot sequential) system using the secondary primaries of yellow, magenta, cyan and black forms the basis of colour printing. In a television system, individual phosphor dots of red, green and blue – provided they are displayed with sufficient spatial frequency – provide the impression of a colour image when viewed from a suitable distance.

Consider the video signal designed to excite such a dot-sequential tube face. When a monochrome scene is being displayed, the television signal does not differ from its black and white counterpart. Each pixel (of red, green and blue) is equally excited, depending on the overall luminosity (or luminance) of a region of the screen. Only when a colour is reproduced does the signal start to manifest a high-frequency component, related to the spatial frequency of the phosphor it is designed successively to stimulate. The exact phase of the high-frequency component depends, of course, on which phosphors are to be stimulated. The more saturated the colour (i.e. the more it departs from grey), the more high-frequency 'colorising' signal is added. This signal is mathematically identical to a black and white television signal whereupon is superimposed a high-frequency colour-information carrier-signal (now known as a colour subcarrier) – a single frequency carrier whose instantaneous value of amplitude and phase respectively determines the saturation and hue of any particular region of the picture. This is the essence of the NTSC[1] colour television system launched in the USA in 1953, although, for practical reasons, the engineers eventually resorted to an electronic dot-sequential signal rather than achieving this in the action of the tube. This technique is considered next.

NTSC and PAL colour systems

If you've ever had to match the colour of a cotton thread or wool, you'll know you have to wind a length of it around a piece of card before you are in a position to judge the colour. That's because the eye is relatively insensitive to coloured detail. This is obviously a phenomenon which is of great relevance to any application of colour picture reproduction and coding; that colour information may be relatively coarse in comparison with luminance information. Artists have known this for thousands of years. From cave paintings to modern animation studios it is possible to see examples of skilled, detailed monochrome drawings being coloured in later by a less skilled hand.

The first step in the electronic coding of an NTSC colour picture is colour-space conversion into a form where brightness information (luminance) is separate from colour information (chrominance) so that the latter can be used to control the high-frequency colour subcarrier. This axis transformation is usually referred to as RGB to YUV conversion and it is achieved by mathematical manipulation of the form:

$$Y = 0.3R + 0.59G + 0.11B$$
$$U = m(B - Y)$$
$$V = n(R - Y)$$

The Y (traditional symbol for luminance) signal is generated in this way so that it as nearly as possible matches the monochrome signal from a black and white camera scanning the same scene. (The colour green is a more luminous colour than either red or blue and red is more luminous than blue.) Of the other two signals, U is generated by subtracting Y from B: for a black and white signal this evidently remains zero for any shade of grey. The same is true of R – Y. These signals therefore denote the amount the colour signal differs from its black and white counterpart. They are therefore dubbed colour difference signals. (Each colour difference signal is scaled by a constant.) These signals may be a much lower bandwidth than the luminance signal because they carry colour information only, to which the eye is relatively insensitive. Once derived, they are low-pass filtered to a bandwidth of 0.5 MHz.[2] These two signals are used to control the amplitude and phase of a high-frequency subcarrier superimposed onto the luminance signal. This chrominance modulation process is implemented with two balanced modulators in an amplitude-modulation-suppressed-carrier configuration – a process which can be thought of as multiplication. A clever technique is employed so that U modulates one carrier signal and V modulates another carrier of identical frequency but phase shifted with respect to the other by 90°. These two carriers are then combined and result in a subcarrier signal which varies its phase and amplitude dependent upon the instantaneous value of U and V. Note the similarity between this and the form of colour information noted in connection with the dot-sequential system: amplitude of high-frequency carrier dependent upon the depth – or saturation – of the colour, and phase dependent upon the hue of the colour. (The difference is that in NTSC, the colour subcarrier signal is coded and decoded using electronic multiplexing and demultiplexing of YUV signals rather than the spatial multiplexing of RGB components attempted in dot-sequential systems.) Figure 15.5 illustrates the chrominance coding process.

While this simple coding technique works well it suffers from a number of important drawbacks. One serious implication is that if the high-frequency colour subcarrier is attenuated (for instance, due to the low pass action of a long coaxial cable) there is a resulting loss of colour saturation. More serious still, if the phase of the signal suffers from progressive phase disturbance the colour in the reproduced colour is likely to change. This remains a problem with NTSC where no means is taken to ameliorate the effects of such a disturbance. The PAL system takes steps to prevent phase distortion having such a disastrous effect by switching the phase of the V subcarrier on alternate lines. This really

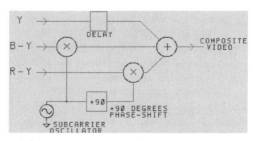

Figure 15.5 *NTSC colour coding process*

involves very little extra circuitry within the coder but has design ramifications which means the design of PAL decoding is a very complicated subject indeed. The idea behind this modification to the NTSC system (for that is all PAL is) is that, should the picture – for argument's sake – take on a red tinge on one line, it is cancelled out on the next when it takes on a complementary blue tinge. The viewer, seeing this from a distance, just continues to see an undisturbed colour picture. In fact, things aren't quite that simple in practice but the concept was important enough to be worth naming the entire system after this one notion – Phase Alternation Line (PAL). Another disadvantage of the coding process illustrated in Figure 15.5 is due to the contamination of luminance information with chrominance and vice versa. Although this can be limited to some degree by complementary band-pass and band-stop filtering, a complete separation is not possible and this results in the swathes of moving coloured bands (cross-colour) which appear across high frequency picture detail on television – herringbone jackets proving especially potent in eliciting this system pathology.

In the colour receiver, synchronous demodulation is used to decode the colour subcarrier. One local oscillator is used and the output is phase shifted to produce the two orthogonal carrier signals for the synchronous demodulators (multipliers). Figure 15.6 illustrates the block schematic of an NTSC colour decoder. A PAL decoder is much more complicated.

Mathematically we can consider the PAL and NTSC coding process thus:

NTSC colour signal $= Y + 0.49(B - Y) \sin \omega t + 0.88 (R - Y) \cos \omega t$
PAL colour signal $\quad = Y + 0.49(B - Y) \sin \omega t \pm 0.88 (R - Y) \cos \omega t$

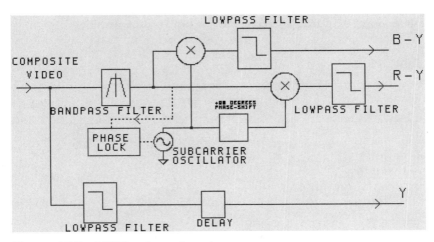

Figure 15.6 *NTSC colour decoder*

Note that following the demodulators, the U and V signals are low-pass filtered to remove the twice frequency component and that the Y signal is delayed to match the processing delay of the demodulation process before being combined with the U and V signals in a reverse colour space conversion. In demodulating the colour subcarrier, the regenerated carriers must not only remain spot-on frequency, but also maintain a precise phase relationship with the incoming signal. For these reasons the local oscillator must be phase locked and for this to happen the oscillator must obviously be fed a reference signal on a regular and frequent basis. This requirement is fulfilled by the colour burst waveform which is shown in the composite colour television signal displayed in Figure 15.7.

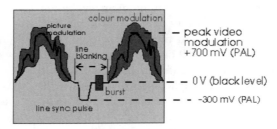

Figure 15.7   *Colour television signal*

The reference colour burst is included on every active TV line at a point in the original black and white signal given over to line retrace. Notice also the high-frequency colour information superimposed on the 'black and white' luminance information. Once the demodulated signals have been through a reverse colour space conversion, and become RGB signals once more, they are applied to the guns of the colour tube.

Table 15.1

	NTSC	PAL
Field frequency	59.94 Hz	50 Hz
Total lines	525	625
Active lines	480	575
Horizontal resolution	440	572
Line frequency	15.75 kHz	15.625 kHz

(Note: Horizontal resolutions calculated for NTSC bandwidth of 4.2 MHz and 52 μs line period; PAL, 5.5 MHz bandwidth and 52 μs period.)

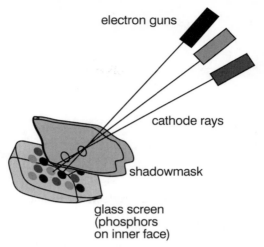

electron guns

cathode rays

shadowmask

glass screen
(phosphors
on inner face)

Figure 15.8 *The shadowmask in action*

As you watch television, three colours are being scanned simultaneously by three parallel electron beams, emitted by three cathodes at the base of the tube and all scanned by a common magnetic deflection system. But how to ensure that each electron gun only excites its appropriate phosphor? The answer is the shadowmask – a perforated, sheet-steel barrier which masks the phosphors from the action of an inappropriate electron gun. The arrangement is illustrated in Figure 15.8. For a colour tube to produce an acceptable picture at reasonable viewing distance there are about half a million phosphor red, green and blue triads on the inner surface of the screen. The electron guns are set at a small angle to each other and aimed so that they converge at the shadowmask. The beams then pass through one hole and diverge a little between the shadowmask and the screen so that each strikes only its corresponding phosphor. Waste of power is one of the very real drawbacks of the shadowmask colour tube. Only about a quarter of the energy in each electron beam reaches the phosphors. Up to 75% of the electrons do nothing but heat up the steel!

Analogue video interfaces

Due to their wide bandwidth, analogue television signals are always distributed via coaxial cables. The technique known as matched termination is universally applied. In this scheme both the sender impedance and the load impedance are set to match the surge impedance

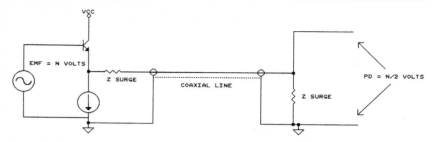

Figure 15.9 *Video interconnection*

of the line itself. This minimises reflections. Standard impedance in television is 75 Ω A typical interconnection is shown in Figure 15.9. Note that matched termination has the one disadvantage, that the voltage signal arriving across the receiver's termination is half that of the signal EMF provided by the sender. Standard voltage levels (referred to above) always relate to voltages measured across the termination impedance.

Digital video

In order to see the forces which have led to the rapid adoption of digital video processing and interfacing throughout the television industry in the 1990s, it is necessary to look at some of the technical innovations in television during the late 1970s and early 1980s. The NTSC and PAL television systems described above were primarily developed as transmission standards, not as television production standards. As we have seen, because of the nature of the NTSC and PAL signal, high-frequency luminance detail can easily translate to erroneous colour information. In fact, this cross-colour effect is an almost constant feature of the broadcast standard television pictures and results in a general 'business' to the picture at all times. That said, these composite TV standards (so named because the colour and luminance information travel in a composite form) became the primary production standard mainly due to the inordinate cost of 'three-level' signal processing equipment (i.e. routing switchers, mixers etc.) which operated on the red, green and blue or luminance and colour-difference signals separately. A further consideration, beyond cost, was that it remained difficult to keep the gain, DC offsets and frequency response (and therefore delay) of such systems constant, or at least consistent, over relatively long periods of time. Systems which did treat the R, G and B components separately suffered particularly from colour shifts throughout the duration of a programme. Nevertheless as analogue technology improved, with the use of integrated circuits as opposed to discrete semiconductor circuits, manufacturers started to produce three-channel, component television equipment which processed the luminance, R – Y and B – Y signals separately. Pressure for this extra quality came particularly from graphics areas which found working with the composite standards resulted in poor quality images which were tiring to work on, and where they wished to use both fine detail textures, which created cross-colour, and heavily saturated colours which do not produce well on a composite system (especially NTSC).

So-called analogue component television equipment had a relatively short stay in the world of high-end production largely because the problems of intercomponent levels, drift and frequency response were never ultimately solved. A digital system, of course, has no such problems. Noise, amplitude response with respect to frequency and time are immutable parameters 'designed into' the equipment – not parameters which shift as currents change by fractions of milliamps in a base-emitter junction somewhere! From the start, digital television offered the only real alternative to analogue composite processing and, as production houses were becoming dissatisfied with the production value obtainable with composite equipment, the death-knell was dealt to analogue processing in television.

4:2:2 protocol description – general

Just as with audio, so with video; as more and more television equipment began to process the signals internally in digital form, so the number of conversions could be kept to a minimum if manufacturers provided a digital interface standard allowing various pieces of digital video hardware to pass digital video information directly without recourse to standard analogue connections. This section is a basic outline of the 4:2:2 protocol (otherwise known as CCIR 601), which has been accepted as the industry standard for digitised component TV signals. The data signals are carried in the form of binary information coded in 8-bit or 10-bit words. These signals comprise the video signals themselves and timing reference signals. Also included in the protocol are ancillary data and identification signals. The video signals are derived by the coding of the analogue video signal components. These components are luminance (Y) and colour difference (Cr and Cb) signals generated from primary signals (R, G, B). The coding parameters are specified in CCIR Recommendation 601 and the main details are reproduced in Table 15.2.

Timing relationships

The digital active line begins at 264 words from the leading edge of the analogue line synchronisation pulse, this time being specified between half amplitude points. This relationship is shown in Figure 15.10. The start

Table 15.2   *Encoding parameter values for the 4:2:2 digital video interface*

Parameters	525-line, 60 field/s systems	625-line, 50 field/s systems
1 Coded signals: Y, Cb, Cr. These signals are obtained from gamma pre-corrected RGB signals.		
2 Number of samples per total line:		
– luminance signal (Y)	858	864
– each colour-difference signal (Cb, Cr)	429	432
3 Sampling structure	Orthogonal line, field and picture repetitive Cr and Cb samples co-sited with odd (1st, 3rd, 5th, etc.) Y samples in each line.	
4 Sampling frequency:		
– luminance signal	13.5 MHz	
– each colour-difference signal	6.75 MHz	
The tolerance for the sampling frequencies should coincide with the tolerance for the line frequency of the relevant colour television standard.		
5 Form of coding	Uniformly quantised PCM, 8 bits per sample, for the luminance signal and each colour-difference signal.	
6 Number of samples per digital active line:		
– luminance signal	720	
– each colour-difference signal	360	
7 Analogue to digital horizontal timing relationship:		
– from end of digital active line to 0 H	16 luminance clock periods (NTSC)	12 luminance clock periods (PAL)

Table 15.2 *(Continued)*

Parameters	525-line, 60 field/s systems	625-line, 50 field/s systems
8 Correspondence between video signal levels and quantisation levels:		
– scale	0 to 255	
– luminance signal	220 quantisation levels with the black level corresponding to level 16 and the peak white level corresponding to level 235. The signal level may occasionally excurse beyond level 235.	
– colour-difference signal	225 quantisation levels in the centre part of the quantisation scale with zero signal corresponding to level 128.	
9 Codeword usage	Codewords corresponding to quantisation levels 0 and 255 are used exclusively for synchronisation. Levels 1 to 254 are available for video.	

(Note that the sampling-frequencies of 13.5 MHz (luminance) and 6.75 MHz (colour difference) are integer multiples of 2.25 MHz, the lowest common multiple of the line frequencies in 525/60 and 625/50 systems, resulting in a static orthogonal sampling pattern for both. The luminance and the colour-difference signals are thus sampled to 8- (or 10-)bit depth with the luminance signal sampled twice as often as each chrominance signal (74 ns as against 148 ns). These values are multiplexed together with the structure as follows:

Cb, Y, Cr, Y, Cb, Y, Cr . . . etc.

where the three words (Cb, Y, Cr) refer to co-sited luminance and colour difference samples and the following word Y corresponds to a neighbouring luminance only sample. The first video data word of each active line is Cb.)

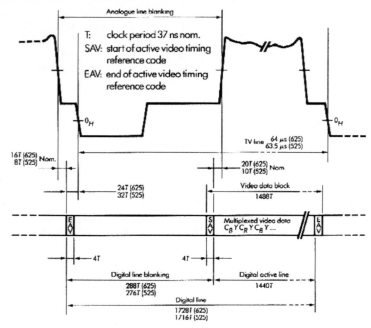

Figure 15.10   *Relationships between timing in digital and analogue TV*

of the first digital field is fixed by the position specified for the start of the digital active line: the first digital field starts at 24 words before the start of the analogue line No. 1. The second digital field starts 24 words before the start of analogue line No. 313.

Video timing reference signals (TRS)

Two video timing reference signals are multiplexed into the data stream on every line, as shown in Figure 15.10, and retain the same format throughout the field blanking interval. Each timing reference signal consists of a four-word sequence, the first three words being a fixed preamble and the fourth containing the information defining:

first and second field blanking;
state of the field blanking;
beginning and end of the line blanking.

This sequence of four words can be represented, using hexadecimal notation, in the following manner:

FF 00 00 XY

in which XY represents a variable word. In binary form this can be represented in the following form:

Data bit number	First word (FF)	Second word(00)	Third word(00)	Fourth word (XY)
7	1	0	0	1
6	1	0	0	F
5	1	0	0	V
4	1	0	0	H
3	1	0	0	P3
2	1	0	0	P2
1	1	0	0	P1
0	1	0	0	P0

The binary values of F, V and H characterise the three items of information listed earlier:

F = 0 for the first field;
V = 1 during the field-blanking interval;
H = 1 at the start of the line-blanking interval.

The binary values P0, P1, P2 and P3 depend on the states of F, V and H in accordance with the following table, and are used for error detection/correction of timing data:

F	V	H	P3	P2	Fl	P0
0	0	0	0	0	0	0
0	0	1	1	1	0	1
0	1	0	1	0	1	1
0	1	1	0	1	1	0
1	0	0	0	1	1	1
1	0	1	1	0	1	0
1	1	0	1	1	0	0
1	1	1	0	0	0	1

Clock signal
The clock signal is at 27 MHz, there being 1728 clock intervals during each horizontal line period (PAL).

Filter templates
The remainder of CCIR Recommendation 601 is concerned with the definition of the frequency response plots for pre-sampling and reconstruction filter. The filters required by Recommendation 601 are practically difficult to achieve and equipment required to meet this

specification has to contain expensive filters in order to obtain the required performance.

Parallel digital interface

The first digital video interface standards were parallel in format. They consisted of 8 or 10 bits of differential data at ECL data levels and a differential clock signal again as an ECL signal. Carried via a multicore cable, the signals terminated at either end in a standard D25 plug and socket. In many ways this was an excellent arrangement and is well suited to connecting two local digital videotape machines together over a short distance. The protocol for the digital video interface is Table 15.3. Clock transitions are specified to take place in the centre of each data-bit cell.

Table 15.3 *The parallel digital video interface*

Pin No.	Function	
1	Clock +	
2	System Ground	
3	Data 7 (MSB) +	
4	Data 6 +	
5	Data 5 +	
6	Data 4 +	
7	Data 3 +	
8	Data 2 +	
9	Data 1 +	
10	Data 0 +	
11	Data −1 +	10-bit systems only
12	Data −2 +	10-bit systems only
13	Cable shield	
14	Clock −	
15	System Ground	
16	Data 7 (MSB) −	
17	Data 6 −	
18	Data 5 −	
19	Data 4 −	
20	Data 3 −	
21	Data 2 −	
22	Data 1 −	
23	Data 0 −	
24	Data −1 −	10-bit systems only
25	Data −2 −	10-bit systems only

Problems arose with the parallel digital video interface over medium/ long distances resulting in mis-clocking of the input data and visual 'sparkles' or 'zits' on the picture. Furthermore the parallel interface required expensive and non-standard multicore cable (although over very short distances it could run over standard ribbon cable) and the D25 plug and socket are very bulky. Today, the parallel interface standard has been largely superseded by the serial digital video standard which is designed to be transmitted over relatively long distances using the same coaxial cable as used for analogue video signals. This makes its adoption and implementation as simple as possible for existing television facilities converting from analogue to digital video standards.

Serial digital video interface

SMPTE 259M specifies the parameters of the serial digital standard. This document specifies that the parallel data in the format given in the previous section be serialised and transmitted at a rate ten times the parallel clock frequency. For component signals this is:

$$27 \, \text{Mbits/s} \times 10 = 270 \, \text{Mbits/s}$$

The serialised data must have a peak to peak amplitude of 800 mV (±10%) across 75 Ω, have a nominal rise time of 1 ns and have a jitter performance of ±250 ps. At the receiving end, the signals must be converted back to parallel in order to present the original parallel data to the internal video processing. (Note that no equipment processes video in its serial form although digital routing switchers and DAs, where there is no necessity to alter the signal, only buffer it or route it, do not decode the serial bit stream.)

Serialisation is achieved by means of a system illustrated in Figure 15.11. Parallel data and a parallel clock are fed into input latches and thence to a parallel to serial conversion circuit. The parallel clock is also fed to a phase-locked loop which performs parallel clock multiplication (by 10 times). A sync detector looks for TRS information and ensures this is encoded correctly irrespective of 8- or 10-bit resolution. The serial data is fed out of the serialiser and into the scrambler and NRZ to NRZI circuit. The scrambler circuit uses a linear feedback shift register which is used to pseudo-randomise the incoming serial data. This has the effect of minimising the DC component of the output serial data stream, the NRZ to NRZI circuit converts long series of ones to a series of transitions. The resulting signal contains enough information at clock rate and is sufficiently DC free that it may be sent down existing video cables. It may be then be reclocked, decoded and converted back to parallel data at the receiving equipment. Due to its very high data rate serial video must be carried by ECL circuits. An illustration of a typical ECL gate is given in

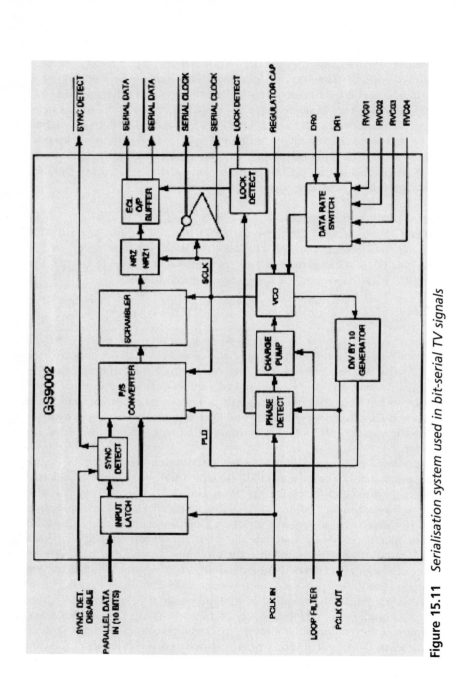

Figure 15.11 Serialisation system used in bit-serial TV signals

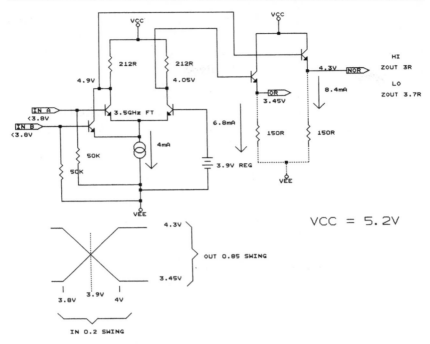

Figure 15.12 *ECL gate*

Figure 15.13 *Gennum Corp. serialiser chip (courtesy of Gennum Corporation)*

Figure 15.12. Note that standard video levels are commensurate with data levels in ECL logic. Clearly the implementation of such a high speed interface is a highly specialised task. Fortunately practical engineers have all the requirements for interface encoders and decoders designed for them by third-party integrated circuit manufacturers (Figure 15.13).

Embedded digital audio in the digital video interface

So far, we have considered the interfacing of digital audio and video separately. Manifestly, there exist many good operational reasons to combine a television picture with its accompanying sound 'down the same wire'. The standard which specifies the embedding of digital audio data, auxiliary data and associated control information into the ancillary data space of the serial digital interconnect conforming to SMPTE 259M in this manner is the proposed SMPTE 272M standard.

The video standard has adequate 'space' for the mapping of a minimum of 1 stereo digital audio signal (or two mono channels) to a maximum of 8 pairs of stereo digital audio signals (or 16 mono channels). The 16 channels are divided into 4 audio signals in 4 'groups'. The standard provides for 10 levels of operation (suffixed A to J) which allow for various different and extended operations over and above the default synchronous 48 kHz/20-bit standard. The audio may appear in any and/or all the line blanking periods and should be distributed evenly throughout the field. Consider the case of one 48 kHz audio signal multiplexed into a 625/50 digital video signal. The number of samples to be transmitted every line is:

$$(48\,000)/(15\,625)$$

which is equivalent to 3.072 samples per line. The sensible approach is taken within the standard of transmitting 3 samples per line most of the time and transmitting 4 samples per line occasionally in order to create this non-integer average data rate. In the case of 625/50 this leads to 1920 samples per complete frame. (Obviously a comparable calculation can be made for other sampling and frame rates.) All that is required to achieve this 'packeting' of audio within each video line is a small amount of buffering either end and a small data overhead to 'tell' the receiver whether it should expect 3 or 4 samples on any given line.

Figure 15.14 illustrates the structure of each digital audio packet as it appears on preferably all, or nearly all, the lines of the field. The packet starts immediately after the TRS word for EAV (end of active line) with the ancillary data header 000,3FF,3FF. This is followed by a unique ancillary data ID which defines which audio group is being transmitted. This is followed with a data-block number byte. This is a free-running counter counting from 1 to 255 on the lowest 8 bits. If this is set to zero, a de-embedder is to assume this option is not active. The 9th bit is even parity for b7 to b0 and the 10th is the inverse of the 9th. It is by means of this data-block number word that a vertical interval switch could be discovered and concealed. The next word is a data count which indicates to a receiver the number of audio data words to follow. Audio subframes

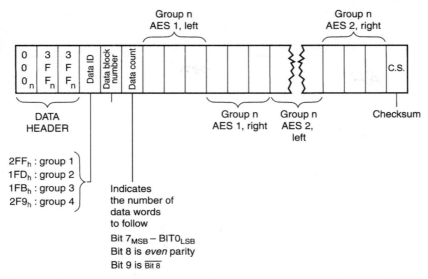

Figure 15.14 *Data format for digital audio packets in SDV bit stream*

then follow as adjacent sets of three contiguous words. The format in which each AES subframe is encoded is illustrated in Figure 15.15. Each audio data packet terminates in a checksum word.

The standard also specifies an optional audio control packet. If the control packet is not transmitted, a receiver defaults to 48 kHz, synchronous operation. For other levels, the control byte must be transmitted in field interval.

Timecode

Longitudinal timecode (LTC)

As we have seen, television (like movie film) gives the impression of continuous motion pictures by the successive, swift presentation of still images, thereby fooling the eye into believing it is perceiving motion. It is probably therefore no surprise that timecode (deriving as it does from television technology) operates by 'tagging' each video frame with a unique identifying number called a timecode address. The address contains information concerning hours, minutes, seconds and frames. This information is formed into a serial digital code which is recorded as a data signal onto one of the audio tracks of a videotape recorder. (Some videotape recorders have a dedicated track for this purpose.)

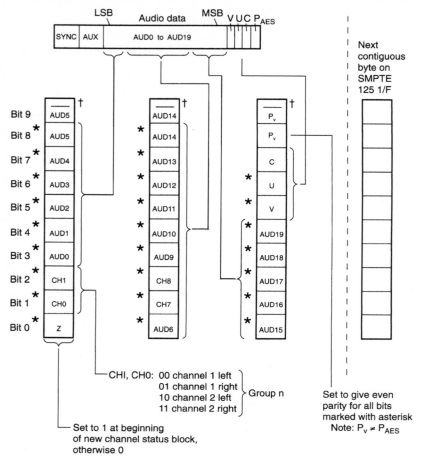

Figure 15.15   *Each EAS subframe is encoded as three contiguous data packets*

Each frame's worth of data is known as a word of timecode and this digital word is formed of 80 bits spaced evenly throughout the frame. Taking EBU timecode 3 as an example, the final data rate therefore turns out to be 80 bits × 25 frames per second = 2000 bits per second, which is equivalent to a fundamental frequency of 1 kHz; easily low enough, therefore, to be treated as a straightforward audio signal. The timecode word data format is illustrated (along with its temporal relationship to a video field) in Figure 15.16. The precise form of the electrical code for timecode is known as Manchester bi-phase modulation. When used in a

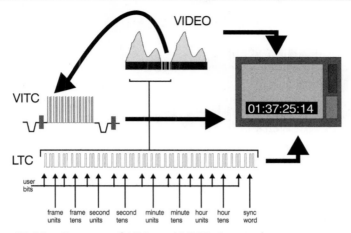

Figure 15.16 *Format of LTC and VITC timecode*

video environment, timecode must be accurately phased to the video signal. As defined in the specification, the leading edge of bit '0' must begin at the start of line 5 of field 1 (±1 line). Time address data is encoded within the 80 bits as 8, 4-bit BCD (binary coded decimal) words (i.e. 1, 4-bit number for tens and 1 for units). Like the clock itself, time address data is only permitted to go from 00 hours, 00 minutes, 00 seconds, 00 frames to 23 hours, 59 minutes, 59 seconds, 24 frames.

However, a 4-bit BCD number can represent any number from 0 to 9, so in principle timecode could be used to represent 99 hours, 99 minutes and so on. But, as there are no hours above 23, no minutes or seconds above 59 and no frames above 24 (in PAL), timecode possesses potential redundancy. In fact some of these extra codes are exploited in other ways. The basic time address data, and these extra bits are assigned their position in the full 80-bit timecode word, are like this:

0–3	Frame units
4–7	First binary group
8–9	Frame tens
10	Drop frame flag
11	Colour frame flag
12–15	Second binary group
16–19	Seconds units
20–23	Third binary group
24–26	Seconds tens
27	Unassigned

28-31	Fourth binary group
32-35	Minutes units
36-39	Fifth binary group
40-42	Minutes tens
43	Unassigned
44-47	Sixth binary group
48-51	Hours units
52-55	Seventh binary group
56-57	Hours tens
58-59	Unassigned
60-63	Eighth binary group
64-79	Synchronising sequence

Vertical interval timecode (VITC)

Longitudinal timecode (LTC) is a quasi-audio signal recorded on an audio track (or hidden audio track dedicated to timecode). VITC, on the other hand, encodes the same information within the vertical interval portion of the video signal in a manner similar to a Teletext signal. Each has advantages and disadvantages; LTC is unable to be read while the player/recorder is in pause, while VITC cannot be read while the machine is in fast forward or rewind modes. It is advantageous that a videotape has both forms of timecode recorded. VITC is illustrated in Figure 15.16 too. Note how timecode is displayed 'burned-in' on the monitor.

PAL and NTSC

Naturally timecode varies according to the television system used and for NTSC (SMPTE) there are two versions of timecode in use to accommodate the slight difference between the nominal frame rate of 30 frames per second and the actual frame rate of NTSC of 29.97 frames per second. While every frame is numbered and no frames are ever actually dropped, the two versions are referred to as 'Drop-' and 'Non-drop-' frame timecode. Non-drop-frame timecode will have every number for every second present, but will drift out of relationship with clock time by 3.6 seconds every hour. Drop-frame timecode drops numbers from the numbering system in a predetermined sequence, so that the timecode-time and clock-time remain in synchronisation. Drop-frame is important in broadcast work, where actual programme time is important.

User bits

Within the timecode word there is provision for the hours, minutes, seconds, frames and field ID that we normally see, and 'user bits' which can be set by the user for additional identification. Use of user bits varies with some organisations using them to identify shoot dates or locations and others ignoring them completely.

Notes

1 NTSC stands for National Television Standards Committee, the government body charged with choosing the American colour system.
2 In NTSC systems or 1 MHz in PAL systems.
3 European Broadcasting Union (EBU) timecode is based on a field frequency of 25 frames per second.

Appendix 1
Music Engineering CD

Part 1 Physics of sound and psychoacoustics

Track 1   Reference level sine wave tone*
1kHz at −18 dB below digital FS (−18 dBFS). The simplest musical sound, the sine-wave, is produced when an object (or in this case, the current in an electrical circuit) vibrates backwards and forwards, exhibiting what physicists call simple harmonic motion.

Track 2   Frequency sweep*
5 Hz to 22 kHz. The pure tone, as illustrated in Track 1, actually sounds rather dull and characterless. But we can vary the number of cycles of oscillation which take place per second. Musicians refer to this variable as pitch – physicists call it frequency. This track also illustrates the possible range of human hearing, which in most adult individuals is limited to about 20 Hz to 15 kHz.

Track 3   Square waves at various frequencies*
A square wave may be synthesised by an infinite number of sine-waves but it's much easier to arrange a circuit known as an astable multivibrator. Just such a circuit was used to generate the sounds on this track.

Track 4   Piano range; 27.5 Hz to 4.19 kHz
The relationship between written musical pitch (on a stave) and frequency (in Hz) is illustrated in Figure 2.5. Remember that the frequency components of the sound produced by each of these instruments extends very much higher than the fundamental tone. Take for instance the highest note on a grand piano. Its fundamental is about 4.19 kHz but the fourth harmonic of this note, which is certainly seen to be present if the sound of this tone is analysed on a spectrum analyser, is well above 16 kHz.

Track 5 The harmonic series, up to the twentieth harmonic
The vibration of the open string produces notes which follow a distinct and repeatable pattern of musical intervals above the note of the open string. These are illustrated in this track. They are termed the harmonic series. A similar pattern is obtainable from exciting the air within a tube as is the case with a pipe organ, recorder, oboe or the clarinet.

Track 6 Even harmonic series (played two times in the form of a musical chord)
Even numbered harmonics are all consonant, their effect is therefore 'benign' musically. The same cannot be said for odd harmonics as demonstrated in the next track.

Track 7 Odd harmonic series
Played two times in the form of a 'dissonant' chord.

Part 2 Microphones and acoustic recording

Track 8 Piano recorded in nearfield. The idea behind near-field microphone technique is to suppress the effect of the acoustics of the room on the signal received by the microphone. This is accomplished by placing the microphone as close as possible to the original sound source.

Track 9 Piano recorded in far-field
Classical microphone technique has 'shied-away' from nearfield method and it is for this reason that recording venues for classical music must be more carefully selected than those for rock and pop recording.

Track 10 Classical session, crossed-pair, stereo recording
Whilst it's possible in principle to 'mike-up' every instrument within an orchestra and then – with a combination of multi-track and electronic panning – create a stereo picture of the orchestra, this is usually not done. Instead, most recordings of orchestras and choirs depend almost exclusively on the application of simple, or 'purist' microphone techniques where the majority of the signal that goes on to the master tape is derived from just two (or possibly three) microphones. In this track, the technique of a coincident crossed pair (sometimes referred to – somewhat incorrectly – as Blumlein Pair) was used.

Track 11 Close-miked recording with artificial reverb
Rock and pop vocalists tend to use smaller working distances in order to capture a more intimate vocal style. This track also illustrates multi-track technique.

Part 3   Electric instruments

Track 12   Electric piano
The most famous electric piano is, without doubt, the Fender Rhodes. This, and its many imitators is actually more of an electronic Glockenspiel (or Vibraphone) than an electronic piano because the sound-producing mechanism is formed from struck metal bars, the hammers being actuated via a conventional keyboard mechanism. An adaptation of the electric guitar type pickup is utilised so that the piano can be amplified.

Track 13   Electric organ and Clavinet
There are two basic types of electronic organ. The divider type uses a digital top-octave generator (one oscillator for each semitone) and chains of divide-by-two bistables to provide the lower octaves. However, this approach tends to produce a 'sterile' tone disliked by musicians. The alternative is known as a free-phase electronic organ. Theoretically the free-phase organ has a different oscillator for each note of the keyboard. In the case of the Hammond each oscillator is mechanical.

The Clavinet was, commercially and artistically, the most successful keyboard produced by German Company Hohner, who designed it to replicate the sound of a Clavichord.

Part 4   Electronic effects

Track 14   Unprocessed guitar. The earliest electric guitars were created by attaching a contact microphone to the top sound-board of a conventional acoustic guitar, the resulting signal being fed to an external amplifier. However, the modern electric guitar was born with the invention of the electromagnetic pick-up and a typical instrument (Fender Stratocaster) is used in this and the following examples.

Track 15   Slap echo effect (50–100 ms)
The Slap (or Slap-back) echo was first heard in Scotty Moore's chiming lead guitar on the early Elvis records.

Track 16   Guitar tape-loop effects
Recordable tape loops originate with the work of Brian Eno and Robert Fripp in the early 1970s, where sounds are recorded over and over onto a loop of magnetic tape on a tape deck which incorporates the crucial modification that the erase head is disabled and an electrical path provided so that sounds may be re-circulated in the manner of a tape-echo device. The sounds are therefore recorded 'on top of one another' and one instrument may create vast, dense, musical structures. Importantly,

subsequent signals do not simply add and from an artistic point of view this is extremely valuable because it means, without continually 'fuelling' the process, the 'sound-scape' gradually dies away. The artist may thereby control the dynamic and tonal 'map' of the piece. Nevertheless, the control of this process is not comprehensive and many of the results are partially random.

Track 17   Fuzz or distorted guitar
In a 'fuzz' circuit, the guitar signal is applied to a circuit at a sufficient amplitude that it drives the circuit beyond its available voltage swing. The waveform is thus 'clipped'. For guitarists this effect is amongst their stock-in-trade. It's now understood, the manner in which the circuit overloads influences the sound timbre.

Track 18   Wah-wah guitar effect
Wah-wah is a dramatic effect derived from passing the signal from the electric guitar's pickup through a high Q band-pass filter, the frequency of which is adjustable usually by means of the position of a foot-pedal. The player may then use a combination of standard guitar techniques together with associated pedal movements to produce a number of instrumental colours.

Track 19   Pitch-shift effect – octave down

Track 20   Octave up
Pitch shifting is used for a number of aesthetic reasons, the most common being the creation of 'instant' harmony. Simple pitch shifters create a constant interval above or below the input signal, like these 'harmonies' at the octave.

Track 21   Pitch-shift, Perfect 4th up; Perfect 5th up
Harmony at a perfect fourth produces only one note which is not present in the original (played) key. However, this extra note is the prominent 'blue' note of the flattened 7th. It is therefore often acceptable in the context of rock music. Harmony at a perfect fifth is usable except for the note a perfect-fifth above the leading-note of the scale which forms a tritone with the root note of the scale.

Track 22   Flanging guitar
George Martin claims the invention of flanging came about due to the slight lack of synchronisation between two 'locked' tape recorders. This effect caused various frequency bands to be alternately reinforced and cancelled; imparting on the captured sound a strange, liquidity – a kind of 'swooshing, swirling' ring. Of course, such an effect is not achieved nowadays using tape recorders; instead, electronic delays are used.

Track 23   Twelve-bar blague
A blend of guitar effects used to create a varied ensemble.

Track 24   Just leave a sample sir!
This track demonstrates a sampler used as an effects (FX) device.

Track 25   Hymn to Aten – for soprano, orchestra and pre-recorded tape
Hymn to Aten was written for the soprano Jane Searle and the Kingston Chamber Orchestra and was first performed in February 1998. Technically the piece is in two halves. The first part (which is the only part recorded here) depicts the primeval and ritualistic elements of a brief, schismatic faith which blossomed in Egypt in about 1345 BC; based on the benevolent, physical presence of the sun. This part is prerecorded and extends the technique used in Birdsong (Track 52) of using effect electronics to process vocal sounds; sometimes creating noises (for instance the bell-like motif) which appear very different from the original material. Harmonically this section is bi-tonal (each group of chords being generated from a MIDI controlled harmoniser) but gradually resolving to a D minor chord for the entry of the orchestra and soprano.

Track 26   Uncompressed guitar – deliberately played with high dynamic contrast
For engineering purposes, it is often desirable to shrink the dynamic range of a signal so as to 'squeeze' or compress it into the available channel capacity. The studio device for accomplishing such a feat is called a compressor. When using a compressor, the peak signal levels are reduced in the manner illustrated in the following track.

Track 27   Compressed guitar
Note that the peaks are reduced but that the gain is not made up. Obviously this would be of little use if the signal (now with compressed dynamic range) was not amplified to ensure the reduced peak values fully exercised the available 'swing' of the following circuits. For this reason, a variable gain amplifier stage is placed after the compression circuit to restore the peak signal values to the system's nominal maximum level. This is demonstrated in the following track.

Track 28   Same effect as Track 27 but with gain made up
Notice that the perceptible effect of the compressor, when adjusted as described, is not so much apparently to reduce the level of the peak signal as to boost the level of the low-level signals; in other words, that the guitar is now apparently louder than in Track 26.

Track 29   Compressed highly distorted guitar
Unfortunately, compression brings with it the attendant disadvantage that low-level noise – both electrical and acoustic – is boosted along with the wanted signal. Notice the noise floor is unacceptably high. The solution is a primitive expansion circuit known as a noise-gate, the effect of which is to suppress all signals below a given threshold and only 'open' in the presence of wanted modulation. The effect of this circuit is illustrated in the next track.

Track 30   Same effect as Track 29, but illustrating the effect of a noise-gate following the compressor

Part 5   Synthetic sound

Track 31   White noise
In white-noise, all frequencies are present (at least stochastically). There is therefore an analogy with white light.

Track 32   Pink noise

Track 33   Red noise
Often composers need a sound which is modified in some way. Examples of this include variations of low-passed filtered noise; so-called pink or red noise because again of an analogy with light.

Track 34   Band-pass filtered noise
Band-pass filtered noise, if generated by a swept band-pass filter can be made to sound like rolling waves or the sound of the wind through trees.

Track 35   Simple near sine-tone patch
The power of the analogue synthesiser lies in its ability to cause each of its individual components to interact in amazingly complex ways. Fundamental to the whole concept is the voltage-controlled oscillator. This may be controlled by a switched ladder of resistances; perhaps by means of a conventional musical keyboard, as in this example, or by means of a constantly variable voltage, thereby providing a sound source with endless portamento like the Ondes Martenot and the Theremin. Alternatively it may be controlled by the output of another oscillator; the resultant being a waveform source frequency modulated by means of another. And perhaps this resultant waveform might be made to modulate a further source! By this means, the generation of very rich waveforms is

possible and herein lies the essential concept behind analogue synthesisers. Some examples are given in the following tracks.

Track 36   Typical analogue bass-synth patch

Track 37   Patch with exaggerated action of LFO

Track 38   Buzzy string patch sound with ASR generator controlling VCF

Track 39   Bass patch, note VCF effect

Track 40   Sampling; used to generate novelty backing!
Digital sampling systems rely on storing high quality, digital recordings of real sounds and replaying these on demand as shown in this simple example.

Track 41   Sampled drums
The tough problem sampling incurs is the sheer amount of memory it requires. Sampling is well suited to repetitive sounds (like drums and other percussion instruments) because the sample is mostly made up of a transient followed by a relatively short on-going (sustain) period. As such, it may be used over and over again so that an entire drum track could be built from as few as half-a-dozen samples.

Track 42   Modern electronic drum samples

Track 43   Gregorian chant voice samples
Sampling is great until long, sustained notes are required; like the sounds generated by the orchestral strings or voices. The memory required to store long sustained notes would be impossibly large, so sampled-synthesis systems rely on 'looping' to overcome the limitation of any non-infinite memory availability.

Track 44   Roland SAS synthesised piano
This particular track demonstrates Roland's proprietary Structured Adaptive Synthesis (SAS) which is an eclectic blend of techniques, honed to give the most realistic piano sound possible.

Track 45   Miller of the Dee
Composite synthesised track used to create slightly frenetic yet, nonetheless, 'classical ensemble' sound: Harpsichord – SAS; Recorder and Bassoon – LS Sound Synthesis/Wavetable; Strings – Yamaha Dynamic Vector Synthesis.

Track 46 Christmas Tree

Another composite ensemble with sampled drums; demonstrating a mix with conventional guitar and vocal track. This is a very cost-effective production technique because the majority of the ensemble can be prepared in advanced in a MIDI programming studio, making the acoustic recording stage very simple and fast.

Part 6 Theremin

One of the earliest electronic instruments, this monophonic (single tone) melodic instrument was originally developed in Russia in about 1920 by Leon Theremin. Magically, the Theremin player does not touch the instrument and has only to bring their hand or body within a small distance of a special aerial to control the pitch produced by the instrument. The Theremin is thus able to produce an endless range of frequencies from the subsonic to the inaudibly high in long sustained glissandi. Despite being very difficult to play, the Theremin has achieved undeniable artistic success. It may be heard in several orchestral pieces and has been used on many film and early TV soundtracks. Furthermore the Theremin remains the emblem of experimental electronic music. A status that it perhaps enjoys because it's one of the very few instruments designed in historical memory to employ a truly novel playing technique.

Track 47 Unprocessed Theremin; some reverb added during recording

The Theremin used on this track has a particularly pure (close to sine-wave) output due to the purity of the RF oscillators and a linear mixer and demodulation stage. The original Theremin had an output nearer to the sound of a violin (i.e. with a large degree of even-harmonic distortion). Waveform distortion is achievable by various means, including reducing the purity of the original RF waveforms or arranging a non-linear detector circuit. However the preferred technique (utilised, for instance, by Bob Moog) is to start with a linear Theremin and to distort the waveform afterwards in the audio domain. Such a technique was used in the following track.

Track 48 Theremin sound subjected to non-linear distortion post demodulation

Track 49 In this track, the Theremin input was used to drive an intelligent harmoniser (Digitech Vocalist Workstation) with the original input suppressed. MIDI data was input to the harmoniser to programme the harmonic progression and the Theremin controlled the arpeggiation.

Track 50   Deep glissando

A short piece for unaccompanied Theremin and effects. Effects include pitch-shift (over two octaves), flange, chorus and reverb as well as non-linear distortion, compression and gating.

Part 7   Music mixing and production

Track 51   Hey Bulldozer

The result of a commission to produce a dance piece for children for the Rainforest Foundation, Hey Bulldozer involves spoken word, a children's choir, sound effects and a standard rock-band arrangement plus amongst other things; pan pipe samples and Spanish guitar. Interestingly, none of the piece's component parts – the choir, the band, the narrators etc. – were able to be at the same place at the same time!

The first step involved making a gash mix from the original MIDI files and dubbing this onto a multi-track tape as a working cue track. The multi-track was then loaded in the back of the car along with an array of microphones, mixer, power amps and fold-back speakers and driven down to the choir. The children recorded several different 'takes' whilst listening to the backing via several fold-back loudspeakers carefully set so that they were loud enough for them to hear them, but not so loud as to cause too much 'spill' of this signal onto the final choir microphone signals. A week later, after the final tweaks to the MIDI sequencer data and samples were complete, the complete vocals track was formed by selecting the best bits from the various 'takes'. The narrated speech and choral speaking excerpts (which were recorded a month earlier in a London dance studio complete with traffic noise!) were then carefully edited and 'topped and tailed' using a digital audio hard disk editing system and spun-in to a track on the multi-track by hand (or do I mean by mouse?). The two acoustic guitar tracks were then added, then the electric guitar track and then a backwards electric guitar which involved turning the multi-track tape over on the deck and having the guitarist play along to the other tracks backwards (something you can only do with an analogue multi-track machine!) And all that was left was to add the sound effects which involved a combination of real jungle sounds (flown direct from South America and on analogue cassette – arrrgh!) and library FX. Hey Bulldozer received several performances during the summer of 1991 and has been performed several times since in different versions.

Track 52   Birdsong

Birdsong demonstrates how sounds may be cut-up and rearranged using digital audio editing software. Here, the technique has been exploited to produce a musical composition based on the minimalist technique of taking tiny sound 'bites' and building them into larger structures. In this case the building 'bricks' are taken from a recording of the renaissance choral composer Victoria's Responses. These are layered, reversed and mixed to produce a collage of sound. The title is from the book by Sebastian Faulks.

Part 8   Stereo techniques and spatial sound processing

Track 53   Classical, crossed-pair stereo recording

When listening to music on a two-channel stereo audio system, a sound 'image' is spread out in the space between the two loudspeakers. The reproduced image thus has some characteristics in common with the way the same music is heard in real-life – that is, with individual instruments or voices each occupying, to a greater or lesser extent, a particular and distinct position in space. Insofar as this process is concerned with creating and re-creating a 'sound-event', it is woefully inadequate. Firstly, the image is flat and occupies only the space bounded by the loudspeakers. Secondly, even this limited image, is distorted with respect to frequency. (There exists an analogy with chromatic aberration in optics.)

Track 54   Track 53, FRANCINSTIEN processed

Happily there exists a simple techniques for both the improvement of existing stereophonic images known as FRANCINSTIEN and for the creation of synthetic sound fields in a 360 degree circle around the listening position (OM processing). These techniques are illustrated in this and the following track.

Track 55   Stand-In for an Echo

Stand-In for an Echo was written during my studies with the Advanced Music Technology Group at Surrey University and was my opportunity to re-write history and re-score some of my favourite moments from a few classic 'Noir' films.

Technically, the music incorporates the discrete phasing technique which Steve Reich introduced in Clapping Music in 1972. All four instrumental parts are derived from the set of patterns played by piano 1. Piano 2 copies these patterns but drops on quaver-beat every 30 quavers, the flute drops a quaver every 20 quavers and the marimba drops a quaver every 9 quavers. The process is extended to incorporate several 'tacit' bars before any instrument starts. In other words, the phasing has 'already begun' before the music starts. Over the instruments are superimposed 'sound-bites' taken from the films of books by Dashiell Hammett and Raymond Chandler. These, monaural, sound sources are treated by means of the OM spatial sound processor to appear outside of the area bounded by the loudspeakers and even behind the listening position – without the use of extra loudspeakers. The title is taken from one a jotting in one of Chandler's note-books; 'I called out and no-one answered, not even a stand-in for an echo.'

* These tones were not digitally generated and are not intended for precision measurement.

Index